MOLECULAR EVOLUTION, PROTEIN POLYMORPHISM AND THE NEUTRAL THEORY

MOLECULAR EVOLUTION, PROTEIN POLYMORPHISM AND THE NEUTRAL THEORY

Edited by

MOTOO KIMURA

JAPAN SCIENTIFIC SOCIETIES PRESS, Tokyo
SPRINGER-VERLAG, Berlin Heidelberg New York

1982

Supported in part by the Ministry of Education, Science and Culture under the Publication Grant-in-Aid.

Published jointly by
Japan Scientific Societies Press Tokyo

ISBN 4-7622-5308-1
and
Springer-Verlag Berlin Heidelberg New York

ISBN 3-540-11466-1 Springer-Verlag Berlin Heidelberg New York
ISBN 0-387-11466-1 Springer-Verlag New York Heidelberg Berlin

Sole distribution rights outside Japan granted to Springer-Verlag Berlin Heidelberg New York

Printed in Japan

To Professor Kozo Imahori
on the occasion of his retirement from Osaka University

Preface

In order to promote molecular evolutionary studies in Japan, a study group was set up in 1978 headed by Dr. Kozo Imahori and sponsored by a grant-in-aid from the Ministry of Education, Science and Culture. The title of the project (freely translated from the Japanese) was: "Elucidation of the mechanism of evolution, with special reference to molecular evolution." The three years of research by the members and collaborators of this study group were extremely fruitful. When the project terminated in 1981, it was proposed that those who had participated would write on topics of their own specialty to produce a volume which would reach a wider audience.

As the editor of this volume, I am very happy that this proposal has now materialized, and I am thankful to those who contributed articles in order to make this possible. I also thank Dr. K. Aoki for his help in editing the book.

It is indeed a pleasure for me to dedicate this book to Professor Kozo Imahori in commemoration of his retirement from Osaka University. As an authority on algal taxonomy, he took a great interest in the phylogeny and evolution of living organisms in general; and as soon as molecular evolutionary studies emerged as a respectable area in evolutionary biology, he felt the need to stimulate younger generations of Japanese scientists to venture into this promising field. In particular, he has done a laudable job of organising the first research group in this field in Japan. His efforts were certainly much appreciated by those of us who participated in this group, and I am sure that they will also be appreciated by those who follow in our path.

January 1982 Motoo KIMURA

Introduction

During the last two decades, the study of evolution has been revolutionized by the introduction of the concepts and the methods of molecular genetics. It has become possible, for example, to construct reliable phylogenetic trees, even when fossil records are missing, by comparative studies of protenis, and, more recently, of DNA sequences. If, in addition, paleontological information is available, we can estimate the rate at which genes are substituted within the species in the course of evolution.

Through the application of electrophoretic methods to proteins, and more recently, the use of restriction enzymes to DNA fragments, it has become possible to study intraspecific variation in molecular terms. Now it is clear that immense genetic variability exists in a sexually reproducing species and our own human species is no exception.

The importance of gene duplication in evolution, as first glimpsed in the early *Drosophila* work, has now been amply demonstrated by directly probing into the genetic material. Existence of gene duplication at various levels of complexity has been revealed in an increasing number of cases, the most dramatic form being the multigene families.

The mathematical theory of population genetics, particularly its stochastic aspect (where the diffusion equation method plays a predominant role) was once considered to be too academic to be of actual use in biology, but the situation has changed. In fact, in conjunction with new molecular data, it has led to the formation of "the neutral theory" of molecular evolution. Unlike the traditional neo-Darwinian approach, this new paradigm emphasizes the importance of mutational pressure and random genetic drift as the major causes of evolution at the molecular level. This has caused the "neutralist-selectionist controversy" in the field of population and evolutionary genetics.

In this connection, I would like to say that although an increasing number of population geneticists have become keenly interested in the development of molecular genetics, very few molecular geneticists still pay attention to population genetics. However, as more and more molecular biologists get involved in comparative studies of DNA sequences, they will inevitably become committed to a discussion of evolution. Thus, a knowledge of population genetics will be needed to guard them from making unrealistic arguments about the mechanism of molecular evolution.

With his characteristic wisdom, Crick (1979) wrote in a recent article (see, Crick, F. *Science* 204: 264–271) in which he reviews the "mini-revolution" in molecular genetics, as follows:

> A molecular biologist who wishes to discuss the evolution of the eukaryotic genome will need not only to know a lot about the way DNA and its transcripts can behave but also something about modern ideas on population genetics.

I sincerely hope that Crick's advice will be followed by those who enter the field of molecular evolution.

The present volume contains theoretical and experimental papers covering a wide range of topics in molecular evolution and protein polymorphism by active Japanese researchers in population and molecular genetics, biochemistry, anthropology, *etc*. In the first chapter, a detailed exposition of the neutral mutation-random drift hypothesis (neutral theory), including new supporting evidence, is presented by M. Kimura. This is intended to form the unifying principle for the remainder of the book. This is also the most comprehensive review of the neutral theory to date.

The subsequent chapters by T. Ohta, N. Takahata, T. Maruyama, and Y. Tateno are theoretical in nature. As mentioned above, multigene families represent extreme forms of gene duplication and recent molecular studies have revealed that they manifest novel phenomena such as coincidental (or concerted) evolution. T. Ohta summarizes her original research on the population genetical treatment of such evolutionary phenomena in chapter 2. N. Takahata investigates the fixation time of recurrent null mutations occurring at duplicated loci by the numerical integration of the Kolmogorov backward equations involved and also by using a simulation method. If null alleles have a pronounced heterozygous effect, then loss of expression is considerably retarded. T. Maruyama introduces a simulation approach to the solution of stochastic equations. In particular, the method is applied to the multiple allele overdominance problem and linkage disequilibrium in finite populations. Y. Tateno investigates the accuracy of three commonly used algorithms for constructing "molecular phylogenetic trees" from gene frequency data. The algorithms are evaluated by creating a tree with known topology and branch lengths and then comparing it to the reconstructed trees.

The chapters by T. Mukai *et al.*, T. Yamazaki, T. Gojobori, T. Miyata, T. Maruyama and E. Soeda, and K. Omoto are statistical analyses of data with interesting conclusions. The paper by Mukai's group summarizes the results of numerous tests for balancing selection conducted on *Drosophila melanogaster* (mainly in his laboratory). These tests provide no evidence for the action of balancing selection in the maintenance of protein and isozyme polymorphisms. Mukai's reputation as a careful worker lends much weight to these results. Genetic variability in a haploid liverwort is examined by T. Yamazaki. As is also true of *Escherichia coli*, the use of a haploid plant is significant since it permits us to rule out the possibility of overdominance as a major cause for the maintenance of protein polymorphism. In the chapter by T. Gojobori, the empirical relationship between the mean and variance of heterozygosity at each locus is compared with the theoretical prediction of the neutral theory in order to detect the presence of functional constraint. T. Miyata's paper deals with functional constraint in gene substitution. The primary structures of various genes among different organisms are compared, revealing that the rates of synonymous change are high and uniform among different genes. Furthermore, the substitution rate for a pseudogene (mouse α-globin) is even higher. These observations suggest functional constraint even against synonymous changes and are consistent with the neutral theory which predicts that the substitution rate should be negatively correlated with the degree of functional constraint. T. Maruyama and E. Soeda provide the first verification, at the molecular level, of coevolution in the papova viruses and their hosts. Furthermore, they give an interesting example of the interaction in substitution rates between overlapping genes of phage. K. Omoto presents hard-won evidence, based on blood genetic markers, that the Philippine negritos are genetically distinct from the African pygmies to whom they bear a superficial resemblance.

The last five chapters by G. Matsuda, T. Tsukihara *et al.*, T. Yamanaka *et al.*, S. Odani and Ikenaka, and T. Oshima are interesting descriptions of the molecular evolution of some specific molecules. G. Matsuda gives a description of the evolution of hemoglobins, which is consistent with the neutral theory although the vocabulary and interpretation sometimes differ. Tsukihara *et al.*'s comparison of plant and halobacterial ferredoxins suggests a common origin, but it also appears that different portions of the amino acid sequence have acquired importance in plants and halobacteria. These portions show little divergence between species. Yamanaka *et al.* present evidence for the coevolution of cytochrome *c* and cytochrome oxidase. Each cytochrome oxidase reacts most rapidly with cytochrome *c* from organisms that are phylogenetically close. S. Odani and Ikenaka use chemical-enzymatic mutation methods to alter the amino acid residues in the active site of proteinase inhibitors. The experiments demonstrate the interchangeability of a variety of residues with respect to the function of proteinase inhibition. Finally, T. Oshima presents a rare example of possible adaptation at the molecular level. The thermostability of a tRNA in a thermophilic bacterium appears to be due to the substitution of a G-C pair for a G-U pair and the modification of another base.

I hope that this book will convey to the readers some idea of the fascination of investigating evolution through the synthesis of population genetics and molecular genetics, a passion for me during the last two decades.

January 1982 Motoo KIMURA

Contents

1

Molecular Evolution, Protein Polymorphism and the Neutral Theory (1982) (ed. M. Kimura), pp. 3–56.
Japan Scientific Societies Press, Tokyo / Springer-Verlag, Berlin

The Neutral Theory as a Basis for Understanding the Mechanism of Evolution and Variation at the Molecular Level

MOTOO KIMURA
National Institute of Genetics, Mishima 411, Japan

The Darwinian theory of evolution by natural selection, which states that organisms become progressively adapted to their environments by accumulating beneficial mutants, is now firmly established. That the Darwinian view pervades our evolutionary thinking is clearly seen throughout contemporary writings in biology. Indeed Darwinian theory is serving as a great unifying principle in the biological sciences.

The storm of opposition and criticism that once raged against the Darwinian view is hard to imagine today, for the theory is now regarded as almost sacrosanct. When Darwin formulated his theory, the mechanism of inheritance and the nature of heritable variations were unknown, preventing him from being over-confident about the role of positive natural selection. Indeed, he gradually came to accept the Lamarckian view that the inheritance of acquired characters also plays some role in evolution.

With the rise of Mendelian genetics in this century, the way was open to elucidate the mechanism of inheritance and the nature of heritable variations. At the beginning there was a temporary disharmony between the "biometricians" who attempted to study natural selection on continuous variations by statistical means, and the "Mendelians" who were preoccupied with the segregation of discontinuous characters (see Provine, 1971).

Eventually, however, the synthesis between Darwinism and Mendelism was achieved by the development of population genetics; it came through the efforts to supply a genetic base to Darwin's theory of natural selection. This was achieved mainly by R. A. Fisher, J.B.S. Haldane, and Sewall Wright, culminating in their definitive writings early in the 1930's. We should add the name of H. J. Muller, who first elucidated the fundamental nature of the gene and its bearing on evolution by natural selection. These four were pre-eminent in their contributions to the "synthetic theory" of evolution.

Upon this foundation, studies of natural populations by T. Dobzhansky and his school, paleontological studies by G. G. Simpson, "ecological genetics" by E. B. Ford and his school, speciation theories of E. Mayr and so on, were added one by one until the synthetic theory had grown into an enormous and impressive edifice. By the early 1960's, consensus seems to have been reached that every biological character can be interpreted in the light of adaptive evolution by natural selection. It was generally agreed, and written that almost no mutant genes are selectively neutral, that is, neither good nor bad for survival and reproduction.

For example in his influential book "*Animal Species and Evolution*," Mayr (1963) claimed " . . . I consider it therefore exceedingly unlikely that any gene will remain selectively neutral for any length of time. . . . Selective neutrality can be excluded almost automatically whenever polymorphism or character clines are found in natural populations." I was greatly influenced by these then prevailing views. If we call the view invoking positive selection to explain all biological characters "panselectionism," this was then a very successful paradigm that still is adhered to by many biologists. All observations and experiments appeared to support it. In retrospect, it is a strange human trait that if a certain doctrine is praised by the majority, endorsed by top authorities in their books and taught in classes, then belief is gradually built up in one's mind, eventually becoming the guiding principle and the basis of value judgement.

At any rate, this was the time when the panselectionist or "neo-Darwinian" position was most secure in the history of biology: the heyday of the traditional "synthetic theory" of evolution. Led by the zeitgeist, a great deal was said on how gene pools of species are organized and how they change in evolution. However, these were inferences based on observations at the phenotypic level, and in reality, there was no way of actually knowing what is going on in evolution at the level of the internal structure of the gene. At that time, much importance was claimed for epistasis (non-additive gene interaction) and overdominance (heterozygote advantage) in fitness. Such terms as *synthetic lethal*, *genetic homeostasis*, *integrated coadapted complex*, *genetic revolution*, *and cohesion of the gene pool* were introduced and accepted. But, there remains a question now as to whether these concepts might not have been more rhetorical than scientific.

Throughout these periods, the mathematical theory of population genetics grew gradually quite sophisticated, which is rather unusual in biology. Particularly noteworthy is the theoretical framework called diffusion models (see Kimura, 1964) which makes use of partial differential equations known as diffusion equations in mathematics. It enables us to treat the behavior of mutant alleles by incorporating random changes due to random sampling of gametes in reproduction as well as deterministic changes due to mutation and selection (for more recent reviews, readers may refer to such books as Crow and Kimura, 1970; Kimura and Ohta, 1971). This powerful method is the out-growth of the great work of Fisher (1930a, b) and Wright (1931, 1945). My own life as a scientist has been largely devoted to the development of diffusion models stimulated by the pioneering work of Wright. Although it involves approximation, the diffusion equation method gives answers to many important but difficult problems which are inaccessible to other methods, such as the problem "what is the probability of a single mutant with a given selective advantage and a given degree of dominance which appears in a finite sized population eventually spreading throughout the whole population?" This probability is often called the *probability of gene fixation* in population genetics.

Despite such a development, its applicability to evolution was quite limited until the era of molecular biology, because population genetics theory is built on the concept of gene frequencies, that is, the proportion of various allelic genes in the population. Conventional studies of evolution are conducted at the phenotypic level, and there is no direct way of unambiguously connecting the two.

Such limitations were removed with the advent of molecular genetics, for evolution could be studied at the molecular level. By comparing proteins or RNA molecules, the direct products of genes, we can investigate the genetic differences between two distantly related organisms. This enables us to estimate the rate by which genes are substituted in evolution.

Furthermore, beginning in the middle 1960's, abundant data have become available regarding genic variability within species at many enzyme loci at the level of electrophoretic mobility. So, at last, the time arrived to apply the mathematical theory of population genetics to find out how genes evolve. It was generally expected then that the principle of Darwinian selection or the survival of the fittest would be clearly shown to prevail at this most fundamental level.

PROPOSAL OF THE NEUTRAL THEORY

The picture of evolutionary change that actually emerged from molecular studies, however, seemed to be quite incompatible with the expectations of neo-Darwinism. Among the salient features that were soon disclosed were the approximate uniformity of the rate of amino acid substitutions per year for each protein among diverse lineages, an apparent randomness in the pattern of substitutions, and a very high overall rate amounting to at least one mutant nucleotide substitution per genome every 2 years when extrapolated to the haploid DNA material of mammals. In fact, the estimate which I obtained for the rate of mutant substitution per genome per generation turned out to be several hundred times higher for mammals than the famous estimate obtained by Haldane several years earlier (Haldane, 1957). His estimate was based on his concept called "cost of natural selection." He considered the situation in which an original gene becomes disadvantageous due to change of environment, while a mutant allele that was originally less fit becomes advantageous. He then computed the amount of selective death (or "cost") needed to carry out gene substitution in the species, and arrived at his estimate that, in the standard-rate evolution, new genes might be substituted in the species at the average rate of once every 300 generations. Since then, Haldane's cost principle has been criticised by various authors (for example, Maynard Smith, 1968) largely on the ground that if advantageous mutants arise constantly without deterioration of the environment, then the cost envisaged by Haldane does not exist, particularly if truncation selection is assumed. I think that such a criticism, even if logically valid, is biologically inappropriate. It is much more realistic to assume that the occurrence of adaptive mutants and their rate of substitution gradually diminish in a stable environment, and that most progressive evolution is generally in response to environmental change.

Turning now to the problem of extant genetic variability within species, the new observations were also remarkable, and once again seemed to be in conflict with conventional neo-Darwinism. Namely, new application of electrophoretic methods suddenly disclosed a wealth of genetic variability, showing molecular polymorphism at a large fraction of loci in diverse organisms. However, these polymorphisms were accompanyed by no visible phenotypic effects and no obvious correlation with environmental conditions. If this variability were generally maintained by overdominance at individual loci, it seemed that the total amount of genetic load, or selective deaths involved, must be intolerably large for higher organisms.

These fascinating but puzzling observations suggested to me during 1967 that a majority of nucleotide substitutions in the course of evolution must be the result of random fixation of selectively neutral or nearly neutral mutants rather than positive Darwinian selection, and that many of the enzyme polymorphisms are selectively neutral and maintained by the balance between mutational input and random extinction. I presented this thesis at the Genetics Club Meeting in Fukuoka in November 1967, and also submitted a short paper to *Nature* in December, which was published in February of the next year (Kimura, 1968a). My theoretical finding, that the rate of neutral mutant substitutions in evolution is independent of population size and equal to the neutral mutation rate, was particularly useful for me to consider problems of molecular evolution.

Then, in 1969 very strong support for my thesis, particularly with respect to molecular evolution, came with the publication of the paper entitled "Non-Darwinian Evolution" written jointly by J. L. King, now at the University of California, Santa Barbara, and T. H. Jukes of the University of California, Berkeley (see King and Jukes, 1969). These authors arrived at the same idea as mine independently and corroborated their claim with a wealth of cogent data from molecular biology. The only difference is that whereas King and Jukes (1969) paid little attention to protein polymorphisms, my contribution to the neutral theory has been equally concerned with the phenomenon of protein polymorphisms.

These papers immediately met strong opposition and severe criticism from those who believed that the new molecular observations could be understood by the orthodox neo-Darwinian view. This led to a controversy (often called "neutralist-selectionist controversy") which has grown a great deal since then. The pros and cons of the neutral theory have been debated in various scientific meetings and publications. Although the debate is still continuing, the neutral theory has survived over a decade, and I believe that supporting evidence has increased, particularly by the recent finding that nucleotide changes within protein-coding DNA regions that cause no amino acid changes (called "synonymous" changes), occur at much higher rates in evolution than those which lead to amino acid changes (Kimura, 1977; Jukes, 1978b; Salser, 1978; Jukes and King, 1979; Nichols and Yanofsky, 1979). Since natural selection acts through phenotype of the organism for which the structure and function of proteins play a decisive role, one should expect that the mutations that do not cause amino acid changes in proteins, other things being equal, are much less subject to natural selection than those which cause amino acid changes. Indeed, undeniable facts have emerged in the last few years suggesting that synonymous changes and "silent" nucleotide changes in the non-coding regions that do not participate in protein formation in DNA are the most prevalent evolutionary changes at the molecular level. It has been discovered very recently that most genes of eukaryotes (higher organisms with a true nucleus) contain regions called "introns" (intervening sequences) which are not included in the mature messenger RNA and therefore do not participate in protein formation. Already there is evidence that evolutionary nucleotide substitutions are very rapid in introns (van Ooyen *et al.*, 1979; Miyata and Yasunaga, 1980; Kimura, 1980).

As I shall examine in more detail later, the rule is that *those molecular changes that are less likely to be subjected to natural selection occur more rapidly in evolution*. I think that King and Jukes had remarkable insight when they wrote in 1969: "if DNA divergence in evolution includes the random fixation of neutral mutations, then the third-position nucleotides should change more rapidly," reasoning that most nucleotide changes at the third position of the codon are synonymous and therefore they are more likely to be neutral, while a majority of changes that produce amino acid substitutions are likely to be harmful and will be weeded out by natural selection. This prediction has now been fully substantiated. Similarly, in one of my papers on the neutral theory (Kimura, 1968b), I wrote: "the recent findings of 'degeneracy' of DNA code, that is, existence of two or more base triplets coding for the same amino acid, seem to suggest that neutral mutations may not be as rare as previously considered." However, I also added a note of caution: "it is important to note that probably not all synonymous mutations are neutral, even if most of them are nearly so." Perhaps, this was a pertinent statement in the light of the recent finding that synonymous codons are often used in "non-random" or unequal fashion (see Kimura, 1981d).

THE NEUTRAL THEORY AS AN EVOLUTIONARY PARADIGM

To explain more fully what the neutral theory (or more precisely, the neutral mutation-random drift hypothesis) purports, and how it differs from the traditional neo-Darwinian (or more precisely, "synthetic") theory, consider an evolutionary process in which mutant genes are substituted one after another within the species. Each such substitution is made up of a sequence of events in which a rare mutant form which appeared, usually singly represented, in the population, finally spreads through the whole population reaching fixation, *i.e.* frequency of 100% (Fig. 1). The neutral theory is concerned with the mechanism by which a mutant form of the gene ("mutant allele") spreads through the population. Nowhere is the contrast between the claims of neutralists and selectionists more clearly expressed than in their interpretations of such spreading of mutants.

The selectionists claim that for a mutant allele to spread through the species it must have some selective advantage, although they admit that occasionally a neutral mutant may be carried along by "hitchhiking" to reach a high frequency. On the other hand, the neutralists claim that mutants do not need a selective advantage for some of them to spread through the population. If mutants are selectively equivalent to the pre-existing forms, their fate is left to chance and their frequencies increase or decrease fortuitously as time goes on. Such a fluctuation arises because of the finite size of populations (Fig. 2); in each generation a relatively small number of gametes are sampled out of a vast number of male and female gametes in order to produce individuals of the next generation.

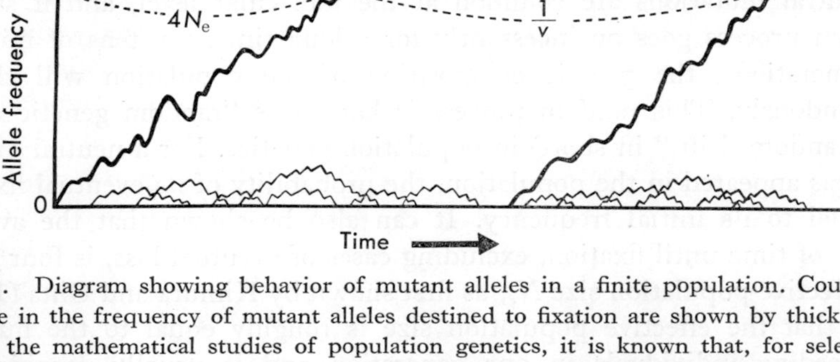

Fig. 1. Diagram showing behavior of mutant alleles in a finite population. Courses of change in the frequency of mutant alleles destined to fixation are shown by thick paths. From the mathematical studies of population genetics, it is known that, for selectively neutral mutations, the average time until fixation of a mutant (excluding the cases in which mutants are lost) is $4N_e$ generations, and the average interval of consecutive fixations is $1/v$, where v is the mutation rate and N_e is the effective population size.

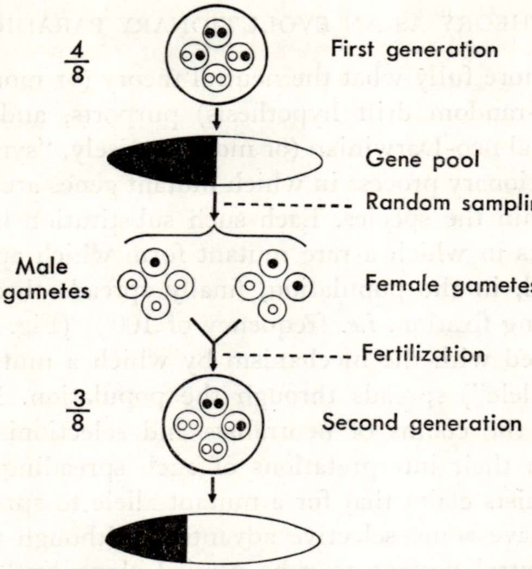

Fig. 2. Diagram illustrating the process of random gene frequency drift, assuming a hypothetical population consisting of four individuals. Each individual has two homologous genes (represented either by a black or white ball) inherited from male and female parents. In the first generation, the frequency of the "black" allele is 4/8 or 50%. In this example, due to random sampling of gametes in reproduction, only one "black" allele happened to be included in the pool of male gametes so that the frequency of the "black" allele changes to 3/8 in the second generation.

Although the overwhelming majority of such mutants are lost by chance, the remaining minority will eventually become fixed in the population. If neutral mutations are common at the molecular level, and if such a random process goes on incessantly for a long time, say tens of millions of generations, the genetic composition of the population will change tremendously. This random process is known as "random genetic drift" (or "random drift," in short) in population genetics. For a neutral mutant that has appeared in the population, the probability of its eventual fixation is equal to its initial frequency. It can also be shown that the average length of time until fixation, excluding cases of eventual loss, is four times the effective population size N_e, as first shown by Kimura and Ohta (1969). Note that the effective population size is roughly equal to the number of breeding individuals in one generation and is usually considerably smaller than the total number of individuals making up the species (for details, see, for example, Kimura and Ohta, 1971, pp. 33–43).

Many of the criticisms that have been directed to the neutral theory have been largely based on misunderstandings. The neutral theory by no means claim that the genes involved are functionless or that most mutations are neutral. What the neutral theory assumes is that mutant forms of each gene participating in molecular evolution may be selectively equivalent, *i.e.*, they can do the job equally well in terms of survival and reproduction of the individual. And these neutral mutations, though they may be a minority of all mutations, constitute the vast majority of the changes that actually occur in evolution and in the creation of genetic polymorphisms. Sometimes, neutral changes are referred to as evolutionary "noise" by anti-neutralists (see, for example, Mayr, 1978), but I think this is a misnomer. Just as synonyms cannot be "noise" in language, it is not proper to regard the substitution of neutral alleles simply as noise or loss of genetic information. If the variants represent amino acid changes in a protein, this means that such changes are equally acceptable for the working of the protein in the body. Furthermore this equality need not be perfect. All one needs is that the resulting difference in fitness is much smaller, such as, less than 10% of the reciprocal of the effective population size (see Fig. 4 to be explained later). Here I call attention to the fact that in organisms, particularly in higher forms, physiological homeostasis is well developed against internal and external environmental disturbances. Thus, fluctuation of environmental conditions by no means automatically implies fluctuation of Darwinian fitness of mutants; a very important point often overlooked by mathematical population geneticists.

Some criticisms of the neutral theory come from a wrong definition of natural selection. Here I must emphasize that the term "natural selection" should be used strictly in the Darwinian sense; it acts through the differential survival and reproduction of the individual. Very often, existence of detectable functional differences between two molecular forms —for example, a shift in the rate constant of an enzyme— is taken as evidence for the existence of natural selection. To prove that natural selection is involved, survival rates and fecundity must be investigated. It is unfortunate that the term "selection" is often used in quite a meaningless way by biochemists. For example, Zuckerkandl (1978) states that "Positive selection does not *per se* confer biological significance upon a mutant." I also think that a clear distinction should be made between positive and negative selection. The latter, which is concerned with the elimination of deleterious mutants, unless the deleterious effect is extremely small, has little to do with gene substitution in evolution. It is known, since the great work of Muller in the early days of *Drosophila* genetics, that negative selection is the most common form of natural selection (see Muller, 1962). "Stabilizing" or "centripetal" selection reflects negative selection, whereas "directional" selection usually involves positive selection. The existence of negative selection does not contradict the neutral theory. Finally, the distinction between gene mutations at the individual level and substitution of mutant forms at the population level must be made clear. Only the substitutions of the mutant forms in the population are directly related to molecular evolution. For advantageous mutants, the rate of evolutionary mutant substitutions is greatly influenced by the population size and the degree of selective advantage (as I shall show later), as well as the mutation rate.

Recently, some authors, such as Zuckerkandl (1976) and Gillespie (1977) claim that they produced "selectionist" theories whose consequences are indistinguishable from the neutral theory. As far as I know, these claims are all based on mistaken theories; some (such as Zuckerkandl) forget the contribution of the population size to the rate of gene substitution by advantageous mutants, others (such as Gillespie) intentionally omit the contribution of mutations. In short, they assume (unnecessarily, I think) that the natural selection acts in such a way that its consequences are indistinguishable from no selection.

TWO FEATURES OF MOLECULAR EVOLUTION

Two of the features revealed by recent studies of molecular evolution are particularly noteworthy, and help to show that the patterns of molecular evolution are quite different from those of phenotypic evolution, and that the laws governing them are also different. (i) For each protein, the rate of evolution in terms of amino acid substitutions is approximately constant per year for various lineages. (ii) Molecules or parts of one molecule that are subject to less functional constraint evolve faster (in terms of mutant substitutions) than those that are subject to stronger constraint. As to the first feature, the constancy (or uniformity) of the evolutionary rate is apparent in hemoglobin. The hemoglobin molecule in bony fishes and higher vertebrates is a tetramer, consisting of two identical α chains and two identical β chains. In mammals, amino acid substitutions in the α chain, consisting of 141 amino acids occur roughly at the rate of one substitution per 7 million years. This corresponds to about 10^{-9} substitution per year per amino acid site, and, surprisingly, this does not seem to depend on such

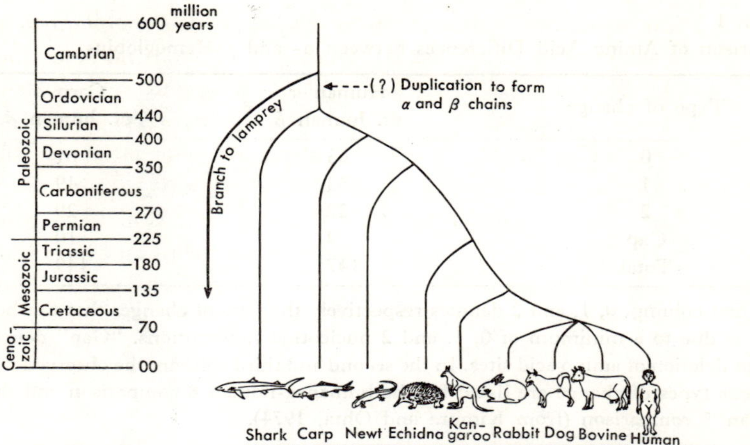

Amino acid differences α chain

	Shark	Carp	Newt	Echidna	Kangaroo	Rabbit	Dog	Bovine	Human
Shark		85	84	84	80	75	80	75	79
Carp			74	75	71	71	67	65	68
Newt				71	67	69	65	64	62
Echidna					49	49	42	43	37
Kangaroo						37	33	26	27
Rabbit							28	25	25
Dog								28	23
Bovine									17
Human									

Fig. 3. A phylogenetic tree of several vertebrates together with their times of divergence. It also shows the numbers of amino acid differences between the hemoglobin α chains of each pair of vertebrates. Note that the numbers of differences are around 20 among the four mammals, while it is about 70 when the carp is compared with any one of the four mammals.

factors as generation time, living conditions and population size. In Fig. 3 the number of amino acid differences between the α chains of several vertebrates including man are given together with their phylogenetic tree and a list of geological epochs to show the times of their divergence.

Uniformity of the evolutionary rate is shown by the divergence between the α and β chains (Table I). The α and β chains have essentially the same tertiary structure, are of about equal length, and show roughly the same evolutionary rate of amino acid substitution. They are derived by gene duplication (possibly at the Ordovician period) followed by differentiation through accumulation of independent mutations. In Table I the divergence between the α and β chains of man is compared with the divergence between the α chain of the carp and β chain of man. Since the human and carp α chains differ from each other at roughly 50% of the

TABLE I

Comparison of Amino Acid Differences between α- and β-Hemoglobins

Type of change	Human α *vs.* human β	Carp α *vs.* human β
0	63	61
1	53	49
2	22	29
Gap	9	10
Total	147	149

In the first column, 0, 1, and 2 denotes respectively the type of changes that can be interpreted as due to a minimum of 0, 1, and 2 nucleotide substitutions. "Gap" denotes addition or deletion of amino acid sites. In the second and third column, the observed numbers of various types of change are listed for the human α-human β comparison and the carp α-human β comparison (from Kimura and Ohta, 1974).

amino acid sites, the table suggests that α chains in two separate lineages, one leading to the carp and the other leading to man, have accumulated mutations independently and at practically the same rate over the span of more than 400 million years. In addition, the rates of amino acid substitutions from these comparisons are very similar to the ones obtained by comparing α chains among mammals.

I pointed out (Kimura, 1969) that if hemoglobins and other molecules of "living fossils" are shown to have undergone as many DNA base (and therefore amino acid) substitutions as corresponding genes (protein) in more rapidly evolving species, this would support my neutral theory. Recently, the amino acid sequences of the β chain and the α chain of the principal hemoglobin of the Port Jackson shark have been determined (Fisher *et al.*, 1977), and the data enables us to test my prediction. According to Romer (1968), this shark is a relict survivor of a type of ancestral shark which had numerous representatives in the late Paleozoic days, notably in Carboniferous period (270–350 million years ago). Therefore, this shark is well entitled to be called a living fossil. We can investigate the divergence between the α and β chains of the Port Jackson shark, and this can be compared with the divergence between the α and β chains of man, similar to what we have done in Table I. The result shows clearly that genes coding for the α and β chains in the shark have diverged to roughly the same extent (or slightly more) as have the corresponding two genes in man by accumulating random mutations since the origin of the α- and β-globin genes by duplication.

My claim of the constancy of the evolutionary rate at the molecular level has been criticized by various authors. Lewontin (1974) in criticizing it, declares, in his book "*The Genetic Basis of Evolutionary Change*" (see p. 229) that "the claimed 'constancy' is simply a confusion between an average and constant." Elsewhere (Stebbins and Lewontin, 1972) it is stated that "the so-called 'constancy' of the rates over millions of years is nothing but the law of large numbers." These statements show a misunderstanding of the nature of molecular evolution. For example, in the hemoglobin α chain which shows a standard rate of molecular evolution, only one amino acid substitution occurs on the average every 7 million years along a lineage! Such events are rare; and entirely different from measuring the temperature each day in Chicago as mentioned by Lewontin (1974). To get a meaningful estimate of the intrinsic evolutionary rate we must observe at least several tens of million of years, preferably much longer. The situation is somewhat similar to accident statistics in actuarial science. In such a situation, "the law of small numbers" (the Poisson law) rather than the law of large numbers is appropriate. We are trying to compare the intrinsic rates of evolution in different lineages. Just as death rates do not become equal between man and an insect by merely averaging over a long period or a large number of individuals, there is no reason to expect that two averages converge toward each other unless the intrinsic parameters in them are the same. My point is that the intrinsic evolutionary rates are essentially determined by the structure and function of the molecules and not by environmental conditions.

The constancy of the evolutionary rate, however, is not as uniform as the rate of radioactive decay. In fact, as shown by Ohta and Kimura (1971a), the observed variance (squared standard deviation) of the evolutionary rates for hemoglobins and cytochrome *c* among mammalian lines is roughly 1.5 to 2.5 times as large as the variance theoretically expected, if the variation is due to purely by chance. Essentially the same result was obtained later by Langley and Fitch (1974) who carried out more elaborate statistical analyses. Using simultaneously the data for α- and β-hemoglobins, cytochrome *c* and fibrinopeptide A, they found that variation of evolutionary rates in terms of mutant substitutions among branches of a phylogenetic tree is about 2.5 times as large as that expected from chance fluctuations. They take this as evidence against the neutral theory. At the same time they show that when the estimated numbers of mutant substitutions are plotted against the corresponding geological dates (in years), they are on a straight line suggesting the uniformity of the evolutionary rates! The only exceptions are three observed points for primates for which there are other uncertainties. In my opinion, emphasizing local fluctuations as evidence against the neutral theory, while neglecting to inquire why the rate is intrinsically so regular or constant is misguided. The constancy of evolutionary rates per year at the molecular level has also been shown by Wilson and his group, using an immunological method involving albumin over a wide range of vertebrate species (for review, see Wilson *et al.*, 1977).

The near-constancy of the amino acid substitution rate in evolution was termed "a molecular evolutionary clock" by Zuckerkandl and Pauling (1965), and has since been one of the most controversial subjects in molecular evolutionary studies. Particularly, it has been debated in relation to the date of human-ape divergence (Wilson *et al.*, 1977), and as evidence for and against the neutral theory (as I have discussed above).

A strong claim of non-constancy has been made by Goodman and his associates in their study of globin evolution (Goodman 1976, 1978; Goodman *et al.*, 1974, 1975). They maintain that mutant substitutions occurred at a very high rate in the early stage of globin evolution, soon after duplication to form myoglobin and α- and β-hemoglobins, and that this was followed by a markedly reduced rate during the last 300 million years from the ancestral amniote to the present. According to Goodman *et al.* (1975), from 500 to 400 million years ago the genes descending from "the basal vertebrate ancestor through the hemoglobin-myoglobin and β–α ancestors to the teleost-tetrapod α ancestor evolved at the rate of 109 NR %," where NR % means the number of nucleotide replacements per 100 codons per 10^8 years. This corresponds to $k_{aa}=10.9\times10^{-9}$ per amino acid per year in my terminology (Kimura, 1969), and it is some 10 times as high as my estimate of about 10^{-9} per amino acid site per year. On the other hand, for the last 300 million years, their estimated rate is only 15 NR % (*i.e.* $k_{aa}=1.5\times10^{-9}$) which is not very different from mine.

Together with their plausible explanation that when new functions emerged in duplicated globin genes rapid mutant substitutions occurred by *positive* Darwinian selection, Goodman *et al.*'s (1975) claim of non-constancy has been quoted widely as evidence against the neutral theory (see, for example, Vogel and Motulsky, 1979). Referring to the discepancy between the evolutionary rates estimated by them and others, Goodman (1976 and elsewhere) writes that both Dickerson (1971) and I (Kimura, 1969) failed to account sufficiently for superimposed mutations in our calculations of distances between contemporary globin sequences, and that therefore we more grossly underestimated the evolutionary distances between the more anciently separated globins.

The statistical method used by Goodman's group is detailed in Goodman *et al.* (1974). They make use of an extensive series of computer programs, termed MMUTD, UWPGM, PSLNG, PSITR, PSANC, *etc.* to assign mutations to various branches of the phylogenetic tree, to determine ancestral sequences at evolutionary forks, *etc.* These are all based on the principle of "maximum parsimony." They determine the geological dates of the branch points from paleontological information and from their own opinions. They correct for undetected mutant substitutions by using their "augmentation" method (considered to be particularly important for unpopulated branches in the "parsimony tree").

Recently I pointed out (Kimura, 1981b) that the relatively high evolutionary rates obtained by Goodman *et al.* (1974, 1975) are not the result of their claimed superior statistical method, but rather of the wrong assignment of geological dates to duplication events in the early history of globin evolution, and that when this is corrected, there is no basis for their claim. I also pointed out that the great majority of globin codons determined by Goodman *et al.* (1974) using their maximum parsimony method are wrong when compared to the actual nucleotide sequences of globins determined by direct sequencing.

If we examine carefully the data used by Goodman *et al.* (1974, 1975), we find that their estimate of the high evolutionary rate (109 NR %) comes almost entirely from their assumption that the gene duplication leading to lamprey globin on the one hand and, on the other, to the ancestral globin that later produced myoglobin and, still later, the α- and β-globins, occurred 500 million years ago (for details, see Kimura, 1981b). I believe that their positioning of gene-duplication points in time contains some errors, of which the most serious is their assumption that the duplication responsible for the myoglobin *versus* hemoglobin divergence occurred about 470 million years ago, after the ancestor of jawed fishes diverged from the ancestor of the lamprey. In other words, they assume that the history of myoglobin is younger than that of the jawless fish (agnatha).

It is much more likely that the gene duplication that led to the lamprey (*Petromyzon marinus*) globin whose sequence was used in their analysis and the ancestral molecule from which myoglobin and α–β globins have descended is much older than they postulate. It is likely that the duplicated globin genes already existed in the ancestral agnatha when they appeared about 500 million years ago. A similar point has been made by Zuckerkandl (1976). According to Hendrickson and Love (1971) the lamprey globin that has been sequenced is only one of six hemoglobins present in *P. marinus* blood; it is hemoglobin V, the most prevalent of the six. This suggests that gene duplication occurs rather frequently in evolution. Much more important is the recent finding by Romero-Herrera *et al.* (1979) that true myoglobin exists in the red muscle of the lamprey heart. Although the amino acid sequence is yet to be determined, this finding completely invalidates Goodman *et al.*'s assumption that myoglobin originated only some 470 million years ago, *after* the jawed-fish diverged from the agnatha, and that therefore myoglobin is newer than hemoglobin V of the lamprey. In this respect, Hunt *et al.* (1978) were more judicious in placing the lamprey globin branch on the opposite side of the phylogenetic tree from the other vertebrate globins, stating that the gene duplication responsible for myoglobin and lamprey globin must have occurred at a very ancient time. In Hunt *et al.*'s figure (see p. 229 of Dayhoff's *Atlas*, 1978) this duplication point is depicted as being one billion or so years old. I would like to emphasize that if the gene duplication that led to myoglobin and lamprey hemoglobin V occurred earlier than postulated by Goodman *et al.*, for example, late in the Pre-Cambrian, some 800 million years ago, the evidence for a very high evolutionary rate that Goodman *et al.* find breaks down completely. In addition, the fallacy of maximum parsimony codons has been revealed in a dramatic way by a recent study of Holmquist (1979). Previously, Tateno and Nei (1978) used a computer to perform Monte Carlo experiments of molecular evolution, simulating the process of divergence of nucleotide sequences by randomly accumulating mutational changes (each hypothetical sequence consisted of 300 nucleotides, corresponding to 100 codons, the first half of which were assumed to be variable and the remaining half constant). Starting from a common ancestor, phylogenetic trees involving 21 contemporary sequences were produced. What is important is that, in these simulation experiments, unlike the evolutionary processes in nature, the exact topology, the numbers of accumulated nucleotide substitutions in each branch of the phylogenetic tree, and also, all the ancestral sequences at the branch points are known. Holmquist obtained the data for these simulation experiments from Nei and Tateno, and sent the 21 contemporary sequences (after being translated into protein sequences) to Goodman asking him to obtain the maximum parsimony solution including both the topology and reconstructed ancestral amino acid sequences (this was called the "inferred topology" solution). Holmquist also asked Goodman to provide him with the ancestral amino acid sequence reconstructions, given the correct topology (the solution thus obtained was designated the "known topology" solution). These amino acid sequences obtained by Goodman were then compared with the correct ancestral sequences. It turned out that there were numerous errors in the amino acid sequences inferred by the method of maximum parsimony. Particularly noteworthy was the high rate at which errors accumulated as the distances of nodal sequences from the contemporary sequences increased. For nodes 200 or more replacements distant, the error was more than 80%. Actually, the errors are so numerous in remote ancestral reconstructions, even when the correct topology is supplied, that the reconstructed sequences are totally worthless (they even contain chain terminating codons). It is clear now that any discussion which involves early stages of globin evolution based on maximum parsimony reconstruction is meaningless.

Furthermore, the "augmentation" procedure that has been used extensively by Goodman and his associates contains various problems as pointed out by Tateno and Nei (1978) and by Kimura (1981c).

All these results show clearly that Goodman *et al.*'s (1974, 1975) work is fraught with errors and uncertainty, and contrary to their claim, there is no evidence that the evolutionary amino acid substitutions were very rapid in the early stage of vertebrate globin evolution. Their claim that the high evolutionary rate is due to positive Darwinian evolution, although plausible sounding, has no real supporting case in molecular evolution. It is much more likely, as we shall discuss in more detail later that the high rate is usually caused by removal of preexisting functional constraint, allowing a large fraction of mutations to become selectively neutral (not harmful) so that they can become fixed by random drift.

This does not mean of course that no adaptive mutant substitutions have occurred in the course of globin evolution. In my opinion, however, such definitely adaptive substitutions are much less frequent that selectively neutral or nearly neutral substitutions caused by random frequency drift. Therefore, I am quite skeptical of Goodman *et al.*'s facile explanations of various evolutionary amino acid substitutions exclusively in adaptive ("selectionist") terms. Although plausible, these explanations entirely lack quantitative analysis based on population genetics theory, and we must take such explanations with a grain of salt.

POPULATION DYNAMICS OF MUTANT SUBSTITUTIONS

Before we proceed to discuss the second feature of the molecular evolution, I shall examine the population genetical implication of the constancy of the evolutionary rate. For this purpose let us again consider how molecular mutants are successively substituted (Fig. 1). To simplify our argument, we consider the entire set of nucleotides making up the genome (haploid chromosome set). For a human, this number is very large, of the order of 3.5 billion (3.5×10^9). Since the mutation rate per site is low (say, 10^{-8} or less), we may assume that whenever a mutant appears, it represents a new site in which mutant forms are not segregating in the population. This assumption is known as the "infinite site model" in population genetics (Kimura, 1971). Let v be the mutation rate per gamete per unit time, such as per generation. Since each individual has two sets of chromosomes (handed down by male and female parents), the total number of new mutants introduced into the population in each generation is $2Nv$ where N is the number of individuals. Let u be the probability of a single mutant ultimately reaching fixation. Then, in the steady state in which the process of substitution has gone on for a very long time, the rate per unit time of mutant substitution in the population is

$$k = 2Nvu, \qquad (1)$$

because $2Nv$ new mutants appear in each generation of which the fraction u eventually reaching fixation. Thus k represents the rate of evolution in terms of mutant substitutions. Although we considered above the genome as a whole, the results can also be applied with good approximation to a single gene locus consisting of at least several hundred nucleotide sites. We may also apply this to a protein coded by a gene. The probability (u) of ultimate fixation of a mutant in Eq. (1) is a well-known quantity in population genetics. In a simple case in which the heterozygote has fitness intermediate between the two homozygotes,

$$u = \frac{1-e^{-2s}}{1-e^{-4Ns}} \qquad (2)$$

where s is the selection coefficient giving the selective advantage of this mutant over the pre-existing allele. To simplify discussion, we tentatively assume that the effective size (N_e) is equal to the actual size (N). For a selectively neutral mutant, $s=0$ and we get $u=1/(2N)$ which we denote as u_0. If the absolute value of s is small, say 0.05 or less, the ratio u/u_0

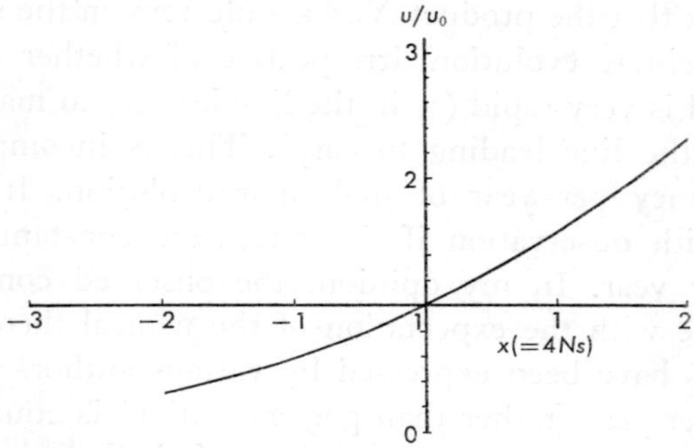

Fig. 4. Relative probability of fixation of a mutant gene in the population as a function of $4Ns$ where N is the population size and s is the coefficient representing its selective advantage or disadvantage as compared with the pre-existing allele.

depends only on $4Ns$, as a good approximation, and this ratio is illustrated in Fig. 4, as a function of $4Ns$ $(=x)$ in the neighborhood of $4Ns=0$. Substituting $u=1/(2N)$ in Eq. (1), we get

$$k = v. \tag{3}$$

Therefore, the rate of evolution (in terms of mutant substitutions in the population) is equal to the mutation rate per gamete independent of population size. This remarkable property is only valid for neutral alleles. However, if a mutant has a much smaller selection coefficient (in absolute value) than $1/N$, then as shown in Fig. 4, u is close to u_0 and Eq. (3) holds approximately. Such a mutant may be called "nearly neutral." On the other hand, if the mutant has a definite selective advantage such that $4Ns \gg 1$, then $u=2s$ with good approximation, so that Eq. (1) gives

$$k = 4Nsv. \tag{4}$$

More generally, if the effective population size (N_e) differs from the actual size (N), it can be shown that

$$k = 4N_e sv. \tag{4a}$$

This means that the rate of evolution depends on the effective population size (N_e), selective advantage (s) as well as the rate at which mutants having such selective advantage (s) are produced each generation. One should expect in this case that the rate of evolution depends strongly on the environment, being high for a species offered a new ecologic opportunity but low for those kept in a stable environment. It is highly unlikely, I think that the product $N_e s v$ should remain the same for diverse lineages of vertebrate evolution, irrespective of whether evolution at the phenotypic level is very rapid (as in the line leading to man) or practically stopped (as in the line leading to carp). This is incompatible with the observed constancy per year of molecular evolution. It would also be incompatible with observation if the rate were constant per generation rather than per year. In my opinion, the observed constancy is much more compatible with the expectation of the neutral theory (see Eq. (3)), although doubts have been expressed by various authors why the neutral mutation rate per year (rather than per generation) is equal among organisms having very different generation times. I shall discuss this problem later using the model of "effective neutral mutations" (Kimura, 1979). Note that the problem does not arise when comparisons are made among organisms with similar generation time but living under very different conditions.

SELECTIVE CONSTRAINT AND THE NEUTRAL SUBSTITUTIONS

The second feature of molecular evolution is even more remarkable than the first which I have discussed. It has become increasingly clear that the weaker the functional constraint which a molecule or a part of one molecule is subject to, the higher the evolutionary rate of mutant substitutions. Among the proteins so far investigated, the highest evolutionary rate is in fibrinopeptides, 8.3×10^{-9} per amino acid site per year (Table II), although a lower estimate 4.5×10^{-9} has also been reported (Barnard *et al.*, 1972). Fibrinopeptides have little known function after they become separated from fibrinogen during blood clotting. A similar example is the C chain of the proinsulin molecule (Fig. 5). The C chain is removed when active insulin is formed. It evolves at a rate several times as fast as that of insulin (for review, see Jukes, 1979). More examples of this sort will be forthcoming because tailoring of a precursor molecule to produce a functionally active form appears to be a common phenomenon. Dickerson (1971) explained the relationship between the functional importance (or more strictly, functional constraint) and the evolutionary rate using four molecules as follows. In fibrinopeptides, virtually any amino acid change (mutant substitution) that permits the peptides to be removed is "acceptable" to the species. Thus, the rate of evolutionary substitution of amino acids may be very near the actual mutation rate. Hemoglobins, because they have a definite function of carrying oxygen and, therefore specifica-

TABLE II
The Rate of Evolution in Terms of Amino Acid Substitution is Listed for Several Proteins

Proteins	Evolutionary rate
Fibrinopeptides	8.3
Pancreatic ribonuclease	2.1
Lysozyme	2.0
α-Hemoglobin	1.1
Myoglobin	1.0
Insulin	0.44
Cytochrome *c*	0.23
Histone H4	0.008

They are expressed taking 10^{-9} per year per amino acid site as the unit of evolutionary rate. These values are either quoted from or computed by using data compiled by Dayhoff (1978).

tions for them are more restrictive than for fibrinopeptides, have a lower evolutionary rate. Cytochrome *c* interacts with cytochrome oxidase and reductase, both of which are much larger than it, and there is more functional constraint in cytochrome *c* than in hemoglobins. Thus cytochrome *c* has a lower evolutionary rate than hemoglobins. Histone H4 binds to

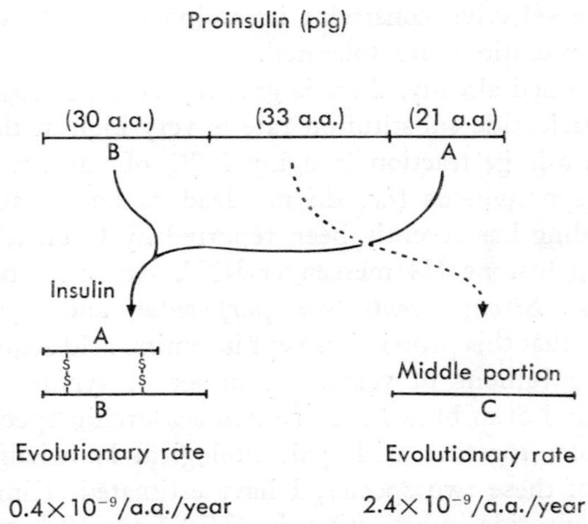

Fig. 5. When the active insulin molecule is formed from its precursor molecule, proinsulin, the middle segment (C peptide) is cut off and discarded. The rate of evolution is several times higher in the C peptide than in A and B which by joining through disulfide bridges form the insulin.

DNA in the nucleus, and is believed to control the expression of genetic information. It is quite probable that a protein so close to the genetic information storage system is highly specified with little evolutionary change over a billion years.

All these examples strongly suggest that natural selection (mostly in the form of negative selection) acts through the function of the folded protein but not the amino acid sequence itself. A similar relationship between functional constraint and evolutionary rate also holds in different parts within hemoglobin molecules. It is known (Perutz and Lehman, 1968) that in hemoglobins the surface part generally has much fewer constraints in maintaining the structure and function of the molecules than the heme pocket which is vitally important. We have shown (Kimura and Ohta, 1973) that in both α- and β-hemoglobins the surface part evolves about ten times as fast as the heme pocket. An additional example is afforded by the report of Barnard *et al.* (1972). According to them, sequence 15–24 of pancreatic ribonucleases evolves at a very high rate comparable to rapidly evolving parts of fibrinopeptides, and this "hypervariability" can be correlated with a lack of any contribution of this part either to the enzymatic activity or to the maintenance of structure required for the activity. Incidentally, their Table III listing frequencies of amino acids in hypervariable segments suggests that in such regions there might still exist some selective constraint in amino acid substitutions, so that not all of the mutations are tolerated.

As I mentioned already, there is growing evidence suggesting that the evolutionary nucleotide substitution rate is very high at the codon's third position where a large fraction (roughly 70%) of random nucleotide substitutions are synonymous (*i.e.* do not lead to amino acid changes). A remarkable finding has recently been reported by Grunstein *et al.* (1976). They compared histone H4 messenger RNA sequences between two sea urchin species, *Strongylocentrotuns purpuratus* and *Lytechinus pictus*. Despite the fact that this protein has kept its amino acid sequence practically unchanged over billions of years, a number of synonymous nucleotide changes are found even between these two sea urchin species (see Fig. 6). Using their data, together with paleontological knowledge on the time of divergence of these two species, I have estimated (Kimura, 1977) that the rate of nucleotide substitution is $(3.7 \pm 1.4) \times 10^{-9}$ per year at the third position of the codon. This is a very high rate, since the amino acid substitution rate of fibrinopeptides when converted into the corresponding nucleotide substitution rate, amount to roughly 4.0×10^{-9}. It is remarkable

L. pictus	Messenger RNA	GA U	AAC	AUC	CAA	GG A	AU A	AC U	AAA	CCG	GC A	AUC
S. purpuratus	Messenger RNA	GA C	AAC	AUC	CAA	GG U	AU C	AC G	?	?	GC U	AUC
Histone H4	Amino acid sequence	Asp	Asn	Ile	Gln	Gly	Ile	Thr	Lys	Pro	Ala	Ile
		24	25	26	27	28	29	30	31	32	33	34

Fig. 6. A part of the comparison between histone H4 messenger RNAs of two sea urchin species *Lytechinus pictus* and *Strongylocentrotus purpuratus*. Although the amino acid sequences of histone H4 of these two species are identical, there are five "synonymous" differences at the third positions of the codon (as indicated by small boxes), even in this short stretch of the messenger RNA (data from Grunstein *et al.*, 1976).

that synonymous mutant substitutions have occurred in this gene almost at the highest known rate despite the fact that amino acid substitutions have occurred in the corresponding protein with the lowest known rate. (I shall treat the problem of estimating the evolutionary rates of base substitutions through comparative studies of nucleotide sequences in the next section.)

The above observations can be explained very simply and consistently by the neutral theory. Suppose that a certain fraction f_0 of the molecular mutants are selectively neutral and that the rest of the mutants are definitely deleterious. Although advantageous mutations will no doubt occur from time to time, we assume that their frequency is negligible in determining the evolutionary rates as estimated by comparative studies of amino acid and nucleotide sequences. If we denote by v_T the total mutation rate, then the mutation rate for neutral alleles is $v=v_T f_0$, so that from Eq. (3) we obtain

$$k = v_T f_0. \tag{5}$$

We now assume that the probability of a mutational change being neutral (*i.e.* not harmful) depends strongly on functional constraints. The weaker the functional constraint, the larger the probability f_0 of a random change being neutral, with the result that k in the above equation increases. According to this explanation, the maximum evolutionary rate is attained when $f_0=1$, that is, when all the mutations are neutral (assuming that v_T per site is the same among different molecules). In my opinion the high evolutionary rates observed at the third position of the codon are rather near this limit. It is sometimes claimed that the amino acid composition of fibrinopeptides do not coincide with the ideal random protein and therefore mutant substitutions involved are unlikely to be selectively neutral. It is possible that there still exist restrictions with regard to which amino acids are used for evolutionary substitution, but, in my opinion the actual restriction (in terms of reducing f_0) may not be as large as it appears. First, the size of the sample examined is very small so that statistical deviation must be very large. Secondly, the genetic code is connected in the sense that a single nucleotide replacement usually leads to similar amino acids so that even if the set of allowable amino acids is somewhat restricted there still remains much room for mutational interchange of nucleotides within the allowable limits. It seems to me that in the case of fibrinopeptide A, the actual value of f_0 might be reasonably high.

Recently, frequencies of four nucleotide bases, A, U, G, and C, at the third position of the codon have been compiled by Jukes (1978a) for a few mammalian messenger RNAs including that of rabbit β-hemoglobin. So far, a marked excess of G and C over A and U is shown to exist among them. This is reflected by the non-equality of frequencies of synonymous codons. For example, GUG predominates over the other three codons (GUU, GUC, and GUA) all of which can code for Valine. Such a deviation from equality (usually referred to as "non-randomness" of codon usage) has started to be mentioned as evidence against the neutral theory (see, for example, Grantham *et al.*, 1980). It is possible that such a deviation from equality is due to negative selection or due to difference of mutation rates among four nucleotide bases. But such inequality *per se* does not prove that mutant substitutions were caused by positive Darwinian selection. For a neutralist solution to the problem of non-random codon usage, see Kimura (1981d).

At any rate, the neutral theory predicts that as functional constraint diminishes the rate of evolution converges to the maximum value set by the total mutation rate (Kimura and Ohta, 1973; Kimura, 1977). If such a convergence (or plateauing) of molecular evolutionary rates is confirmed by future studies, it will turn out to be very strong evidence supporting the neutral theory.

All these interpretations will not make sense to selectionists. In their view the rapidly evolving molecules or parts of one molecule (in terms of mutant substitutions) have some important but unknown functions, and are undergoing rapid adaptive improvements by accumulating beneficial mutants. Also, there will be no reason to believe that the upper limit of evolutionary rate is related to the total mutation rate.

ESTIMATION OF EVOLUTIONARY DISTANCES THROUGH COMPARATIVE STUDIES OF NUCLEOTIDE SEQUENCES

Although comparative studies of protein sequences have been a major tool for the studies of molecular evolution during the last 15 years or so, we are now experiencing a new change in the type of data being obtained, with outburst of reports on nucleotide sequences. This is a result of the revolutionary technology of amplifying gene copies in bacterial plasmids and then rapidly sequencing the DNA by Maxam-Gilbert method (Maxam and Gilbert, 1977). It is likely that comparative studies of amino acid sequences will soon be superseded by those of nucleotide sequences. Already, it has become evident that the preponderance of synonymous and other silent base substitutions is a general but remarkable feature of molecular evolution and that this is consistent with the neutral theory (Kimura, 1977, 1980, 1981a; Jukes, 1980a, b; Miyata *et al.*, 1980, Miyata and Yasunaga, 1981).

In estimating the evolutionary distances between homologous sequences in terms of the number of base substitutions, corrections for multiple and revertant changes at homologous sites are essential. This is because there are only four kinds of bases in nucleotide sequences and even two random sequences show 25% match on the average at individual sites. Therefore, we need suitable statistical methods which incorporate such corrections in the estimation of evolutionary distances (and, also, the evolutionary rate when the divergence time is known).

Let us consider a homologous site in two nucleotide sequences compared (denoted by sequences 1 and 2). In the following I use RNA (rather than DNA) codes so that the four bases are expressed by letters U, C, A, and G. This fascilitates translating back and forth between codon (triplet) and amino acid, when needed, using a standard code table. We investigate how the homologous sites differentiate from each other in the course of evolution starting from a common ancestor T years back (Fig. 7). At individual sites, bases are substituted one after another in the course of evolution. To formulate this process, we assume a model of evolutionary base substitutions as shown in Fig. 8. In this figure, α is the rate of transition type substitutions, and 2β is that of transversion type substitutions ("transitions" are substitutions of purines by purines, and pyrimidines by pyrimidines, while "transversions" are substitutions in which a purine is replaced by a pyrimidine, or a pyrimidine is replaced by a purine). The total rate of substitutions per site per year is then $k=\alpha+2\beta$. Note

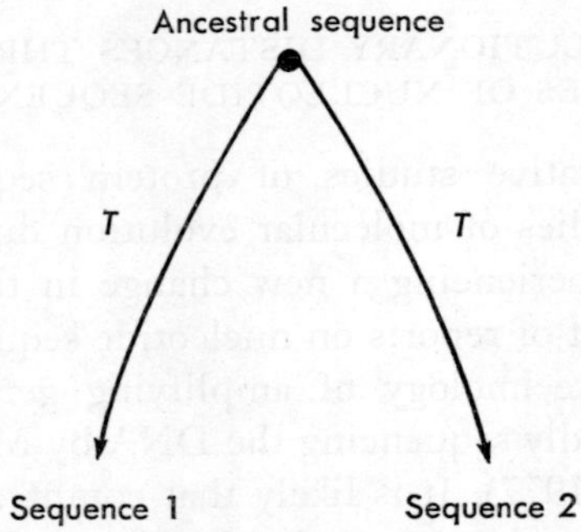

Fig. 7. Divergence of two homologous sequences from a common ancestral sequence T years back.

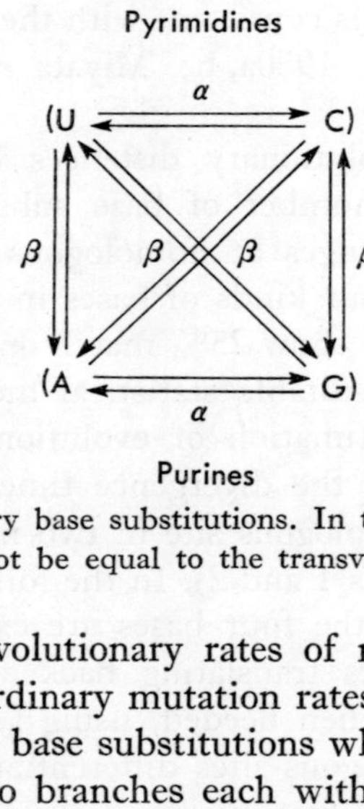

Fig. 8. Model of evolutionary base substitutions. In this model, the rate of transition type substitutions (α) need not be equal to the transversion type substitutions (β).

that α and β refer to evolutionary rates of mutant substitutions in the species rather than the ordinary mutation rates at the level of individuals.

The total number of base substitutions which separate two sequences and therefore involve two branches each with length T is given by $2Tk$ which we denote by K.

$$K = 2Tk = 2\alpha T + 4\beta T \tag{6}$$

In comparing homologous sites of sequences 1 and 2, we note that there are 12 combinations of different bases as shown in Table III. Let P be the probability (relative frequency) that, at time T, homologous sites are occupied by UC, CU, AG, or GA, that is, the probability of homologous sites showing the transition type base differences. Similarly, let

TABLE III
Various Types of Different Base Pairs at Homologous Nucleotide Sites in Two Sequences Compared

Type of difference	Transition type				Transversion type							
Sequence 1	U	C	A	G	U	A	C	G	U	G	C	A
Sequence 2	C	U	G	A	A	U	G	C	G	U	A	C
(Frequency)	(P)				(Q)							

Q be the probability of homologous sites showing the transversion type differences.

Then, it can be shown that P and Q satisfy the following set of differential equations (for derivation, see Kimura, 1980).

$$\left.\begin{aligned} \frac{dP}{dT} &= 2\alpha-4(\alpha+\beta)P-2(\alpha-\beta)Q \\ \frac{dQ}{dT} &= 4\alpha-8\beta Q \end{aligned}\right\} \quad (7)$$

The solution of this set of equations which satisfies the condition $P=Q=0$ at $T=0$, that is, the two sequences are identical at the start, is as follows.

$$\left.\begin{aligned} P &= \frac{1}{4}-\frac{1}{2}e^{-4(\alpha+\beta)T}+\frac{1}{4}e^{-8\beta T} \\ Q &= \frac{1}{2}-\frac{1}{2}e^{-8\beta T} \end{aligned}\right\} \quad (8)$$

From these equations we get

$$4(\alpha+\beta)T = -\log_e(1-2P-Q) \quad (9)$$

and

$$8\beta T = -\log_e(1-2Q). \quad (10)$$

Since the evolutionary distance in terms of the number of base substitutions between the two sequences is given by Eq. (6), we obtain

$$K = -\frac{1}{2}\log_e\{(1-2P-Q)\sqrt{1-2Q}\} \quad (11)$$

(Kimura, 1980). This equation may be used to estimate the evolutionary distance between two sequences in terms of the number of base substitutions per site that have occurred in the course of evolution extending over T years. In this equation, P and Q are respectively the fraction of sites for which two sequences differ from each other with respect to transition type and transversion type substitutions (see Table III).

In the special case of $\alpha=\beta$, Eq. (8) reduce to $P=Q/2=[1-\exp(-8\alpha T)]/4$, so that Eq. (11) reduces to

$$K=-\frac{3}{4}\log_e\left(1-\frac{4}{3}\lambda\right), \tag{12}$$

where $\lambda=P+Q=3Q/2$ is the fraction of sites for which two sequences differ from each other (Jukes and Cantor, 1969; Kimura and Ohta, 1972). In actual situations, however, particularly when the third positions of codons are compared, P is often larger than Q, and therefore the assumption of $\alpha=\beta$ or $P=Q/2$ is not always realistic. This is one reason why formula (11) is better than (12). Eq. (11) also has a desirable property that as P and Q get small, it converges to $K=P+Q$ independent of α and β. The base substitution rate per year is given by $k_{nuc}=K/(2T)$ if the divergence time T is known, where the subscript *nuc* means that the estimate refers to the rate per nucleotide site.

Since a large fraction of substitutions at the third positions of codons are synonymous, that is, they do not cause amino acid changes, it would be of interest to estimate the synonymous component of the substitution rate at this position. From the standard RNA code table we note that for a given pair of bases in the first and the second codon positions, roughly speaking there are two situations; either the third position is completely synonymous (4-fold degeneracy) or synonymy is restricted within purines or pyrimidines (2-fold degeneracy). These two situations occur roughly in equal numbers. Thus, the synonymous component of substitutions at the third position which we denote by k_S' may be estimated by $0.5(\alpha+2\beta)+0.5\alpha$ or $\alpha+\beta$. Let $K_S'=2Tk_S'=2(\alpha+\beta)T$ be the synonymous component of the distance, then, we get

$$K_S'=-\frac{1}{2}\log_e(1-2P-Q), \tag{13}$$

if we apply Eq. (9).

It is desirable to have formulae for the error variance (due to sampling) of the estimated values of K and K_S'. Let n be the number of nucleotide sites for which the two sequences are compared with each other. Then the large sample standard errors of K and K_S' are

$$\sigma_K=\frac{1}{\sqrt{n}}\{\sqrt{(a^2P+b^2Q)-(aP+bQ)^2}\} \tag{14}$$

and

$$\sigma_{K_S'} = \frac{\sqrt{4P+Q-(2P+Q)^2}}{2(1-2P-Q)\sqrt{n}}, \tag{15}$$

where $a=1/(1-2P-Q)$ and $b=(1/2)\ [1/(1-2P-Q)+1/(1-2Q)]$.

As an example, we compare the nucleotide sequence of the rabbit β-globin (Efstratiadis *et al.*, 1977) with that of chicken β-globin (Richards *et al.*, 1979). There are 438 nucleotide sites which can be compared, corresponding to 146 amino acid sites (codons). Among these sites, we find that there are 58 sites for which these two sequences show transition type differences, and 63 sites with transversion type differences. Thus, $P=0.132$, $Q=0.144$ and we obtain $K=0.348$. Mammals and birds probably diverged during the Carboniferous period (see Romer, 1968), so we tentatively take $T=300\times10^6$ years. The evolutionary rate per site is then $k_{nuc}=K/(2T)=0.58\times10^{-9}$ per year. This is the overall rate per site, but it is much more interesting to estimate separately the evolutionary rates for the three codon positions. For the first position, there are 146 nucleotide sites compared, and we find $P=15/146$ and $Q=21/146$, giving $K_1=0.300$, where subscript 1 denotes that it refers to the first codon position. Similarly, for the second position, we find $P=7/146$ and $Q=18/146$ so that $K_2=0.195$. Finally, for the third position, $P=36/146$ and $Q=24/146$, and we get $K_3=0.635$, which is much higher than the corresponding estimates for the first and second positions. We can also estimate the synonymous component of the evolutionary distance per third codon position. From Eq. (13), this turns out to be $K_S'=0.535$.

In Table IV, results of similar calculations are listed for various comparisons involving the human β-(Marotta *et al.*, 1977), the mouse β-(Konkel *et al.*, 1978), and rabbit α-globin sequences (Heindell *et al.*, 1978), in addition to the chicken and rabbit β-globin sequences. Except for the last two comparisons involving the abnormal, globin-like α-3 gene (Nishioka *et al.*, 1980), the table suggests that the relationship $K_2<K_1<K_3$ holds generally; the evolutionary mutant substitutions are most rapid at the third position, and this is followed by the first position, and then, at the second position, the substitutions are the slowest. This can readily be interpreted by the neutral theory as follows. Among the three codon positions, base substitutions at the second positions tend to produce more drastic changes in the physico-chemical properties of amino acids than those at the first positions. Take for example the codon for proline. The code table shows that it is CCN, where N is any one of four bases. Substitutions of U,

TABLE IV
Evolutionary Distances in Terms of the Number of Base Substitutions Estimated for Several Comparisons of Globin Sequences

Comparison	Evolutionary distances per nucleotide site			
	K_1	K_2	K_3	K_S'
Human β *vs.* mouse β	0.17±0.04	0.13±0.03	0.34±0.06	0.28±0.05
Rabbit β *vs.* mouse β	0.16±0.04	0.13±0.03	0.43±0.07	0.36±0.07
Human β *vs.* rabbit β	0.06±0.02	0.06±0.02	0.28±0.06	0.25±0.05
Chicken β *vs.* rabbit β	0.30±0.05	0.19±0.04	0.64±0.11	0.53±0.10
Rabbit α *vs.* rabbit β	0.54±0.09	0.44±0.07	0.90±0.15	0.69±0.13
Rabbit α *vs.* mouse α-1	0.12±0.03	0.11±0.03	0.54±0.09	0.47±0.09
Rabbit α *vs.* mouse α-3	0.27±0.06	0.28±0.06	0.69±0.13	0.56±0.12
Mouse α-1 *vs.* mouse α-3	0.16±0.04	0.20±0.05	0.30±0.06	0.22±0.05

K_1, K_2, and K_3 denote the number of base substitutions at the first, second, and third positions of codons, while K_S' stands for the estimated number of substitutions due to synonymous changes at the third position. Estimated values of these parameters together with their standard errors are listed (from Kimura, 1980).

A, and G for base C at the first position lead respectively to serine, threonine, and alanine. In terms of Miyata's distance (based on polarity and volume differences between an amino acid pair; see Miyata *et al.*, 1979), they are 0.56, 0.87, and 0.06 units apart from proline, with the average distance of about 0.5. On the other hand, the corresponding average distance resulting from substitutions for C at the second position of the codon turns out to be about 2.5.

This means that mutational changes at the first position have a higher chance of not being harmful (*i.e.*, selectively neutral or equivalent) than those at the second position, and therefore, have a higher chance of being fixed in the species by random drift (Kimura and Ohta, 1974). This type of reasoning applies more forcibly to the third position (as compared with the first and the second positions), since a majority of mutational changes at this position do not cause amino acid changes.

As an index for the relative frequencies of synonymous and amino acid-altering substitutions, we may take the ratio $K_S' : 1/2(K_1+K_2)$. This is 4.17 for the human β-rabbit β comparison ($T=8\times10^7$ years), 2.16 for the chicken β-rabbit β comparison ($T=3\times10^8$ years) but only 1.41 for the rabbit α-rabbit β comparison ($T=5\times10^8$ years). Perier *et al.* (1980) claim, based on a similar analysis of preproinsulin and globin evolution, that initially the silent sites accumulate changes some seven times more rapidly than replacement sites (*i.e.*, amino acid altering changes), but that accumulation saturates at 85–100 million years. Such a "saturation" phenomenon of silent substitutions is probably an artefact (they are more apparent than real), coming from the difficulty of detecting all the synonymous substitutions, as more and more of them accumulate in the codons. One reason why I suspect this is that the higher rates are obtained from more recent comparisons (such as between mammals), and that because evolution proceeds from the remote past toward the more recent past, but not in the reverse direction, the "saturation" phenomenon can only be superficial. If evolutionary change really decreased, the more recent comparisons should show more slow changes, but what is observed is the other way round. Furthermore, as compared with the first and the second positions, the third position shows marked deviation of the base composition from equality (*e.g.*, G 36%, C 30%, U 27%, A 7% in human β) and this complicates the problem of estimation for a very large value of T. For additional discussion on the estimation of synonymous substitution at position 3 of codons readers may refer to Kimura (1981a).

Much more fascinating are the last two comparisons in Table IV which involve the globin-like α–3 gene recently sequenced in the mouse (Nishioka *et al.*, 1980). This gene completely lacks two intervening sequences normally present in all the α- and β-globin genes, and it does not encode globin. In other words, it is inactive in the production of stable messenger RNA. However, from a comparison of this sequence with the normal, adult mouse α-globin (α–1) and the rabbit α-globin nucleotide sequences, it is evident that this gene evolved from a normal ancestral α-globin gene through duplication and subsequent loss of its intervening sequences. This must have occurred after the mouse and the rabbit diverged from their common ancestor some 80 million years ago. This α-globin-like gene acquired in its coding region a number of insertions and deletions of nucleotides. In making sequence comparisons, therefore, I chose only those codons of the α–3 gene which do not contain such changes and which are either identical with or differ only through base substitutions from the corresponding (homologous) codons of the mouse α–1 and the rabbit α genes.

As seen from the last three lines of Table IV, this "dead gene" (α–3) evolved at a much faster rate than its normal counterpart (mouse α–1 gene), particularly with respect to the first and the second codon positions. This is easy to understand from the neutral theory. Under a normal situation, each gene is subject to a selective constraint coming from the requirement that the protein which it produces must function normally.

Evolutionary changes are restricted to such a set of base substitutions. However, once a gene is freed from this constraint, as is the case for this globin-like α–3 gene, practically all the base substitutions in it become indifferent to Darwinian fitness, and the rate of base substitutions should approach the upper limit set by the mutation rate (this holds only if the neutral theory is valid, but not if the majority of base substitutions are driven by positive selection). If the rates of synonymous substitutions are not very far from this limit (Kimura, 1977) we may expect that the rates of evolution of a "dead gene" are roughly equal to those of synonymous substitutions. Recently Miyata and Yasunaga (1981) made a more detailed study of the evolutionary rate of a mouse pseudo α-globin gene ($\psi\alpha$ 30.5) which was sequenced by Vanin *et al.* (1980) and which appears to be essentially equivalent to the α–3 gene. They also obtained a result supporting this prediction.

In addition to coding regions, it would be interesting to estimate base substitution rates in non-coding regions such as introns. I use data presented by van Ooyen *et al.* (1979) who investigated the similarity between the nucleotide sequences of rabbit and mouse β-globin genes. They list (see their Table II) separately the numbers of "transition" and "transversion" type differences between the homologous parts of these sequences. For the small introns, excluding 5 gaps that amount to 6 nucleotides, $P=27/113$, $Q=18/133$, and $n=113$. Using Eqs. (11) and (14), we get $K=0.60\pm0.12$. This is not significantly different from the substitution rate at the third position $K_3=0.43\pm0.07$. The large introns of rabbit and mouse β-globin genes differ considerably in length, and when the alignment of the sequences is optimized there are 14 gaps which amount to 109 nucleotides. Excluding these parts, $P=113/557$ and $Q=179/557$, from which we get $K=0.90\pm0.07$. This value is significantly larger than K_3. It is likely, as pointed out by van Ooyen *et al.*, that insertions and deletions occur rather frequently in this part in addition to point mutations, and that they inflate the estimated value of the "nucleotide substitution rate," since a majority of these changes may also be selectively neutral and subject to random fixation by genetic drift.

MAINTENANCE OF GENETIC VARIABILITY AT THE MOLECULAR LEVEL

The neutralists and selectionists also give diametrically opposed interpretations regarding the mechanism by which genetic variability is maintained within the species, particularly in the form of protein polymorphisms. Here, polymorphism means coexistence of two or more allelic forms in a species, usually excluding the situation where the frequency of the most prevalent allele is higher than 99%.

A short history on the concept of polymorphism will help to see the problem in proper perspective. Before electrophoretic surveys became popular, nobody knew whether the so-called wild type or normal allele was typically a single entity or a mixture of chemically different forms. If it were a mixture forming "cryptic polymorphism," one would have naturally reasoned that such a variation could only be maintained by natural selection, for the prevailing view then was that selectively neutral genes practically do not exist and that recurrent mutations are insufficient for a mutant allele to reach any appreciable frequency. At that time, Ford's definition of "genetic polymorphism" had been widely adopted in the textbooks and very few geneticists questioned its adequacy. According to his definition, genetic polymorphism is the occurrence together in the same habitat of two or more discontinuous forms of a species in such proportions that the rarest of them cannot be maintained merely by recurrent mutation (see Ford, 1965). In my opinion, this was a peculiar definition, for it tacitly presupposes the mechanism involved; mutation-selection balance was automatically excluded from it. It was also often claimed that the selective forces involved should generally be very large for genetic variants in natural populations, and that overdominance (*i.e.*, heterozygote superiority in fitness) was responsible for genetic difference between individuals.

Returning to the neutralist-selectionist controversy, neutralists claim that those polymorphisms are selectively neutral and maintained in the population through mutational input and random extinction (Kimura, 1968a, b; Kimura and Ohta, 1971). In each generation, a number of neutral or nearly neutral mutants arise in the population, and they will either be lost from the population or fixed in it. In the process, they contribute to the genetic variability in the form of polymorphisms (Fig. 9). Thus, polymorphism and molecular evolution are not two distinct phenomena; polymorphism is simply a phase of molecular evolution. On the other hand, selectionists claim that polymorphisms are actively maintained by some form of "balancing selection." Although the details of "balancing selection" are often not specified, the two most popular explanations of selection are (i) overdominance (heterozygote advantage), and (ii) "frequency-dependent selection" that favors the minority. At one time, over-

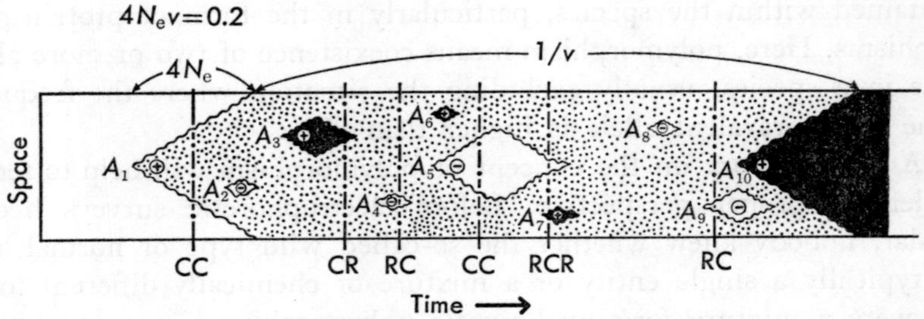

Fig. 9. A typical pattern of extinction and multiplication of selectively neutral (*i.e.*, selectively equivalent) mutants in a finite population when they occur at the rate of one mutation in every ten generations. The allele A_1 whose protein has a positive electric charge happens to spread through the whole population by chance, followed by alleles A_2 (negative charge), A_3 (more positive charge than A_1), A_4, *etc.*, which increase to various intermediate frequencies but are eventually lost on the way. Then, A_{10}, because of concatenation of good luck, finally spreads through the whole population to replace A_1. A large number of small dots represent innumerable mutant alleles that are lost in a few generations after their appearance in the population. If the mutation rate (v) per gamete per generation for neutral mutants is one in a million, the average interval between A_1 and A_{10} is $1/v$ or one million generations. At any given moment, the population contain various alleles. For example, at the phase marked as CR (second from the left), the population happens to contain a pair of alleles, one common (A_1) and one rare (A_3). In other words, the population is polymorphic for A_1 and A_3 (from Kimura and Ohta, 1973, slightly modified).

dominance was enthusiastically proposed as the main cause, but it has become unpopular after abundant protein polymorphisms were found to exist in *Escherichia coli* by Milkman (1973). Since *E. coli* is a haploid organism, heterozygotes do not exist, and therefore the mechanism assuming heterozygote advantage does not work.

Nowadays, many selectionists resort to frequency-dependent selection as the most likely mechanism for the maintenance of protein polymorphism (see, for example, Clarke, 1979). This was first proposed by the late Ken-Ichi Kojima and his group who reported experimental results showing marked frequency-dependent selection at the esterase-6 and alcohol dehydrogenase loci in the fruit fly *Drosophila melanogaster* (Kojima and Yarbrough, 1967). He also claimed that under this type of natural selection (assuming that various genotypes become equally fit at equilibrium), the problem of genetic load does not exist. Later, very carefully designed, large scale experiments were conducted by Yamazaki (1971) using the esterase-5 locus in *Drosophila pseudoobscura*. But, Yamazaki could not detect any selection. Similarly, a group led by Mukai of Kyushu University carried out extensive studies of selection at several enzyme loci in *D. melanogaster*, but no evidence was obtained for fitness differences between these "isozyme" genes (see Mukai's article in this book). Yet, a number of reports have been published by various authors claiming that they demonstrated experimentally the occurrence of natural selection between alleles at some enzyme loci. In most cases, however, it is likely that the experimental results do not reflect the effects of the single locus in question; rather they are caused by surrounding linked loci as pointed out by Mukai and Yamazaki (1980).

So far, the case for natural selection appears to be strongest in alcohol dehydrogenase (ADH) locus in fruit flies. Clarke who has been particularly active in criticizing the neutral theory, claims that his group showed clearly that natural selection discriminates between two allelic forms F and S (corresponding to "fast" and "slow" in mobility on a starch gel) in the fruit fly *D. melanogaster* (Clarke, 1975a), and that they identified the agents for selection. Namely, F is more active in detoxifying ethanol in rotting foods but S is more registant to heat. He also claims that in his laboratory, frequency-dependent selection was confirmed at the ADH locus (Clarke, 1975b). He assumes that frequency-dependent selection is at work in most of enzyme loci. He goes so far as to say that one can tackle the problem of protein polymorphisms directly using the method of ecological genetics, so that mathematical treatments are no longer necessary or desirable to account for the observed properties of evolving populations. Such a claim, however, cannot go unchallenged. In order to check the existence of frequency-dependent selection at ADH locus in *D. melanogaster*, Mukai and his associates recently have done very careful, large scale experiments. Contrary to the claim of Clarke's group, they completely failed to find such selection (Yoshimaru and Mukai, 1979). So, the existence of marked frequency-dependent selection at ADH locus is now much in doubt. It is likely that selection involved in protein polymorphism, if any, is extremely weak as shown by the study of Mukai *et al.* (1980). These authors extracted more than 500 second chromosomes from an isolated population of *D. melanogaster* in Okinawa, and compared the homozygous load of the standard chromosomes and that of the inversion *In(2L)t* carrying chromosomes. Taking advantage of the fact that the isozyme loci are nearly monomorphic in the inversion-carrying chromosomes, but are highly polymorphic in the standard chromosomes, they obtained the estimate $\bar{s} = -0.00013 \pm 0.000375$ for the average selection coefficient per locus. This estimate is not significantly different from zero.

During the last few years, interesting new observations have started to emerge that are either favorable to the neutral theory or can readily be interpreted by it. I shall mention some of them. Several years ago, the possibility of marked non-random association of polymorphic variants between different gene loci on one chromosome (called "linkage disequilibrium") was suggested based on simulation experiments assuming balancing selection.

Of particular interest is the work by Franklin and Lewontin (1970). They investigated a model in which a large number of overdominant loci are tightly packed within a chromosome with multiplicative fitness effect among loci. They disclosed a remarkable property of the model that, under a suitable condition, an extremely strong linkage disequilibrium develops in such a way that a population consists mainly of two complementary chromosome types. In one series of simulation experiments, for example, they assumed 36 overdominant loci on a chromosome with recombination fraction $r=0.0025$ between adjacent loci and with 10% fitness decrease in homozygous condition at individual loci. The effective population size was assumed to be 400. Starting from the condition that gene frequencies are equal (*i.e.*, 0.5) at all the loci and combinations of genes between loci are completely at random, they observed, for instance, in one of replicate trials, that, in generation 300, the population consisted mainly of two chromosome types, denoted by

$$011 \quad 010 \quad 110 \cdots 101$$

and

$$100 \quad 101 \quad 001 \cdots 010,$$

in roughly equal frequencies. These two chromosomes amounted to some 83% while the remaining types were only 17% in all. Once such an equilibrium is reached, the average fitness of the population becomes about $\bar{w}=0.4$ which is much higher than $\bar{w}=0.16$ expected under random combination of genes between loci. No such remarkable polarization effect occurred if the recombination fraction between adjacent loci was $r=0.005$ or more. Thus, linkage must be tight for the effect to appear. However, no appreciable increase of fitness occurred for $r=0.0$ (complete linkage). What makes this model attractive is that $r=0.0025$ can be a reasonable map distance between two adjacent segregating loci in *Drosophila*. They also found a very interesting property of the model: by increasing the number of segregating loci within a chromosome while holding the total map length and the inbreeding effect (*i.e.*, the decrease of fitness in the homozygous state) constant, they found that the average correlation between genes on the chromosome remained essentially unchanged (for analytical treatments, see Slatkin, 1972). In other words, the property of the system does not depend on the number of genes. Thus, what matters is the chromosome as a whole but not individual loci. In an extreme situation, allelic combinations are polarized into only two groups so that the situation becomes somewhat analogous to the balanced lethal system.

If such a polarization really occurs in natural populations, the argument of "heavy segregational load" against the assumption of a large number of overdominant loci breaks down; the load under random mating does not necessarily become much larger than 0.5.

Franklin and Lewontin (1970) also investigated various cases such as asymmetrical fitness between two homozygotes at each loci, unequally spaced loci on a chromosome and truncation selection. They confirmed that their new finding is robust to such a change.

Although interesting, whether such marked linkage disequilibrium really exists in nature is another matter. In fact, as more and more actual observations accumulate, it has become increasingly clear that there is virtually no linkage disequilibrium among enzyme loci in *Drosophila*, unless gene loci are associated with inversions (a structural change with a chromosome segment inverted).

That linkage disequilibrium between isozyme loci seldom exists in natural populations of *Drosophila* has been shown by extensive studies of Mukai and his associates (Mukai *et al.*, 1971, 1974; Mukai and Voelker, 1977), and also by Langley *et al.* (1978) and Charlesworth *et al.* (1979). Even between such closely linked loci as *Odh* (octanol dehydrogenase) and *Est-C* (esterase-C) with recombination fraction 0.0058 between them, no linkage disequilibrium was observed by Mukai and Voelker (1977). On the other hand, significant linkage disequilibria were found in some cases between isozyme genes and polymorphic inversions, but these may most appropriately be explained by founder effect (*i.e.*, histrical accident and random drift).

The problem of non-random association between "electromorphs" (isozyme alleles) and inversion chromosomes in finite populations was investigated mathematically by Nei and Li (1980). Using the diffusion equation method, they estimated the probable magnitude of non-random association, that is, the difference in the frequencies of a given allele between inversion and non-inversion chromosomes.

According to their analysis, significant non-random association will develop if $4N_I r$ is small (say, less than 1), where N_I is the effective number of the inversion chromosome and r is the rate by which the allele in the inversion chromosome is exchanged with the allele in the non-inversion chromosome. They revealed that the standard error of non-random association often becomes larger than the mean. Previously, Prakash and Lewontin (1968) noted that, at the P_t*–10* and *amylase* loci, there is a strong association of different but characteristic alleles with different inversion types (such as "Standard" and "Santa Cruz") in the third chromosome of *D. pseudoobscura*, irrespective of the geographical origin of the strains. Furthermore, in the sibling species, *D. persimilis* whose third chromosome gene arrangements belong to (or are directly derived from) Standard type, they found that the allelic constitution is very similar to Standard type of *D. pseudoobscura*. Considering the possibility that the separation of different inversion types dates back at least several million generations, they conclude that the gene contents of the inversions are "coadapted," in agreement with Dobzhansky (1951), so that any introduction of foreign alleles is rejected. However, Nei and Li (1980), through a more meticulous reexamination of the problem, point out that Prakash and Lewontin's (1968) observation can easily be accomodated by the neutral theory.

If the alleles involved in enzyme polymorphisms are selectively equivalent as assumed by the neutral theory, what kind of observations are pertinent to further test the theory.

Recently, Koehn and Eanes (1977, 1978), and Nei *et al.* (1978) have shown that in *Drosophila* there is a significant positive correlation between the genetic variability of proteins and their subunit molecular weight. This is easy to understand from the neutral theory since the larger the subunit size, the higher the mutation rate. Such a correlation was not found in human polymorphisms, as investigated by Harris and his collaborators (Harris *et al.*, 1977), but they found that monomeric enzymes are more variable (*i.e.*, more polymorphic) than multimeric enzymes (with more than one subunits) in agreement with an earlier finding of the same relationship in *Drosophila* by Zouros (1976). Particularly the finding that multimeric enzymes which form interlocus hybrid molecules have a markedly reduced level of polymorphism is consistent with the neutral theory, because subunit interaction in such enzymes will increase the functional constraint and reduce the possibility of a mutation being harmless (*i.e.*, neutral). At the same time this finding is inconsistent with the hypothesis (Fincham, 1966) that hybrid molecules can be the basis of overdominance.

Whereas neutralists regard the molecular structure and function as the major determinants of protein polymorphisms, selectionists consider that ecological conditions are the major determinants. In fact, many selectionists claimed, in various versions, that a paralellism should exist between environmental variability (or heterogeneity) and genetic variability. They predicted that organisms living deep at the bottom of the sea would generally have very low genetic variability since the environment there is the most stable and homogeneous imaginable, while organisms living in the inter-tidal zone would have high genetic variability because the environment there is ever changing. Although logical and plausible, this prediction was a failure as observations came in completely the other way round; genetic variability is generally extremely high among organisms living at the bottom of the oceans (Gooch and Schopf, 1972; Ayala *et al.*, 1975), while it is low among those living in the inter-tidal zone (Somero and Soulé, 1974).

MODELS OF MUTATION IN MOLECULAR POPULATION GENETICS

To carry out quantitative studies on genetic variability at the molecular level based on population genetics, specific models on mutational production of new alleles are required. The first model of this kind, which was proposed by Kimura and Crow (1964), is based on the consideration that each gene consists of a large number of nucleotide sites, so that practically an infinite number of alleles can be produced. This model (now known as "the infinite allele model") assumes that whenever a new mutant occurs it represents a new, not pre-existing allele. It can then be shown that at steady state in which mutational production of neutral alleles is balanced by random extinction, the average heterozygosity per gene locus is

$$\bar{H}_e = \frac{4N_e v}{4N_e v+1}, \tag{16}$$

where N_e is the effective population size and v is the mutation rate per locus per generation. For example, if $v=10^{-6}$ and $N_e=10^5$, then $\bar{H}_e \approx 0.286$. Note that the average heterozygosity is determined by the product $N_e v$ rather than by N_e and v separately. This model is appropriate if all the alleles can be identified at the level of nucleotide sequences. At the moment,

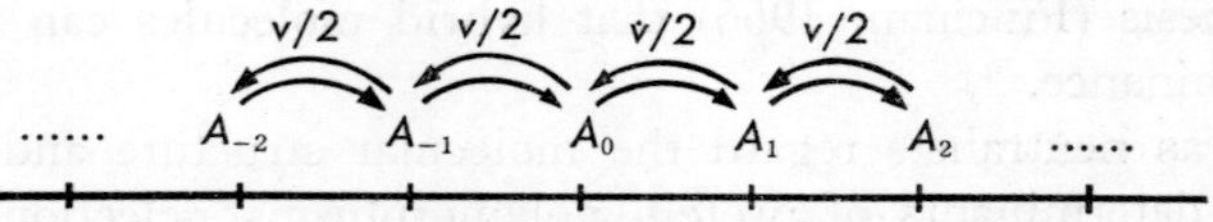

Fig. 10. A mathematical model called "stepwise mutation model" or "Ohta-Kimura model" in population genetics. This model assumes that all allelic states are expressed by integers (. . . , −1, 0, 1, 2, . . .) and that every member of the set of homologous genes making up a population changes its allelic state according to the following rule: whenever a mutation occurs it moves one step in either the positive direction or the negative direction along the integer axis. Each gene is also subject to random sampling in each generation.

however, we resort to more crude methods of detecting allelic differences by electrophoresis. Such electrophoretic variants or "electromorphs" (using the term coined by King and Ohta, 1975) usually correspond to amino acid changes that cause charge differences at the surface of the protein molecule. To accommodate such a limited resoving power, the "stepwise mutation model" was proposed (Ohta and Kimura, 1973). It is assumed that the entire space of allelic states can be represented by integers on the horizontal axis, and that whenever a mutation occurs to a gene, its allelic states moves either one step in the positive direction or one step in the negative direction (see Fig. 10). With this model the average heterozygosity at equilibrium is

$$\bar{H}_e = 1 - \frac{1}{\sqrt{1+8N_e v}} \tag{17}$$

which again depends on the product $N_e v$. These two models give similar results if $N_e v$ is much smaller than unity, but as $N_e v$ gets larger the results diverge; $\bar{H}_e$ increases much more slowly in the stepwise mutation model. What then are the observed values of $\bar{H}_e$ in natural populations? Since the pioneering studies by Harris (1966) on man and Lewontin and Hubby (1966) on fruit flies, the average heterozygosity has been investigated for a large number of species including vertebrates, invertebrates, flowering plants, fungi, and bacteria (in haploid organisms, instead of heterozygosity, one minus the sum of squares of allelic frequencies, which may be termed "virtual heterozygosity," can be used to represent genetic variability). A remarkable fact that emerged from these investigations is that in the majority of species the average heterozygosity ($\bar{H}_e$) per enzyme locus comes in the range 0–20%. In other words, although $\bar{H}_e$ is zero or nearly so for some species which have very small effective population sizes or have apparently experienced severe population bottlenecks, $\bar{H}_e$ seldom exceeds 20%.

For mammalian species with large body size, $N_e=10^5$ is a reasonable value, so that $\bar{H}_e=0.1$ is attained if $v=3\times10^{-7}$ (in Eq. (17)), which is also a plausible value for the mutation rate. The problem, however, is much more difficult than appears from this example. Using the above two equations we can see that for $\bar{H}_e$ to stay in the observed range (0.0–0.2), $N_e v$ has to remain very roughly in the range 0.00–0.07. If $N_e v$ gets larger, $\bar{H}_e$ rapidly approaches unity. In nature, the apparent population sizes of various species differ by several orders of magnitude, and there is a possibility that $N_e v$ changes enormously among them. Then, why do species having very large population sizes do not have nearly 100% heterozygosity? Ayala *et al.* (1972) criticised the neutral theory on the ground that in the neotropical fruit fly *D. willistoni* which has an immense population size (which they think has at least the effective size $N_e=10^9$, with geographic distribution encompassing several million square kilometers), the observed heterozygosity is roughly 18%. They point out that even if we assume a very low neutral mutation rate such as $v=10^{-7}$ (per generation), we still have $4N_e v=400$, and from the neutral theory the predicted heterozygosity should be practically 100%, contrary to their observation.

There are at least two ways out of this dilemma. First, it is possible that the effective population size of this species is not as large as 10^9 even if the apparent size is enormous. It can be shown mathematically that if a population bottleneck occurs from time to time, the genetic variability due to neutral alleles is greatly reduced, and then it takes millions of generations to build up to the theoretical level that is expected if the population size is constant and remains enormous all the time. In this sense, the effect of the last glaciation still remains in neotropical fruit flies.

A much more important and likely cause for reducing the effective population size is the frequent local extinction of colonies that must occur fairly frequently in *Drosophila*. It has been shown mathematically (Maruyama and Kimura, 1980) that if a species consists of a large number of subpopulations or colonies and if local extinction and recolonization occur frequently, not only the effective size of the total population (species) is much reduced but also divergence of subpopulations is largely prevented. This explanation is also pertinent to meet a similar criticism made by Milkman (1976) based on his observations on *E. coli* (Milkman, 1973, 1975). He claims (Milkman, 1976) that *E. coli* has been at a population size well over 10^{10} for at least 4×10^{10} generations, while the mutation rate v could hardly be less than 10^{-8}, so that $4N_ev$ must be at least 400. This is not consistent with his observations suggesting that $4N_ev$ must be less than 1 if the neutral theory is true. Maruyama and Kimura (1980) showed that if a population (species) consists of n haploid lines (subpopulations) which reproduce asexually and each of which is subject to random extinction and subsequent replacement, then the virtual heterozygosity (one minus the sum of squares of allelic frequencies) at equilibrium under the infinite allele model is $\bar{H}_e=2N_ev/(1+2N_ev)$, where

$$N_e=\tilde{N}+\frac{n}{2\lambda}+\frac{n\tilde{N}v}{\lambda} \tag{18}$$

in which $\tilde{N}$ is the harmonic mean of the population size per line, n is the number of lines (assumed to be large), and λ is the rate of line extinction. The corresponding formula for the diploid population (species) consisting of n subpopulations (colonies) is

$$N_e=\tilde{N}+\frac{n}{4(v+\lambda+m)}+\frac{n\tilde{N}(v+m)}{v+\lambda+m}, \tag{19}$$

where m is the rate by which each colony exchanges individuals with the species as a whole in each generation. In this case the heterozygosity is given by Eq. (16).

As an example of the haploid case, let us consider a bacterial species consisting of 100,000 lines ($n=10^5$). Let us assume that each line starts from a single individual and, although it may grow into a line comprising an immense number of individuals, it then becomes extinct on the average in 1,000 generations ($\lambda=10^{-3}$) with the result that the harmonic mean of the population size is only $\tilde{N}=100$. Let us also assume that the mutation rate per generation is $v=10^{-8}$. Then, from Eq. (18), we get, as the effective size of the population $N_e\approx5\times10^7$, and therefore $H_e\approx0.5$. This means that the virtual heterozygosity is only 50%. Yet at any moment, this species may comprise an immense number, say 10^{20}, of individuals. Under such a situation, the genetic variability is almost entirely due to line differences. Such an example will help us understand the genetic structure of *E. coli*. Recently, Selander and Levin (1980) surveyed electrophoretic variation at 20 enzyme loci in 109 clones of *E. coli* from natural populations and found that the number of distinctive *E. coli* genotypes is rather limited; electrophoretically identical clones were obtained from unassociated hosts. The mean genetic diversity (*i.e.*, virtual heterozygosity) turned out to be

0.4718. The genetic structure of the *E. coli* population can most easily be understood by regarding it as a collection of asexual lines. Random sampling of such lines occurs frequently through periodic selection. Thus, the effective size of *E. coli* is not really very large, and Milkman's criticism against the neutral theory is not warranted. Also, in the case of *D. willistoni*, it is possible that the effective population size of this species is two orders of magnitude less than what Ayala *et al.* claim and that their criticism of the neutral theory is also unwarranted.

In addition to the possibility of the effective population size being much smaller than the apparent population size, there is a second, probably supplementary, explanation for the observation that the average heterozygosity seldom exceeds 0.2. As first seriously proposed by Ohta (1973, 1974), the majority of "neutral alleles" may not be strictly neutral but very slightly deleterious with their selection coefficients not excessively large as compared with the mutation rate. These very slightly deleterious alleles will behave as if selectively neutral in relatively small populations such as found in many mammalian species. On the other hand, a majority of them will be effectively selected against in very large populations. It can be shown that if the product $N_e s'$ (where s' is the selection coefficient against the mutant) is large, mutation-selection balance is established in the population, and this will prevent the genetic variability reaching a high level. This explanation, although it invokes natural selection, differs totally from the "selectionist" explanation. More recently, Ohta (1977) refined her hypothesis by assuming that the selection coefficient s' follows an exponential distribution so that the frequency of occurrence of mutants with selective disadvantage s' is proportional to $e^{-cs'}$ where c is a positive constant. As an extension of Ohta's model, and especially to accomodate more neutral mutations, I considered the situation in which s' follows a γ-distribution as shown in Fig. 11 (Kimura, 1979). The mutation rate for variants whose s' value is smaller than the reciprocal of $2N$ may be called the "effectively neutral mutation rate" (denoted by v_e). It can then be shown that v_e decreases as N increases; $v_e \propto 1/\sqrt{N}$ in the case illustrated in Fig. 11. In this model the level of heterozygosity increases only very slowly as N increases. Also we can show that under a reasonable assumption (*i.e.*, generation length is proportional to $1/\sqrt{N}$ as in the case of Fig. 11), the rate of evolution in terms of mutant substitutions is roughly constant per year among lineages if the mutation rate per generation is constant.

The neutral theory of molecular evolution and polymorphism, at least the one I have been developing in collaboration with my colleagues,

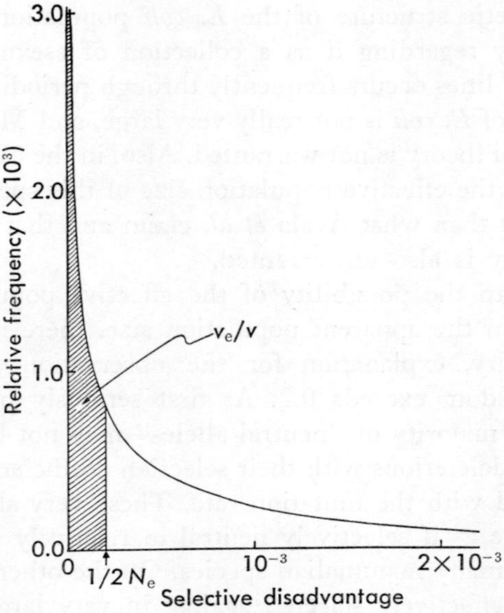

Fig. 11. In this figure, a theoretical model is illustrated which shows the frequency distribution of selective disadvantage among mutants at different sites within a gene. The shaded area represents the fraction of "effectively neutral mutations" whose selective disadvantage is less than $1/(2N_e)$ where N_e is the effective population size. In this example, the relative frequency y is proportional to $e^{-500s'}/\sqrt{s'}$, where s' is the selective disadvantage. The mean selective disadvantage is assumed to be 0.001 in this figure.

has one important feature which distinguishes it from most of the "selectionist" approach, particularly the one adopted by people with an "ecological genetics" bent. My aim is to make studies of molecular evolution *quantitative* by using the diffusion equation method. Such a venture may be called "molecular population genetics" (Kimura, 1971), to which Nei and his associates also contributed greatly, notably through their efforts to connect theoretical predictions with actual observations. For example, they showed (Nei *et al.*, 1976; Fuerst *et al.*, 1977) that the variance of heterozygosity among enzyme loci within a species can be predicted fairly accurately by the neutral theory using observations on the mean heterozygosity within the species (Fig. 12).

In our approach, various parameters such as mutation rates, selection coefficients, population sizes and migration rates (or sometimes, only one parameter such as N_ev as in the case of Fig. 12) must be estimated in order to check theoretical predictions with observations and experiments. Be-

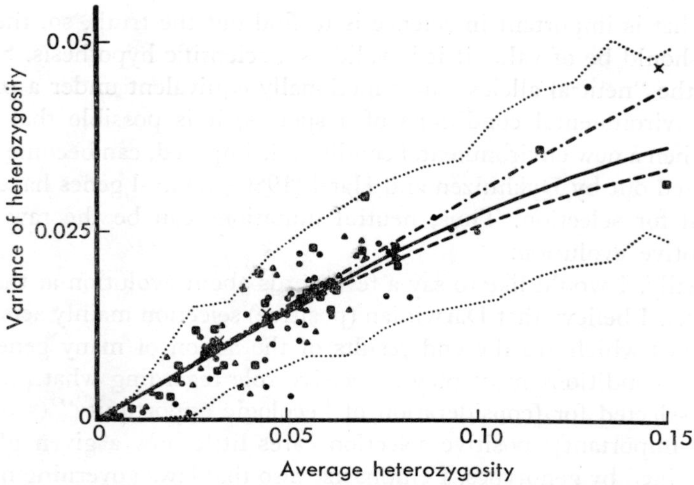

Fig. 12. The relationship between the average and the variance of the heterozygosity among enzyme loci within vertebrate species. The observed values are plotted by various dots; ■ represents mammals; ● reptiles; ▲ fishes; and × amphibians. Three central lines represent the theoretical values based on the neutral theory using various models; —— "infinite allele model"; – – – "stepwise mutation model"; and ——•—— infinite allele model with varying mutation rate. The dotted lines (. . .) represent the 95% significance intervals (from Fuerst *et al.*, 1977).

cause of this testability, the neutral theory is much more liable to be killed, if wrong, than the "selectionist" theories that can invoke all sorts of selection to fit the circumstances, usually not taking any trouble to make quantitative predictions. Among the students of evolutionary biology, there has been a strong tendency to claim that these population genetical parameters will never be known accurately and therefore theories which contain such parameters are of little use. I have the opposite view that these parameters have to be investigated and measured if we really want to understand the mechanism of evolution. Can astronomers and cosmologists claim that theories which contain various astronomical parameters should be avoided because such parameters are difficult to estimate accurately?

"NEUTRAL GENES" AND PHENOTYPIC EVOLUTION

Many people, directly and indirectly, have told me that the neutral theory is not biologically important, since neutral genes are by definition not concerned with adaptation. I believe that this is a too narrow view.

First, what is important in science is to find out the truth, so, the neutral theory should be of value if it is valid as a scientific hypothesis. Secondly, even if the "neutral alleles" are functionally equivalent under a prevailing set of environmental conditions of a species, it is possible that some of them, when a new environmental condition is imposed, can become selected. As pointed out by Dykhuizen and Hartl (1980), neutral genes have a latent potential for selection. Thus, neutral mutations can be the raw material for adaptive evolution.

Finally, I would like to say a few words about evolution at the phenotypic level. I believe that Darwinian (positive) selection mainly acts through phenotypes which are the end results of the action of many genes. Here, ecological conditions must play a decisive role regarding what phenotypes will be selected for [consideration of "ecologic opportunity" (see Wright, 1977) is important]; positive selection cares little how a given phenotype is determined by genotypes. I emphasize also that laws governing molecular evolution are clearly different from those governing phenotypic evolution.

It may be asked, then, why natural selection is so prevalent at the phenotypic level and yet random fixation of selectively neutral or nearly neutral alleles prevails at the molecular level. The answer to this question, I think, comes from the fact that the most common type of natural selection at the phenotypic level is "stabilizing" selection. Unlike the type of natural selection which Darwin had in mind when he tried to explain evolutionary change, stabilizing selection eliminates phenotypically extreme individuals and preserves those that are near the population mean. Since the early work of Bumpus on the house sparrow and Weldon on the land snail, many examples of stabilizing selection have been reported. Probably the best example in man is the relationship between the birth-weights of babies and their neo-natal mortality studied by Karn and Penrose (for review, see, for example, Mather, 1973; Parkin, 1979). It has been shown (Kimura, 1981d) that if a large number of segregating loci or sites are involved in a quantitative character, the average selection coefficient per mutant under stabilizing selection may be exceedingly small. It has also been shown that mutants are very slightly deleterious but nearly neutral, so that mutant substitutions are mainly controlled by random drift, although the rate of "neutral evolution" may be lower as compared with the situation in which all the mutations are strictly neutral. This model was applied to the problem of "non-random" or unequal usage of synonymous codons (Grantham *et al.*, 1980; Ikemura, 1981), and it was shown (Kimura, 1981d) that such non-randomness could readily be understood within the framework of the neutral theory.

The picture of evolution that emerges from this analysis is as follows. From time to time, the position of the optimum of a phenotypic character shifts due to change of environment, and the species tracks such a change rapidly by altering its mean. During this short period of change, extensive shift of gene frequencies is expected to occur at many loci, but this process itself will seldom cause gene substitutions. But, most of the time, stabilizing selection predominates, under which "neutral evolution" or random fixation of mutant alleles occurs extensively, transforming all genes including those of living fossils.

Thus, even if Darwin's principle of natural selection prevails in determining the evolutionary change at the phenotypic level, deep down at the level of the internal structure of the genetic material, there is a great deal of evolutionary change propelled by random drift. Although slow and insignificant for our ephemeral existence, at the level of geological time the random process is of enormous scale. In this way, the footprints of time should be evident in all genetic material.

SUMMARY

The neutral theory (or more precisely, the neutral mutation-random drift hypothesis) asserts that the great majority of evolutionary changes at the molecular level are caused by the random fixation of selectively neutral or nearly neutral mutants in the species, rather than by positive Darwinian selection. It also claims that much of the intraspecific genetic variability at the molecular level, such as is demonstrated by protein polymorphisms, is selectively neutral so that polymorphic alleles are maintained in the species by mutational input and random extinction. Thus the neutral theory regards protein and DNA polymorphisms as a transient phase of molecular evolution. The essential aspect of the neutral theory is not that the alleles involved are selectively neutral in the strict sense. Rather, the emphasis is on mutation and random drift as explanatory factors in molecular evolution because the selection intensity involved is exceedingly small. The theory does not deny the occurrence of deleterious mutations. In fact, selective constraint due to such negative selection is a very important part of the neutralist explanation of some important features of molecular evolution. The neutral theory does not deny the possibility that some changes are adaptive. Thus it is by no means antagonistic to the Darwinian theory of evolution by natural selection. However, because of its emphasis on mutation and random drift, and also because of its concentration on negative selection rather than positive Darwinian selection, the theory clearly differs in its theoretical framework from the traditional neo-Darwinian or "synthetic" theory of evolution.

The neutral theory is accompanied by a well-developed mathematical theory based on the stochastic theory of population genetics. The latter enables us to treat evolution and variation quantitatively, and therefore to check the theory by observations and experiments. The neutral theory in the above sense does not rule out the possibility that the mutant alleles involved have different effects on form and function. These mutants can become selected under a special circumstance, even if they are selectively neutral (*i.e.*, selectively equivalent) under the prevailing environmental conditions. This mean that "neutral" variations can be the raw material for adaptive evolution. In this paper the following topics are discussed: (i) the historical development of the neutral theory; (ii) how it explains the main features of molecular evolution; (iii) what kind of mathematical models are employed to treat the mechanism for the maintenance of protein polymorphisms; (iv) how it differs from rival "selectionist" hypotheses in the face of hard facts; and (v) why neutral mutations are so prevalent at the molecular level, and how they relate to natural selection at the phenotypic level.

REFERENCES

Ayala, F. J., Powell, J. R., Tracey, M. L., Mourão, C. A., and Pérez-Salas, S. (1972). Enzyme variability in the *Drosophila willistoni* group. IV. Genetic variation in natural populations of *Drosophila willistoni*. *Genetics* **70**, 113–139.

Ayala, F. J., Valentine, J. W., Hedgecock, D., and Barr, L. G. (1975). Deep-sea asteroids: high genetic variability in a stable environment. *Evolution* **29**, 203–212.

Barnard, E. A., Cohen, M. S., Gold, M. H., and Kim, J. K. (1972). Evolution of ribonuclease in relation to polypeptide folding mechanisms. *Nature* **240**, 395–398.

Charlesworth, B., Charlesworth, D., Loukas, M., and Morgan, K. (1979). A study of linkage disequilibrium in British populations of *Drosophila subobscura*. *Genetics* **92**, 983–994.

Clarke, B. (1975a). The causes of biological diversity. *Sci. Am.* **233** (2), 50–60.

Clarke, B. (1975b). The contribution of ecological genetics to evolutionary theory: detecting the direct effects of natural selection on particular polymorphic loci. *Genetics* **79** (Suppl.), 101–113.

Clarke, B. (1979). The evolution of genetic diversity. *Proc. Roy Soc. Lond. B.* **205**, 453–474.

Crow, J. F. and Kimura, M. (1970). *An Introduction to Population Genetics Theory.* Harper & Row, New York, reprint by Burgess, Minneapolis.

Dayhoff, M. O., ed. (1978). *Atlas of Protein Sequence and Structure*, Vol. 5, Supplement 3. National Biomedical Research Foundation, Washington, D.C.

Dickerson, R. E. (1971). The structure of cytochrome *c* and the rates of molecular evolution. *J. Mol. Evol.* **1**, 26–45.

Dobzhansky, Th. (1951). *Genetics and the Origin of Species*, 3rd ed. Columbia Univ. Press, New York.

Dykhuizen, D. and Hartl, D. L. (1980). Selective neutrality of 6PGD allozymes in *E. coli* and the effects of genetic background. *Genetics* **96**, 801–817.

Efstratiadis, A., Kafatos, F. C., and Maniatis, T. (1977). The primary structure of rabbit β-globin mRNA as determined from cloned DNA. *Cell* **10**, 571–585.

Fincham, J. R. S. (1966). *Gene Complementation.* Benjamin, Amsterdam.

Fisher, R. A. (1930a). The distribution of gene ratios for rare mutations. *Proc. Roy. Soc. Edinburgh* **50**, 205–220.

Fisher, R. A. (1930b). *The Genetical Theory of Natural Selection.* Oxford Univ. Press, 2nd revised edition (1958), Dover Publ.

Fisher, W. K., Nash, A. R., and Thompson, E. O. P. (1977). Haemoglobins of the shark, *Heterodontus portusjacksoni.* III. Amino acid sequence of the β-chain. *Aust. J. Biol. Sci.* **30**, 487–506.

Ford, E. B. (1965). *Genetic Polymorphism.* Faber & Faber, London.

Franklin, I. and Lewontin, R. C. (1970). Is the gene the unit of selection? *Genetics* **65**, 707–734.

Fuerst, P. A., Chakraborty, R., and Nei, M. (1977). Statistical studies on protein polymorphism in natural populations. I. Distribution of single locus heterozygosity. *Genetics* **86**, 455–483.

Gillespie, J. H. (1977). Sampling theory for alleles in a random environment. *Nature* **266**, 443–445.

Gooch, J. L. and Schopf, T. J. M. (1972). Genetic variability in the deep sea: relation to environmental variability. *Evolution* **26**, 545–552.

Goodman, M. (1976). Protein sequences in phylogeny. In *Molecular Evolution* (ed. F. J. Ayala), pp. 141–159. Sinauer, Sunderland.

Goodman, M. (1978). Substitutional trends and non-random changes in rates during protein evolution. In *Evolution of Protein Molecules* (eds. H. Matsubara and T. Yamanaka), pp. 17–32. Japan Sci. Soc. Press, Tokyo.

Goodman, M., Moore, G. W., Barnabas, J., and Matsuda, G. (1974). The phylogeny of human globin genes investigated by the maximum parsimony method. *J. Mol. Evol.* **3**, 1–48.

Goodman, M., Moore, G. W., and Matsuda, G. (1975). Darwinian evolution in the genealogy of hemoglobin. *Nature* **253**, 603–608.

Grantham, R., Gautier, C., Gouy, M., Mercier, R., and Pavé, A. (1980). Codon catalog usage and the genome hypothesis. *Nucleic Acids Res.* **8**, r49–r62.

Grunstein, M., Schedl, P., and Kedes, L. (1976). Sequence analysis and evolution of sea urchin (*Lytechinus pictus* and *Strongylocentrotus purpuratus*) histone H4 messenger RNAs. *J. Mol. Biol.* **104**, 351–369.

Haldane, J. B. S. (1957). The cost of natural selection. *J. Genet.* **55**, 511–524.

Harris, H. (1966). Enzyme polymorphisms in man. *Proc. Roy. Soc. Lond. B* **164**, 298–310.

Harris, H., Hopkinson, D. A., and Edwards, Y. H. (1977). Polymorphism and the subunit structure of enzymes: a contribution to the neutralist-selectionist controversy. *Proc. Natl. Acad. Sci. U.S.* **74**, 698–701.

Heindell, H. C., Liu, A., Paddock, G. V., Studnicka, G. M., and Salser, W. A. (1978). The primary sequence of rabbit α-globin mRNA. *Cell* **15**, 43–54.

Hendrickson, W. A. and Love, W. E. (1971). Structure of lamprey haemoglobin. *Nature New Biol.* **232**, 197–203.

Holmquist, R. (1979). The method of parsimony: an experimental test and theoretical analysis of the adequacy of molecular restoration studies. *J. Mol. Biol.* **135**, 939–958.

Hunt, L. T., Hurst-Calderone, S., and Dayhoff, M. O. (1978). Globins. In *Atlas of Protein Sequence and Structure* (ed. M. O. Dayhoff), Vol. 5, Supplement 3, p. 229. National Biomedical Research Foundation, Washington, D.C.

Ikemura, T. (1981). Correlation between the abundance of *Escherichia coli* transfer RNAs and the occurrence of the respective codons in its protein genes. *J. Mol. Biol.* **146**, 1–21.

Jukes, T. H. (1978a). Codons and nearest-neighbor nucleotide pairs in mammalian messenger RNA. *J. Mol. Evol.* **11**, 121–127.

Jukes, T. H. (1978b). Neutral changes during divergent evolution of hemoglobins. *J. Mol. Evol.* **11**, 267–269.

Jukes, T. H. (1979). Dr. Best, insulin, and molecular evolution. *Can. J. Biochem.* **57**, 455–458.

Jukes, T. H. (1980a). Silent nucleotide substitutions and molecular evolutionary clock. *Science* **210**, 973–978.

Jukes, T. H. (1980b). Silent nucleotide substitutions during evolution. *Naturwissenschaften* **67**, 534–539.

Jukes, T. H. and Cantor, C. R. (1969). Evolution of protein molecules. In *Mammalian Protein Metabolism* (ed. H. N. Munro), Vol. III, pp. 21–132. Academic Press, New York.

Jukes, T. H. and King, J. L. (1979). Evolutionary nucleotide replacement in DNA. *Nature* **281**, 605–606.

Kimura, M. (1964). Diffusion models in population genetics. *J. Appl. Probab.* **1**, 177–232.

Kimura, M. (1968a). Evolutionary rate at the molecular level. *Nature* **217**, 624–626.

Kimura, M. (1968b). Genetic variability maintained in a finite population due to mutational production of neutral and nearly neutral isoalleles. *Genet. Res. Camb.* **11**, 247–269.

Kimura, M. (1969). The rate of molecular evolution considered from the standpoint of population genetics. *Proc. Natl. Acad. Sci. U.S.* **63**, 1181–1188.

Kimura, M. (1971). Theoretical foundation of population genetics at the molecular level. *Theor. Popul. Biol.* **2**, 174–208.

Kimura, M. (1977). Preponderance of synonymous changes as evidence for the neutral theory of molecular evolution. *Nature* **267**, 275–276.

Kimura, M. (1979). Model of effectively neutral mutations in which selective constraint is incorporated. *Proc. Natl. Acad. Sci. U.S.* **76**, 3440–3444.

Kimura, M. (1980). A simple method for estimating evolutionary rate of base substitutions through comparative studies of nucleotide sequences. *J. Mol. Evol.* **16**, 111–120.

Kimura, M. (1981a). Estimation of evolutionary distances between homologous nucleotide sequences. *Proc. Natl. Acad. Sci. U.S.* **78**, 454–458.

Kimura, M. (1981b). Was globin evolution very rapid in its early stages?: a dubious case against the rate-constancy hypothesis. *J. Mol. Evol.* **17**, 110–113.

Kimura, M. (1981c). Doubt about studies of globin evolution based on maximum parsimony codons and the augmentation procedure. *J. Mol. Evol.* **17**, 121–122.

Kimura, M. (1981d). Possibility of extensive neutral evolution under stabilizing selection with special reference to non-random usage of synonymous codons. *Proc. Natl. Acad. Sci. U.S.* **78**, 5773–5777.

Kimura, M. and Crow, J. F. (1964). The number of alleles that can be maintained in a finite population. *Genetics* **49**, 725–738.

Kimura, M. and Ohta, T. (1969). The average number of generations until fixation of a mutant gene in a finite population. *Genetics* **61**, 763–771.

Kimura, M. and Ohta, T. (1971). *Theoretical Aspects of Population Genetics*. Princeton Univ. Press, Princeton.

Kimura, M. and Ohta, T. (1972). On the stochastic model for estimation of mutational distance between homologous proteins. *J. Mol. Evol.* **2**, 87–90.

Kimura, M. and Ohta, T. (1973). Mutation and evolution at the molecular level. *Genetics* **73** (Suppl.), 19–35.

Kimura, M. and Ohta, T. (1974). On some principles governing molecular evolution. *Proc. Natl. Acad. Sci. U.S.* **71**, 2848–2852.

King, J. L. and Jukes, T. H. (1969). Non-Darwinian evolution. *Science* **164**, 788–798.

King, J. L. and Ohta, T. (1975). Polyallelic mutational equilibria. *Genetics* **79**, 681–691.

Koehn, R. K. and Eanes, W. F. (1977). Subunit size and genetic variation of enzymes in natural populations of *Drosophila*. *Theor. Popul. Biol.* **11**, 330–341.

Koehn, R. K. and Eanes, W. F. (1978). Molecular structure and protein variation within and among populations. In *Evolutionary Biology* (eds. M. K. Hecht, W. C. Steere, and B. Wallace), Vol. 11, pp. 39–100. Plenum Press, New York and London.

Kojima, K. and Yarbrough, K. M. (1967). Frequency-dependent selection at the esterase 6 locus in *Drosophila melanogaster*. *Proc. Natl. Acad. Sci. U.S.* **57**, 645–649.

Konkel, D. A., Tilghman, S. M., and Leder, P. (1978). The sequence of the chromosomal mouse β-globin major gene: homologies in capping, splicing and poly (A) sites. *Cell* **15**, 1125–1132.

Langley, C. H. and Fitch, W. M. (1974). An examination of the constancy of the rate of molecular evolution. *J. Mol. Evol.* **3**, 161–177.

Langley, C. H., Smith, D. B., and Johnson, F. M. (1978). Analysis of linkage disequilibria between allozyme loci in natural populations of *Drosophila melanogaster*. *Genet. Res. Camb.* **32**, 215–229.

Lewontin, R. C. (1974). *The Genetics Basis of Evolutionary Change*. Columbia Univ. Press, New York and London.

Lewontin, R. C. and Hubby, J. L. (1966). A molecular approach to the study of genic heterozygosity in natural populations. II. Amount of variation and degree of heterozygosity in natural populations of *Drosophila pseudoobscura*. *Genetics* **54**, 595–609.

Marotta, C. A., Wilson, J. T., Forget, B. G., and Weissman, S. M. (1977). Human β-globin messenger RNA. III. Nucleotide sequences derived from complementary DNA. *J. Biol. Chem.* **252**, 5040–5053.

Maruyama, T. and Kimura, M. (1980). Genetic variability and effective population size when local extinction and recolonization of subpopulations are frequent. *Proc. Natl. Acad. Sci. U.S.* **77**, 6710–6714.

Mather, K. (1973). *Genetical Structure of Populations*. Chapman and Hall, London.

Maxam, A. M. and Gilbert, W. (1977). A new method for sequencing DNA. *Proc. Natl. Acad. Sci. U.S.* **74**, 560–564.

Maynard Smith, J. (1968). " Haldane's dilemma " and the rate of evolution. *Nature* **219**, 1114–1116.

Mayr, E. (1963). *Animal Species and Evolution*. Harvard Univ. Press, Cambridge.

Mayr, E. (1978). Evolution. *Sci. Am.* **239** (3), 39–47.

Milkman, R. (1973). Electrophoretic variation in *E. coli* from natural sources. *Science* **182**, 1024–1026.

Milkman, R. (1975). Allozyme variation in *E. coli* of diverse natural origins. In *Isozymes* (ed. C. L. Markert), Vol. IV, pp. 273–285. Academic Press, New York.

Milkman, R. (1976). Molecular evolution-selection is the major determinant. *Trends Biochem. Sci.* **1** (7), N152–N154.

Miyata, T., Miyazawa, S., and Yasunaga, T. (1979). Two types of amino acid substitutions in protein evolution. *J. Mol. Evol.* **12**, 219–236.

Miyata, T. and Yasunaga, T. (1980). Molecular evolution of mRNA: a method for estimating evolutionary rates of synonymous and amino acid substitutions from homologous nucleotide sequences and its application. *J. Mol. Evol.* **16**, 23–36.

Miyata, T., Yasunaga, T., and Nishida, T. (1980). Nucleotide sequence divergence and functional constraint in mRNA evolution. *Proc. Natl. Acad. Sci. U.S.* **77**, 7328–7332.

Miyata, T. and Yasunaga, T. (1981). Rapidly evolving mouse α globin-related pseudo gene and its evolutionary history. *Proc. Natl. Acad. Sci. U.S.* **78**, 450–453.

Mukai, T., Mettler, L. E., and Chigusa, S. I. (1971). Linkage disequilibrium in a local population of *Drosophila melanogaster*. *Proc. Natl. Acad. Sci. U.S.* **68**, 1065–1069.

Mukai, T., Watanabe, T. K., and Yamaguchi, O. (1974). The genetic structure of natural populations of *Drosophila melanogaster*. XII. Linkage disequilibrium in a large local population. *Genetics* **77**, 771–793.

Mukai, T. and Voelker, R. A. (1977). The genetic structure of natural populations of *Drosophila melanogaster* XIII. Further studies on linkage disequilibrium. *Genetics* **86**, 175–185.

Mukai, T., Tachida, H., and Ichinose, M. (1980). Selection for viability at loci controlling protein polymorphisms in *Drosophila melanogaster* is very weak at most. *Proc. Natl. Acad. Sci. U.S.* **77**, 4857–4860.

Mukai, T. and Yamazaki, T. (1980). Test for selection on polymorphic isozyme genes using the population cage method. *Genetics* **96**, 537–542.

Muller, H. J. (1962). *Studies in Genetics*. Indiana Univ. Press, Bloomington.

Nei, M., Fuerst, P. A., and Chakraborty, R. (1976). Testing the neutral mutation hypothesis by distribution of single locus heterozygosity. *Nature* **262**, 491–493.

Nei, M., Fuerst, P. A., and Chakraborty, R. (1978). Subunit molecular weight and genetic variability of proteins in natural populations. *Proc. Natl. Acad. Sci. U.S.* **75**, 3359–3362.

Nei, M. and Li, W.-H. (1980). Non-random association between electromorphs and inversion chromosomes in finite populations. *Genet. Res. Camb.* **35**, 65–83.

Nichols, B. P. and Yanofsky, Ch. (1979). Nucleotide sequences of *trpA* of *S. typhimurium* and *E. coli*: An evolutionary comparison. *Proc. Natl. Acad. Sci. U.S.* **76**, 5244–5248.

Nishioka, Y., Leder, A., and Leder, P. (1980). Unusual α-globin-like gene that has cleanly lost both globin intervening sequences. *Proc. Natl. Acad. Sci. U.S.* **77**, 2806–2809.

Ohta, T. (1973). Slightly deleterious mutant substitutions in evolution. *Nature* **246**, 96–98.

Ohta, T. (1974). Mutational pressure as the main cause of molecular evolution and polymorphism. *Nature* **252**, 351–354.

Ohta, T. (1977). Extension to the neutral mutation random drift hypothesis. In *Molecular Evolution and Polymorphism* (ed. M. Kimura), pp. 148–167. National Institute of Genetics, Mishima.

Ohta, T. and Kimura, M. (1971a). On the constancy of the evolutionary rate of cistrons. *J. Mol. Evol.* **1**, 18–25.

Ohta, T. and Kimura, M. (1973). A model of mutation appropriate to estimate the number of electrophoretically detectable alleles in a finite population. *Genet. Res. Camb.* **22**, 201–204.

Parkin, D. T. (1979). *An Introduction to Evolutionary Genetics*. Edward Arnold, London.

Perier, F., Efstratiadis, A., Lomedico, P., Gilbert, W., Kolodner, R., and Dodgson, J. (1980). The evolution of genes: the chicken preproinsulin gene. *Cell* **20**, 555–566.

Perutz, M. F. and Lehman, H. (1968). Molecular pathology of human haemoglobin. *Nature* **219**, 902–909.

Prakash, S. and Lewontin, R. C. (1968). A molecular approach to the study of genic heterozygosity in natural populations. III. Direct evidence of coadaptation in gene arrangements of *Drosophila*. *Proc. Natl. Acad. Sci. U.S.* **59**, 398–405.

Provine, W. B. (1971). *The Origins of Theoretical Population Genetics*. Univ. Chicago Press, Chicago.

Richards, R. I., Shine, J., Ulrich, A., Wells, J. R. I., and Goodman, H. M. (1979). Molecular cloning and sequence analysis of adult chicken β globin cDNA. *Nucleic Acids Res.* **7**, 1137–1146.

Romer, A. S. (1968). *The Procession of Life*. Weidenfeld and Nicolson, London.

Romero-Herrera, A. E., Lieska, N., and Nasser, S. (1979). Characterization of the myoglobin of the lamprey *Petromyzon marinus*. *J. Mol. Evol.* **14**, 259–266.

Salser, W. (1978). Globin mRNA sequences: analysis of base pairing and evolutionary implications. *Cold Spring Harbor Symp. Quant. Biol.* **12** (part 2), 985–1002.

Selander, R. K. and Levin, B. R. (1980). Genetic diversity and structure in *Escherichia coli* populations. *Science* **210**, 545–547.

Slatkin, M. (1972). On treating the chromosome as the unit of selection. *Genetics* **72**, 157–168.

Somero, G. N. and Soulé, M. (1974). Genetic variation in marine fishes as a test of the niche-variation hypothesis. *Nature* **249**, 670–672.

Stebbins, G. L. and Lewontin, R. C. (1972). Comparative evolution at the levels of molecules, organisms and populations. In *Proc. of the Sixth Berkeley Symp. on Math. Stat. and Prob. Vol. V: Darwinian, Neo-Darwinian, and Non-Darwinian Evolution* (eds. L. M. LeCam, J. Neyman, and E. L. Scott), pp. 23–42. Univ. California Press, California.

Tateno, Y. and Nei, M. (1978). Goodman *et al.*'s method for augmenting the number of nucleotide substitutions. *J. Mol. Evol.* **11**, 67–73.

van Ooyen, A., van den Berg, J., Mantel, N., and Weissmann, C. (1979). Comparison of total sequence of a cloned rabbit β-globin gene and its flanking regions with a homologous mouse sequence. *Science* **206**, 337–344.

Vanin, E. F., Goldberg, G. I., Tucker, P. W., and Smithies, O. (1980). A mouse α-globin-related pseudogene lacking intervening sequences. *Nature* **286**, 222–226.

Vogel, F. and Motulsky, A. G. (1979). *Human Genetics*. Springer-Verlag, Berlin.

Wilson, A., Carlson, S. S., and White, T. J. (1977). Biochemical evolution. *Annu. Rev. Biochem.* **46**, 573–639.

Wright, S. (1931). Evolution in Mendelian populations. *Genetics* **16**, 97–159.

Wright, S. (1945). The differential equation of the distribution of gene frequencies. *Proc. Natl. Acad. Sci. U.S.* **31**, 382–389.

Wright, S. (1977). *Evolution and the Genetics of Populations*, Vol. 3. Univ. Chicago Press, Chicago.

Yamazaki, T. (1971). Measurement of fitness at the esterase-5 locus in *Drosophila pseudoobscura. Genetics* **67**, 579–603.

Yoshimaru, H. and Mukai, T. (1979). Lack of experimental evidence for frequency-dependent selection at the alcohol dehydrogenase locus in *Drosophila melanogaster. Proc. Natl. Acad. Sci. U.S.* **76**, 876–878.

Zouros, E. (1976). Hybrid molecules and the superiority of the heterozygote. *Nature* **262**, 227–229.

Zuckerkandl, E. (1976). Evolutionary processes and evolutionary noise at the molecular level. II. A selectionist model for random fixations in proteins. *J. Mol. Evol.* **7**, 269–311.

Zuckerkandl, E. (1978). Multilocus enzymes, gene regulation, and gene sufficiency. *J. Mol. Evol.* **12**, 57–89.

Zuckerkandl, E. and Pauling, L. (1965). Evolutionary divergence and convergence in proteins. In *Evolving Genes and Proteins* (eds. V. Bryson and H. J. Vogel), pp. 97–166. Academic Press, New York.

2

Molecular Evolution, Protein Polymorphism and the Neutral Theory (1982) (ea. M. Kimura), pp. 59–77.
Japan Scientific Societies Press, Tokyo / Springer-Verlag, Berlin

Some Evolutionary Problems of Multigene Families Considered from the Standpoint of Population Genetics

TOMOKO OHTA

National Institute of Genetics, Mishima 411, Japan

Recent developments in eukaryote molecular biology have revealed several new features of gene organization, thereby providing new subjects of reasearch in various fields. Evolution of repeated genes which form a multigene family is one such subject. It is now known that the globin genes are duplicated on the chromosome (Jeffreys, 1979; Goossens *et al.*, 1980; Zimmer *et al.*, 1980) and that genes of transfer RNA, ribosomal RNA, histone, and the variable region of immunoglobulins are repeated many times (for review, see Hood *et al.*, 1975). An interesting property of such gene families is that genes in one family evolve coincidentally, *i.e.*, there are species specific amino acids at certain sites of gene products even if these amino acids are different between the species (Hood *et al.*, 1975). A most probable hypothesis to account for this phenomenon is the model of unequal crossing-over which states that the continued occurrence of unequal crossing-over results in spreading of mutant genes on the chromosome (Smith, 1974). See Fig. 1 for a hypothetical case where a gene spreads on the chromosome by continued unequal crossing-over. The hypothesis has subsequently been verified for the ribosomal DNA

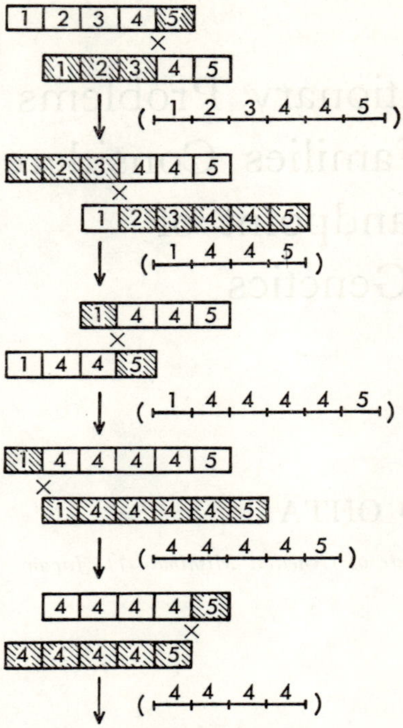

Fig. 1. A model of gene fixation by unequal crossing-over.

gene family of yeast by introducing a marker in the region of this gene family (Szostak and Wu, 1980; Petes, 1980).

As compared with the ordinary single locus case, the mutant dynamics should be considerably different for such gene systems. For the last several years, I have studied the evolution of multigene families from the standpoint of population genetics (Ohta, 1978, 1980). The evolution of large multigene families is treated as a double diffusion process; diffusion of gene lineages within a chromosome through unequal crossing-over and diffusion of chromosome types within a population by random genetic drift. An interesting application of the theory is statistical analyses on amino acid diversity of variable region sequences of immunoglobulins. Analysis is extended to the small multigene family such as globin gene families (Ohta, 1981), for which data on gene structure and organization are rapidly accumulating. In this report, I shall briefly review such studies. The problem of accumulation of pseudogenes among the repeated members is also discussed.

GENETIC VARIATION IN LARGE MULTIGENE FAMILIES

As reviewed by Hood *et al.* (1975), genes of several important proteins and nucleic acid families belong to the large multigene family; transfer RNA, ribosomal RNA, histon, and the variable region of immunoglobulins. Recently, genes of some structural proteins such as chorion, collagen, and actin, were reported to form multigene families, although not so large as in the previous examples (Jones *et al.*, 1979; Solomon *et al.*, 1980). Most of the large families seem to be clustered on a single chromosome. However, some families such as actin genes are scattered throughout the genome. As stated, gene families are characterized by coincidental evolution and the model of unequal crossing-over most satisfactorily explains this phenomenon. The hypothesis presumably applies to the clustered multigene families, and other mechanisms should be responsible for the coincidental evolution of dispersed ones. Following my report (Ohta, 1980), I shall here present the analyses on the evolution of clustered gene families.

We consider a randomly mating population with an effective size N. It is assumed that a multigene family is evolving by the means of mutations, inter-chromosomal crossing-overs at meiosis, intra-chromosomal unequal crossing-overs at somatic division of germ cell lines as well as by random genetic drift. Thus one generation consists of the following processes; mutation $\rightarrow$ random genetic drift $\rightarrow$ intra-chromosomal crossing-over $\rightarrow$ inter-chromosomal crossing-over. In the following of this report, I shall use the term family to mean one cluster of homologous repeated genes on a chromosome. Let n_g be the mean gene number per family and be constant. The gene number in an individual family changes through unequal crossing-overs, and it may differ from family to family. The problem of gene family size is itself an important one and needs to be investigated (Hood *et al.*, 1980; Ohta and Kimura, 1981); however, I shall concentrate here on gene composition. In particular, the simplest and most convenient quantity called "identity coefficient" will be studied which is defined as the probability of two genes belonging to the family being identical by descent. It is equivalent to the measure of homozygosity at enzyme loci.

The parameters are: v is the mutation rate per generation, which may be per gene unit or per smaller unit inside of a gene such as amino acid site; γ is the rate of intra-chromosomal unequal crossing-over with mean shift of m gene units; γ' is the rate of inter-chromosomal crossing-over with mean shift of m' genes. In my previous study (Ohta, 1976, 1978b,

1980), it is assumed that unequal crossing-over occurs in cycles so that n_g remains unchanged. Such an assumption is unrealistic but the model may approximate the real situation. It should be noticed, however, that when family size may differ in each genome, exact (equal) pairing between the families at inter-chromosomal crossing-over is unlikely to occur and practically all inter-chromosomal crossing-overs are unequal. Thus, no *equal* inter-chromosomal crossing-over is assumed here, though it is incorporated in the previous study (Ohta, 1980). In the future, a more precise formulation of the process is desirable which considers both family size distribution in the population and probability distribution of gene shift at chromosome pairing. Such an investigation should properly take account of the effect of natural selection on gene family size. In other words, the crossover products with too many or too few gene copies would be selectively eliminated from the population and, therefore, the effect of unequal crossing-over correspondingly becomes smaller. The situation is somewhat analogous to the case of small multigene families that will be discussed in the later section.

Next, I shall present the transitional and equilibrium properties of the identity coefficient under our model. Let us define the following two measures of identity coefficient; the identity coefficient of genes on one chromosomes *versus* that between members of the same family residing on different chromosomes in the population. Let us denote the average identity coefficient of one chromosomes by C_{w1} and that of two different chromosomes by C_{w2}, where the subscripts w1 and w2 mean one chromosome and two chromosomes within the population. The identity coefficient increases by means of two mechanisms; within a gene family through unequal crossing-over and within the population through random genetic drift. It decreases by mutation. Inter-chromosomal crossing-over also contributes to decreasing gene identity in an ordinary situation. When the mean shift of genes (m or m') at unequal crossing-over is sufficiently large, say larger than 10% of family size (n_g), the amount of change of C_{w1} and C_{w2} in one generation can be shown to become roughly as follows (Ohta, 1980),

$$\begin{bmatrix} C_{w1} \\ C_{w2} \end{bmatrix}_{t+1} = (1-2v)\begin{bmatrix} 1 & 0 \\ \dfrac{1}{2N} & 1-\dfrac{1}{2N} \end{bmatrix}\begin{bmatrix} 1-\dfrac{\beta'}{3} & \dfrac{\beta'}{3} \\ 0 & 1 \end{bmatrix}$$

$$\times\begin{bmatrix}1-\alpha & 0\\ 0 & 1\end{bmatrix}\begin{bmatrix}1-n_g\alpha' & (n_g-1)\alpha'\\ 0 & 1\end{bmatrix}\begin{bmatrix}C_{w2}\\ C_{w2}\end{bmatrix}_t+\begin{bmatrix}\alpha+\alpha'\\ 0\end{bmatrix} \qquad (1)$$

in which t denotes t-th generation, $\beta'=\gamma'\{1-3(m'/n_g)\}$, $\alpha=m\gamma/n_g^2$, $\alpha'=m'\gamma'/n_g^2$ and parameters, v, $1/(2N)$, β', α, and $n_g^{\alpha'}$ are assumed to be much less than unity. Of particular interest is the equilibrium value. At equilibrium, the identity coefficients remain unchanged. Therefore $(C_{w1}, C_{w2})_t =(C_{w1}, C_{w2})_{t+1}$, and the solutions become,

$$\hat{C}_{w1}=\frac{\alpha+\alpha'}{2v+\frac{\beta'}{3}\frac{4Nv}{1+4Nv}+\alpha+\alpha'\frac{1+4n_gNv}{1+4Nv}} \qquad (2)$$

and

$$\hat{C}_{w2}=\frac{\hat{C}_{w1}}{1+4Nv}, \qquad (3)$$

where the hat over C denotes the equilibrium value. These formulas give the average identity coefficients, when various forces balance each other, which are equivalent to the identity probability of two genes randomly chosen from the family.

Next consider a situation where a population was split into two daughter populations and they are isolated thereafter (speciation). When there is no gene exchange between the populations, identity coefficients decrease with time. The average rate of decrease is twice the mutation rate, because neither crossing-over nor sampling drift influences the mean identity coefficients. Let us define the identity coefficient of two genes each taken from the isolated populations, and denote by C_b its average value. The subscript b means "between-populations" to distinguish from the within-population values. For simplicity, I assume that the population expanded at the time of splitting and the effective size of the two daughter population is equal to that of the original one. The average values of the within-population identity coefficients (C_{w1} and C_{w2}) do not change under such an assumption, whereas the between-population identity coefficient decreases by the following formula,

$$C_b(t)=\hat{C}_{w2}e^{-2vt} \qquad (4)$$

in which t denotes the time since splitting measured in generations.

It should be noticed that the above Eqs. (2)–(4) give merely average values and an individual case may deviate much from the expectation. The situation is analogous to homozygosity and genetic distance at selectively neutral enzyme loci (Nei, 1975). The magnitude of such deviation may be known if variance and covariance of the identity coefficients are obtained. I have carried out the analyses which enables one to calculate the variance and covariance of identity coefficients numerically (Ohta, 1978, 1979). The analyses are very complicated and not given here. Together with the results presented here, they provide the theoretical basis to understand various observed facts on repeated gene families in a unified way. In the next section, some applications of the theory to real examples like sequence variability of immunoglobulins will be given.

APPLICATION TO AMINO ACID DIVERSITY OF IMMUNOGLOBULINS

Variable region genes of immunoglobulins represent a most interesting example to apply the results in previous section. I shall first briefly review the organization of this gene family. There are three major classes of immunoglobulins: IgG, IgM, and IgA. The most common class in adult mammals is IgG and an individual molecule of this class is composed of two identical light chains and two identical heavy chains. A light chain has two homologous domains, each having about 110 amino acids, and a heavy chain has four domains. The amino terminal domain is different from chain to chain and is called the *variable* region. The other domains are stable and called *constant* regions. Various classes of immunoglobulins are defined by the kind of constant region of heavy chain. The light chains occur in two distinct forms, κ and λ. Genes for the κ chain, the λ chain and the heavy chain are each located on different chromosomes and thus constitute different gene families. Figure 2 illustrates the situation.

We now know the fine details of gene arrangements within these gene families. Seidman *et al.* (1978) and Tonegawa *et al.* (1978) have shown that the variable region gene family of the light chain is divided into two gene families; *V* (variable) region and *J* (junction) region families and the latter contains only about 15 amino acid sites. Figure 3 illustrates the situation. The heavy chain family has an even more complicated structure, *i.e.*, between the *V* and the *J* families there apparently exist *D* (diversity) segment family, which comprises only a few amino acid sites (Early *et al.*, 1980; Sakano *et al.*, 1980). Figure 3 also depicts the diagram of heavy chain family structure. Note that the constant region family also has a

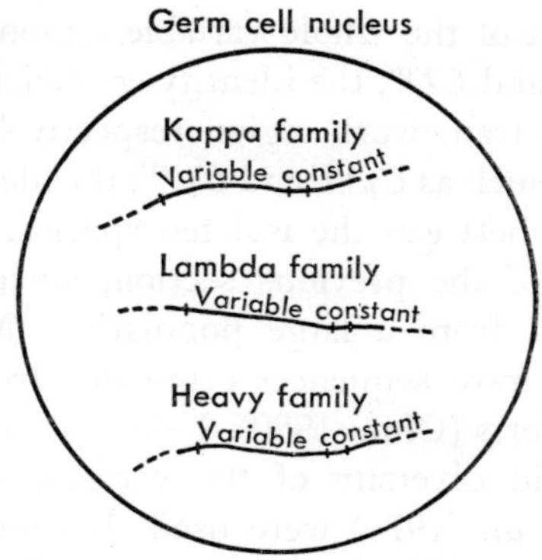

Fig. 2. A model showing the chromosomal locations of the three immunoglobulin gene families.

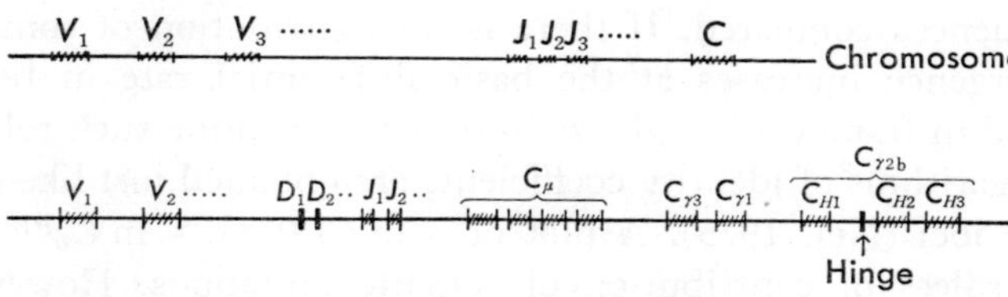

Fig. 3. Models illustrating the organization of immunoglobulin gene families. The upper figure is for the light chain and the lower one, for the heavy chain.

more complicated structure in the heavy chain than in the light chain, consisting of several C_H genes which define the different classes of antibody molecule (Yaoita and Honjo, 1980). Yet the diagram of constant region genes above is incomplete and lacks a few genes on the right side. The antibody diversity is generated by combinatorial use of each of these segments of the variable region during ontogeny. It is also suggested that during rearrangement of various segments, further diversity may be generated at the joining point.

The variable region sequence, on the other hand, can be divided into two sets of amino acid positions according to whether they show high or low degrees of variability and these are called the hypervariable and framework regions (Wu and Kabat, 1970). The former constitutes the antigen combining sites. A controversy has existed on whether or not somatic mutation in terms of amino acid replacement is responsible for hypervariability. Thus, the identity coefficient presented in the previous section should be measured as a function of individual amino acid site rather than at the level of the whole variable region of immunoglobulins. Let us denote by C^{HV} and C^{FW}, the identity coefficient at the hypervariable region and that at the framework region respectively. Also we denote by the subscripts w and b such as C_w^{HV} and C_b^{HV}, the identity coefficient within a population and that between the isolated species. Here C_w corresponds to C_{w2} (formula (3)) of the previous section, since immunoglobulin sequences were sampled from a large population. Mouse sequences may be exceptional in that most sequences originated from a few inbred lines.

Following my reports (Ohta, 1980), I shall present results of statistical analyses of amino acid diversity of the variable region. Sequence data compiled by Kabat *et al.* (1976) were used. If somatic mutations are responsible in lowering the identity coefficient in the hypervariable region, the lower value of C_b^{HV} compared to C_b^{FW} depends upon the added contribution of the somatic changes which does not depend upon remoteness of the sequences compared. If there is no contribution of somatic mutations, divergence increases at the basic differential rate in both hypervariable and in framework regions. In order to examine such relationships, negative logarithms of identity coefficients are obtained just like in analyses on enzyme loci (Nei, 1975). A plot of $-\ln C_b^{HV}$ *vs.* $-\ln C_b^{FW}$ should be linear regardless of contribution of somatic mutations. However, under the germline hypothesis, it should give a regression coefficient larger than one, whereas the plot under the somatic mutation theory should give a regression line of slope one. By denoting the (germ-line) mutation rate per amino acid site per year by v_{aa}, the average value of the identity coefficient between the populations decreases by the constant rate $2v_{aa}$ (see Eq. (4)). This applies both for the hypervariable and framework regions. Therefore the slope of the plot of $Y=-\ln C_b^{HV}$ *vs.* $X=-\ln C_b^{FW}$ should be v_{aa}^{HV}/v_{aa}^{FW} in which the superscripts HV and FW again represent the hypervariable and framework regions.

Actually, the within-species identity coefficient, C_w, varies considerably from species to species. The main cause would be the change in family size in evolution. It is expected theoretically that, when the family size differs, C_w is lowered by v_{aa} with a coefficient positively correlated with the family size (Eqs. (2) and (3)). Thus, one would expect a similar relationship between $-\ln C_w^{HV}$ and $-\ln C_w^{FW}$, and both C_w and C_b values are used for the analysis.

Figure 4 shows the regression line of $Y=-\ln C^{HV}$ on $X=-\ln C^{FW}$ which is $Y=2.59X+0.01$. Each point represents an observed value. As to the data source of these points, and the details of calculation, see

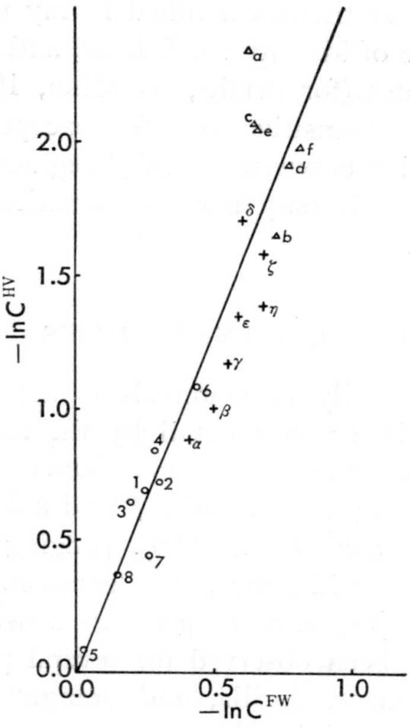

Fig. 4. Relationship between $-\ln C^{FW}$ and $-\ln C^{HV}$. The straight line represents the regression line, and each dot represents an observed value.

Ohta (1980). The regression coefficient, 2.59, is significantly larger than unity, therefore $v_{aa}{}^{HV}$ is significantly larger than $v_{aa}{}^{FW}$, and the ratio $v_{aa}{}^{HV}/v_{aa}{}^{FW}$ is estimated to be 2.59. This coefficient is likely to be an underestimate, because the Poisson correction by taking negative logarithm is often an underestimation of the true divergence when the comparison is remote (Dayhoff, 1972; Fitch, 1973). If another statistical analysis is carried out based on amino acid diversity of various subgroups of immunoglobulins, $v_{aa}{}^{HV}$ is estimated to be about 3.5 times larger than $v_{aa}{}^{FW}$ (Ohta, 1980). Thus it may be concluded that the rate of mutant substitution is roughly three times higher in the hypervariable region than in the framework region, and no somatic mutation is required to explain the observed pattern of sequence variability. With these estimated values of $v_{aa}{}^{HV}$ and $v_{da}{}^{FW}$, and with pertinent values for other parameters, various other statistics fit the prediction of the simple theory remarkably well. In addition, $v_{aa}{}^{HV}$ can be shown to be about the same as the rate of amino acid substitution of fibrinopeptides which is known as the protein with highest evolutionary rate, and $v_{aa}{}^{FW}$ roughly equals the average evolutionary rate of various proteins. Higher order statistics examined in my previous study are the variance and covariance of identity coefficients, and linkage disequilibrium between amino acid sites (for details, see Ohta, 1980). Of course, these statistical tests may be insensitive to a few exceptional amino acid sites such as the joining point between *V* and *J* segments, or between *V* and *D*, at which additional diversity may be somatically generated through recombination process.

EVOLUTION OF SMALL MULTIGENE FAMILIES

The globin gene family in mammals is the genetic system which has been investigated in greatest detail by the molecular cloning technique. It is now known that the normal human genome contains two hemoglobin α genes, two γ genes, and a β and a δ gene (Hardison *et al.*, 1979; Jeffreys, 1979; Lauer *et al.*, 1980). Judging from the amino acid sequence data (Dayhoff, 1972), the two α genes must be very similar to each other, whereas the β gene and δ gene are considerably differentiated. A similar situation has been observed for several primate species: chimpanzee, pygmy chimpanzee, gorilla, and orangutan, by Zimmer *et al.* (1980). These authors emphasize that the same mechanism as the coincidental evolution of large multigene family should be responsible for the evolution of two tandem genes and suggest a phrase "concerted evolution." In order to understand the evolution of such genetic systems, a population genetic approach is again required. Following my report (Ohta, 1981), I shall present my analyses on genetic variation in small gene families again with special reference to the probability of gene identity.

As in the previous section, we consider a finite population with effective size *N*. A small gene family (two tandem genes) is present on a chromosome and is evolving under mutation, random genetic drift, unequal intra-chromosomal crossing-over between sister-chromatids at somatic division of germ cell lines, and inter-chromosomal recombination at meiosis. In order to make the treatment manageable, let us consider the following two models and examine the correspondence between the two. In the first model, the unequal crossing-over is assumed to occur in cycles as in large gene families, *i.e.*, the duplication phase of unequal crossing-over is followed by the deletion phase and thereby the number of genes remains unchanged. In the second model, natural selection is assumed to be responsible for keeping the gene number per family constant, *i.e.*, the crossover products with three genes or with a single gene are selected against and their frequencies in the population are in balance between crossing-over and selection. Here the first model is biologically unrealistic, but can be easily handled, whereas the second one is based on biological phenomena yet very complicated to formulate.

We assume that a chromosome contains two tandem genes, and cycles of two unequal intra-chromosomal crossing-overs of duplication and deletion as in Fig. 5 occur with the rate γ per chromosome per generation. Also, at meiosis, inter-chromosomal crossing-over takes place. No unequal inter-chromosomal crossing-over is assumed. Let β be the rate of such crossing-over per chromosome per generation. Let v be the mutation rate per gene per generation as before. We shall define a set of identity coefficients as in Fig. 6 which are analogous to the coefficients of Kimura

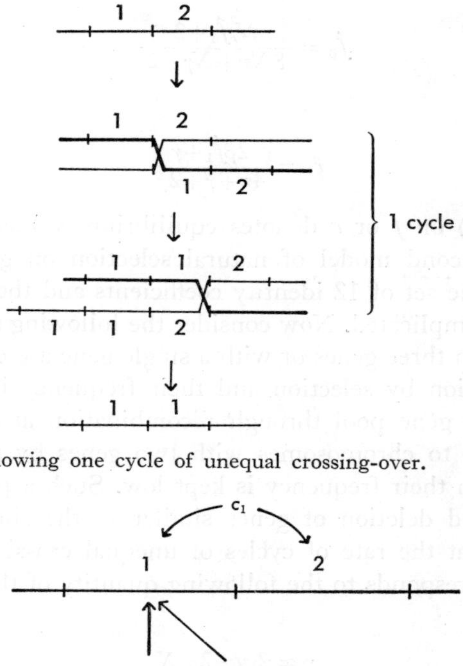

Fig. 5. Diagram showing one cycle of unequal crossing-over.

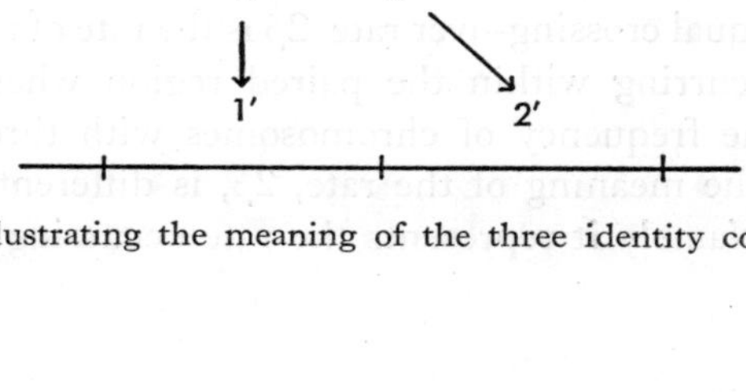

Fig. 6. Diagram illustrating the meaning of the three identity coefficients.

and Ohta (1979) for treating a large gene family. As shown in the figure, c_1 is the identity probability between the two genes residing on the same chromosome, f_0 is that between two genes located at homologous sites, but on different chromosomes, and f_1 is that between two genes at non-homologous sites, and on different chromosomes.

It is possible to formulate the change of these identity coefficients by various forces in one generation (Ohta, 1981). Their equilibrium values are of special interest and can be shown to be as follows, by assuming all parameters, γ, β, $1/(2N)$, and v are much less than unity.

$$\hat{f}_1 = \frac{\gamma(6Nv+N\gamma+N\beta+1)}{\left(4Nv+\frac{N\gamma}{2}+1\right)^2(4v+\gamma+2\beta)-\left(\frac{N\gamma}{2}\right)^2(4v+\gamma+2\beta)-2\beta\left(4Nv+\frac{N\gamma}{2}+1\right)} \tag{5}$$

$$\hat{f}_0 = \frac{N\gamma\hat{f}_1+2}{8Nv+N\gamma+2} \tag{6}$$

and

$$\hat{c}_1 = \frac{2\beta\hat{f}_1+\gamma}{4v+\gamma+2\beta} \tag{7}$$

where the hat (ˆ) on f or c denotes equilibrium values.

As to the second model of natural selection on gene number, one needs to define the set of 12 identity coefficients and therefore calculation becomes much complicated. Now consider the following situation in which chromosomes with three genes or with a single gene are quickly eliminated from the population by selection and their frequency is very low. They contribute to the gene pool through recombination at meiosis, and also when they revert to chromosomes with two genes by unequal crossing-over, even though their frequency is kept low. Such a process may result in duplication and deletion of genes similar to the simple cycle model. I have shown that the rate of cycles of unequal crossing-over (γ) of the simple model corresponds to the following quantity of the selection model (Ohta, 1981),

$$\gamma \approx 2(u+2\beta)X \tag{8}$$

where u is the unequal crossing-over rate, 2β is the rate of inter-chromosomal recombination occurring within the paired region when two gene units pair, and X is the frequency of chromosomes with three genes or with one gene. Here, the meaning of the rate, 2β, is different from that of the previous model. Namely it represents the rate occurring within the entire paired region, whereas the rate β of the previous model is the recombination value occurring between the units to compare identity. Therefore the expected rate of actual recombination taking place between the units in the selection model is $(1/2)\times 2\beta=\beta$ which is the same as in the previous model, by assuming that recombination occurs uniformly in the paired region. Also, if the selective disadvantage of chromosomes with three genes or that with one gene is s, X becomes at equilibrium between selection and unequal crossing-over,

$$X=\frac{u}{2s}. \tag{9}$$

By using Eqs. (5)–(7), (8), and (9), it is possible to estimate the identity coefficients under various sets of parameter values. It would be expected that, in the ordinary case of $Nv\ll 1$, $N\gamma\ll 1$, $4v+\gamma\ll 2\beta$ and $4v<\gamma$, all three identity coefficients are close to unity but $\hat{c}_1$ and $\hat{f}_1$ are slightly less than $\hat{f}_0$. In the following I shall try to apply the results. From Table 1 of Zimmer *et al.* (1980), the two duplicated loci of α-hemoglobin of primates are quite similar, *i.e.*, only one species (orangutan) out of five produce two kinds of α-hemoglobin which differ by one amino acid. Then, the identity coefficients approximately become, $c_1\approx f_1\approx 0.8$ and $f_0\approx 1$, when measured by the amino acid identity of the total polypeptide. Thus, the above condition (Nv, $N\gamma\ll 1$, $4v+\gamma\ll 2\beta$ and $4v<\gamma$) would be satisfied. If the rate of amino acid substitution for α-hemoglobin loci is roughly 1.5×10^{-7} per year per locus, and under the neutral mutation-random drift hypothesis (see Kimura, this volume), v in the above equations takes this value when measured per year. Then the rate of duplication and deletion (γ) would be 10^{-5}–10^{-6} per year. If X (frequency of chromosomes with three genes or with one gene) is about 10^{-2} (Goosens *et al.*, 1980), the above value of γ implies that $2(u+2\beta)$ is 10^{-3}–10^{-4} from Eq. (8); thus the rate of unequal crossing-over is probably about 10^{-4} per generation, since it is likely that $u<2\beta$. The selection coefficient, s, would be roughly 10^{-2} from Eq. (9). Such an estimation is tentative and may require revisions in the future.

Let us consider the case in which the parameter values estimated in α-hemoglobin genes can be applied to the γ-hemoglobin gene family. An interesting observation concerning this gene family is that there is a polymorphism with respect to a restriction enzyme map in the intervening sequence of human γ gene (Jeffreys, 1979). Noteworthy is the fact that this polymorphism exists in both of the tandem genes. Thus the identity coefficient, f_0, is not unity and may be the same as f_1 when measured by the restriction enzyme. It is expected that f_0 (zero step identity, see Fig. 6) is larger than c_1 and f_1 (one step identity) when $4v<\gamma$. However the identity coefficients as expressed by Eqs. (5) and (6), merely represent average values and an individual case may differ much even if the parameters take similar values. By chance, we may observe $f_0 \approx f_1$ or even $f_0<f_1$. The situation is similar to the case of enzyme loci and identity coefficients may be associated with a large variance. In the case of the β-hemoglobin gene family, the two tandem genes are considerably differentiated (Zimmer *et al.*, 1980). Therefore the rate of duplication and loss or cycles of unequal crossing-over (γ) is much smaller in this family than in γ or α genes. Thus, in some cases, this rate is fairly high, whereas it is very low in other families. An important aspect is that natural selection seems to be working mainly on the quantity of gene products and therefore on number of genes, and the majority of amino acid substitutions are selectively neutral as Kimura shows (for review, see Kimura, this volume). If the evolution of globin gene families is representative of a fairly common phenomenon, duplication and deletion of genetic material may be partly responsible for Mukai's observation on slightly deleterious mutations (for review, see Mukai, this volume), that they occur with too high a frequency in *Drosophila*. Also, the globin gene families represent a most interesting example of progressive evolution through gene duplication as proposed by Ohno (1970). As suggested above, the unequal crossing-over rate may be as high as 10^{-4} per generation, and can be an important evolutionary force. One has to realize that the present well-organized gene families are the endproduct of natural selection, such that every species has experienced numerous trials and errors in terms of duplication and deletion of genes in the long course of evolution.

DISCUSSION

The analyses presented in previous sections are based on several overly simple assumptions. In particular, it is assumed that the gene family size stays more or less constant. Actually, however, it is likely that the gene family size, especially of large families, varies from time to time. Furthermore, when there exist many copies of similar genes, accumulation of defective genes (pseudogenes) among them is unavoidable due to mutational pressure. Several notable examples are already reported; alternate arrangement of gene and pseudogene of 5S ribosomal DNA of *Xenopus* (Jacq *et al.*, 1977; Fedoroff, 1979) and the presence of one to two pseudogenes in one cluster of globin genes of human and some other mammals (Hardison *et al.*, 1979; Nishioka *et al.*, 1980; Vanin *et al.*, 1980). The study of gene family size is not independent of that of accumulation of defective genes, since total gene product or the number of functional genes in a genome is subject to natural selection.

The dynamics of spreading of pseudogenes in the population, under the condition that the genomes with extra-gene constitute majority of the population, is itself an important subject of population genetics, and extensive research has been carried out (for review, see Takahata, this volume). The main result is that, by mutational pressure, disappearance of one of the duplicate genes will occur under probable assumptions within a reasonable time period. Now our interest here lies in the joint process of spreading of duplicated genes through unequal crossing-over and random drift, and degeneration of extra-gene copies. Such a process is rather complicated and its clarification is left to future study, together with the problem of gene family size.

Next, let us consider the long term average of the rate of accumulation of pseudogenes, *i.e.*, a question on the number of pseudogenes expected to accumulate in a long period such as 10^7 or 10^8 years. The problem is different from the previous one in that it is independent of time of spreading of pseudogenes. The situation is analogous to the case of mutant dynamics; the problem of the rate of mutant substitution in a population *versus* that of the time until fixation of a mutant gene. In case of small multigene families, I have argued (Ohta, 1981) that, if the rate of occurrence of defective mutation is v_D per generation, and if a chromosome with extra-gene, which is defective, is selectively neutral, then the long term average of the rate of accumulation of defective gene in the population would be v_D times the frequency of chromosomes with extra-gene, which is equal to $v_D u/(2s)$ where u is the unequal crossing-over rate and s is the selective disadvantage of chromosomes with an extra-gene (see Eq. (9)). Then the number of pseudogenes would increase with time at this rate. Actually, selection may eliminate chromosomes with too many pseudogenes or those with pseudogenes in inappropriate positions. An important implication of the above argument is that accumulation of pseudogenes is proportional to the product of the rate of occurrence of defective mutations and that of unequal crossing-over.

As to the large multigene families, the situation is even more complicated. Crow and Kimura (1970, pp. 294–296) presented a model in which the gene family size is in equilibrium between natural selection and unequal crossing-over. They did not consider accumulation of pseudogenes. In their model, it is assumed that the fitness is maximum for an intermediate family size and decreases with the square of the deviation from this optimum value. Also, the difference of the family size between the new crossover product and the parental family is small and independent of parental size. Then the mean ($\bar{n}$) and variance (V_n) of family size in the population become at equilibrium (Crow and Kimura, 1970),

$$\bar{n} = n_{opt} \tag{10}$$

and

$$V_n = \sqrt{\frac{u\overline{x^2}}{2s}}, \tag{11}$$

in which n_{opt} is the optimum n, u is the rate of unequal crossing-over, $\overline{x^2}$ is the mean square of genes expanded or deleted, and s is the selection coefficient such that the fitness of a family with size n is $1-s(n-n_{opt})^2$. I equated the following parameters in connecting the Crow-Kimura model with my theory on genetic variation as in section 2 (Ohta, 1980, p. 109),

$$u = \gamma \text{ and } \overline{x^2} = m^2. \tag{12}$$

Here γ and m are defined as the rate of unequal crossing-over and the mean number of genes expanded or deleted *under the condition that all crossover products are selectively neutral.* When selection is incorporated as in the Crow-Kimura model, the above Eq. (12) would no longer hold. The situation is similar to the small multigene family (pp. 66–67) in that γ (rate of cycle of unequal crossing-overs in small family case) is equal to $2(u+2\beta)X$. The effects of unequal crossing-over (γ and m) for the large family case should become correspondingly smaller as selective elimination of chromosomes with unbalanced gene number becomes more intense. For a quantitative assessment, future study is necessary.

Hood *et al.* (1980) performed Monte Carlo simulation on gene family size and on accumulation of pseudogenes. They incorporated natural selection working at two levels, *i.e.*, to eliminate pseudogenes and to increase the number of functional genes, and concluded that the dynamic equilibrium is achieved by the balance of these two types of selection. Their figure indicates, however, that the proportion of pseudogenes increases with time, and an extremely long time would be required to reach equilibrium. The more detailed population genetical investigation would be desirable in the future.

In finishing this report, I emphasize that duplication and deletion of genes by unequal crossing-over are much more frequent, and important for the evolution of higher organisms than previously thought. In addition to unequal crossing-over, gene conversion may be playing an important role as recently suggested (Baltimore, 1981; Dover and Coen, 1981), and it would have the same effect as one cycle of unequal crossing-over, in resulting duplication of a DNA segment and deletion of another segment. The two mechanisms may provide various possibilities for the progressive evolution of gene families. At the same time, in terms of base sequences, genes are evolving with some steady rate, and the neutral theory of Kimura (1968) is essentially correct.

SUMMARY

For the last several years, population genetics theory on repeated genes which form multigene families has been investigated and this report summarizes such studies. Genetic variation is measured by an identity coefficient which is the probability of identity of two genes belonging to the same family. Amounts of change of identity coefficients by various evolutionary forces in unit time are formulated. An important force is unequal crossing-over, by means of which mutant genes may spread on the chromosome and coincidental evolution occurs. Equilibrium and transient values of identity coefficients are obtained under simplifying assumptions and are applied to the statistical analyses of amino acid diversity of variable region of immunoglobulins, genes of which represent a most interesting example of an evolving multigene family. Theoretical study is extended to the small multigene family in which only two or three genes are tandemly duplicated, such as in various globin gene families of mammals.

Acknowledgment

I thank Dr. N. Takahata for his many useful comments on the first draft of this paper, and Dr. K. Aoki for carefully going over the manuscript and making many helpful suggestions.

REFERENCES

Baltimore, D. (1981). Gene conversion: some implications for immunoglobulin genes. *Cell* **24**, 592–594.

Crow, J. F. and Kimura, M. (1970). *An Introduction to Population Genetics Theory*. Harper & Row, New York.

Dayhoff, M. O. (1972). *Atlas of Protein Sequence and Structure 1972*. National Biomedical Research Foundation, Silver Spring, Maryland.

Dover, G. A. and Coen, E. S. (1981). Springcleaning ribosomal DNA: a model for multigene evolution? *Nature* **291**, 731–732.

Early, P., Huang, H., Davis, M., Calame, K., and Hood, L. (1980). An immunoglobulin heavy chain variable region gene is generated from three segments of DNA: V_H, D and J_H. *Cell* **19**, 981–992.

Fedoroff, N. V. (1979). On spacers. *Cell* **16**, 697–710.

Fitch, W. M. (1973). Aspects of molecular evolution. *Annu. Rev. Genet.* **7**, 343–380.

Goossens, M., Dozy, A. N., Embury, S. H., Zachariades, Z., Hadjiminas, M. G., Stamatoyannopoulos, G., and Kan Y. W. (1980). Triplicated α-globin loci in humans. *Proc. Natl. Acad. Sci. U.S.* **77**, 518–521.

Hardison, R. C., Butler III, E. T., Lacy, E., Maniatis, T., Rosenthal, N., and Efstratiadis, A. (1979). The structure and transcription of four linked rabbit β-like globin genes. *Cell* **18**, 1285–1297.

Hood, J. M., Huang, H. V., and Hood, L. (1980). A computer simulation of evolutionary forces controlling the size of a multigene family. *J. Mol. Evol.* **15**, 181–196.

Hood, L., Campbell, J. H., and Elgin, S. C. R. (1975). The organization, expression, and evolution of antibody genes and other multigene families. *Annu. Rev. Genet.* **9**, 305–353.

Jacq, C., Miller, J. R., and Brownlee, G. G. (1977). A pseudogene structure in 5S DNA of *Xenopus laevis*. *Cell* **12**, 109–120.

Jeffreys, A. J. (1979). DNA sequence variants in the ${}^{G}\gamma$-, ${}^{A}\gamma$-, δ- and β-globin genes of man. *Cell* **18**, 1–10.

Jones, C. W., Rosenthal, N., Rodakis, G. C., and Kafatos, F. C. (1979). Evolution of two major chorion multigene families as inferred from cloned cDNA and protein sequences. *Cell* **18**, 1317–1332.

Kabat, E. A., Wu, T. T., and Bilofsky, H. (1976). *Variable Regions of Immunoglobulin Chains*. Med. Computer Systems, Bolt, Beranek and Newman, Cambridge, Mass.

Kimura, M. (1968). Evolutionary rate at the molecular level. *Nature* **217**, 624–626.

Kimura, M. (1982). The neutral theory as a basis for understanding the mechanism of evolution and variation at the molecular level. This volume pp. 3–56.

Kimura, M. and Ohta, T. (1979). Population genetics of multigene family with special reference to decrease of genetic correlation with distance between gene members on a chromosome. *Proc. Natl. Acad. Sci. U.S.* **76**, 4001–4005.

Mukai, T. Yamaguchi, O., Kusakabe, S., Tachida, H., Matsuda, M., Ichinose, M., and Yoshimaru, H. (1982). Lack of balancing selection for protein polymorphism. This volume pp. 81–120.

Lauer, J., Shen, C. J., and Maniatis, T. (1980). The chromosomal arrangement of human α-like globin genes: sequence homology and α-globin gene deletions. *Cell* **20**, 119–130.

Nei, M. (1975). *Molecular Population Genetics and Evolution*. Elsevier/North-Holland, Amsterdam.

Nishioka, Y., Leder, A., and Leder, P. (1980). An unusual alpha globin-like gene that has cleanly lost both globin intervening sequences. *Proc. Natl. Acad. Sci. U.S.* **77**, 2806–2809.

Ohno, S. (1970). *Evolution by Gene Duplication*. Springer-Verlag, Berlin.

Ohta, T. (1976). A simple model for treating the evolution of multigene families. *Nature* **263**, 74–76.

Ohta, T. (1978a). Theoretical study on genetic variation in multigene families. *Genet. Res.* **31**, 13–28.

Ohta, T. (1978b). Theoretical population genetics of repeated genes forming a multigene family. *Genetics* **88**, 845–861.

Ohta, T. (1979). An extension of a model for the evolution of multigene families by unequal crossing over. *Genetics* **91**, 591–607.

Ohta, T. (1980). Evolution and variation of multigene families. In *Lecture Notes in Biomathematics* (ed. S. Levin), Vol. 37. Springer-Verlag, Berlin.

Ohta, T. (1981). Genetic variation in small multigene families. *Genet. Res.* **37**, 133–149.

Ohta, T. and Kimura, M. (1981). Some calculations on the amount of selfish DNA. *Proc. Natl. Acad. Sci. U.S.* **78**, 1129–1132.

Petes, T. D. (1980). Unequal meiotic recombination within tandem arrays of yeast ribosomal DNA genes. *Cell* **19**, 765–774.

Sakano, H., Maki, R., Kurosawa, Y., Roeder, W., and Tonegawa, S. (1980). Two types of somatic recombination are necessary for the generation of complete immunoglobulin heavy-chain genes. *Nature* **286**, 676–683.

Seidman, J. G., Leder, A., Norman, M. N. B., and Leder, P. (1978). Antibody diversity. *Science* **202**, 11–17.

Smith, G. P. (1974). Unequal crossover and the evolution of multigene families. *Cold Spring Harbor Symp. Quant. Biol.* **38**, 507–513.

Solomon, E. (1980). The collagen gene family. *Nature* **286**, 656–657.

Szostak, J. W. and Wu, R. (1980). Unequal crossing over in the ribosomal DNA of *Saccharomyces cerevisiae*. *Nature* **284**, 426–430.

Takahata, N. (1982). The disappearance of duplicate gene expression. This volume pp. 169–190.

Tonegawa, S., Maxam, A. M., Tizard, R., Bernard, O., and Gilbert, W. (1978). Sequence of a mouse germ line gene for a variable region of an immunoglobulin light chain. *Proc. Natl. Acad. Sci. U.S.* **75**, 1485–1489.

Vanin, E. F., Goldberg, G. I., Tucker, P. W., and Smithies, O. (1980). A mouse α-golbin-related pseudogene lacking intervening sequences. *Nature* **286**, 222–226.

Wu, T. T. and Kabat, E. A. (1970). An analysis of the sequences of the variable regions of Bence Jones proteins and myeloma light chains and their implications for antibody complementarity. *J. Exp. Med.* **132**, 211–250.

Yaoita, Y. and Honjo, T. (1980). Deletion of immunoglobulin heavy chain genes from expressed allelic chromosome. *Nature* **286**, 850–853.

Zimmer, E. A., Martin, S. L., Beverley, S. M., Kan, Y. W., and Wilson, A. C. (1980). Rapid duplication and loss of genes coding for the α chains of hemoglobin. *Proc. Natl. Acad. Sci. U.S.* **77**, 2158–2162.

3

Molecular Evolution, Protein Polymorphism and the Neutral Theory (1982) (ed. M. Kimura), pp. 81–120.
Japan Scientific Societies Press, Tokyo / Springer-Verlag, Berlin

Lack of Balancing Selection for Protein Polymorphisms*

TERUMI MUKAI, OSAMU YAMAGUCHI,
SHIN-ICHI KUSAKABE, HIDENORI TACHIDA,
MUNEO MATSUDA, MOTOSHI ICHINOSE,
AND HIROSHI YOSHIMARU

Department of Biology, Kyushu University, Fukuoka 812, Japan

The astonishingly rapid development of molecular genetics has caused a revolution in the study of population genetics and evolution, and it is now possible to discuss the evolution of organisms at the molecular level. Since Lewontin and Hubby (1966) reported a large number of protein polymorphisms in *Drosophila pseudoobscura*, a tremendous amount of data has been accumulated in many organisms, including humans (Harris, 1966). It is now accepted that polymorphisms are the general rule in wild populations of organisms.

There are two almost diametrically opposed theories for the explanation of the above phenomenon: the selection theory and the neutral theory. The former says that some form of balancing selection is responsible for the maintenance of these polymorphisms (Prakash *et al.*, 1969), while the latter states that the polymorphism is one phase of the spreading of selectively neutral or effectively neutral genes by random genetic drift in a population (Kimura, 1968; Kimura and Ohta, 1971).

* This is paper No. 13 from the Laboratory of Population Genetics, Department of Biology, Kyushu University.

It should be pointed out here that the selectionists do not believe that balancing selection is operating on all polymorphic loci whereas the neutralists accept the operation of selection at a minority of polymorphic loci but claim that the majority of polymorphisms is due to random genetic drift of selectively neutral or near-neutral alleles.

A number of investigators have attempted to solve the above problem. Some made a census of gene frequencies over many populations and attempted to relate the results to the ecological conditions of these populations (Schaffer and Johnson, 1974); some investigated the nature of isozymes and tried to explain natural selection on the basis of the difference in characteristics between isozymes (Vigue and Johnson, 1973); and some applied a highly sophisticated mathematical method to the measurement of selection (Yamazaki and Maruyama, 1972). However, there is still no concensus among biologists as to which mechanism is the correct one. We have attempted to check experimentally in the laboratory whether or not selection is operating on individual polymorphic loci. The principal candidates for balancing selection are overdominance, frequency-dependent selection, and diversifying selection, and these will be discussed in the present chapter.

OVERDOMINANCE

1. *Linkage Disequilibrium*

Franklin and Lewontin (1970) have shown on the basis of computer simulation that if overdominance with respect to fitness is universal, strong linkage disequilibrium is formed under random mating conditions even when multiplicative (or independent) gene action is the general rule. We have conducted extensive studies to test linkage disequilibrium for the isozyme loci together with polymorphic inversions of *D. melanogaster* for the Raleigh, North Carolina (N.C.) population, and significant nonrandom associations between isozyme genes were not detected except in a few cases which can be understood as due to sampling error. However, highly significant linkage disequilibria were detected between some isozyme genes and polymorphic inversions (Mukai *et al.*, 1974), and these disequilibria are considered to be due to historical accidents (Mukai and Voelker, 1977). The effective size of the above population must be extremely large for neutral or near-neutral genes since populations appear to be continuous over a very wide area (the continent of North America) and, even for lethal genes, it has been estimated to be larger than 20,000 (Mukai and Yamaguchi, 1974).

Although the finding of Franklin and Lewontin (1970) described above was confirmed analytically by Slatkin (1972), his results together with the findings of Yamazaki (1977) indicates that if the population size increases the formation of linkage disequilibrium becomes gradually more difficult or that the intensity of linkage disequilibrium decreases with an increase in the size of the population. Thus, there was a possibility that linkage disequilibrium was not formed in the Raleigh, N. C. population since the effective population size was too large, although slight overdominance was manifested at many loci. In order to test the above hypothesis, we have examined the linkage disequilibrium for isozyme genes and polymorphic inversions of *D. melanogaster* on two isolated islands and in one marginal population. One of the characteristics of these populations is that their effective sizes are small. Thus, if overdominance and multiplicative gene action are common, there is a possibility of linkage disequilibria.

The second chromosomes were extracted from the following populations: Ishigaki-jima Island, Okinawa, in May 1976 (abbreviated as IG), the same island, in May 1977 (IW), Chichi-jima Island, Ogasawara, in December 1975 (OG), and Hirosaki, Aomori, in October 1976 (HS). With the *Cy-Pm* inversion method, 504 second chromosomes from the IG population, 498 from IW, 539 from OG, and 477 from HS were extracted. The following four enzyme loci on the second chromosome were studied: α-glycerol-3-phosphate dehydrogenase-1 (α*Gpdh-1*, map position 2-17.8), malate dehydrogenase-1 (*Mdh-1*, map position 2-35.3), alcohol dehydrogenase (*Adh*, map position 2-50.1), and hexokinase-C (*Hex-C*, map position 2-74.5). The salivary gland chromosomes of all chromosome lines were studied. For the identification of alleles of the loci, the starch-gel electrophoresis method was used (*cf.* Stone *et al.*, 1968). Linkage relationships of the loci are presented in Fig. 1 together with the two polymorphic inversions, *In(2L)t* and *In(2R)NS*. The data on IG (1976) are shown in Table I as representative of the four populations investigated. In this population, two alleles were found at each locus (*F* and *S*), but the gene frequency of the *F* allele at the *Hex-C* locus was extremely low (0.002 ± 0.002) and this locus was not analyzed. For the sake of simplicity, *In(2L)t*, α*Gpdh-1*, *Mdh-1*, *Adh*, and *In(2R)NS* are designated as I, A, B, C, and J, respectively. Analyses for linkage disequilibria were only conducted

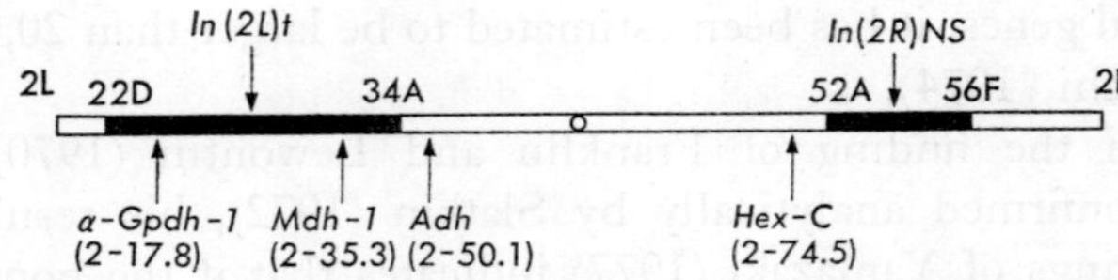

Fig. 1. The locations of isozyme genes and polymorphic inversions in the second chromosome of *D. melanogaster*.

TABLE I
Frequency of Second Chromosome Genotype for Four Isozyme Loci and Two Polymorphic Inversions in the IG population (1976)

Isozyme genotype	Chromosome structure				Total	Expected
	Standard	*In(2L)t*	*In(2R)NS*	*In(2L)t*+ *In(2R)NS*		
FFFF[a]	0	0	0	0	0	0.00
FFFS	0	0	0	0	0	2.05
FFSF	0	0	0	0	0	0.03
FFSS	5	5	5	1	16	13.62
FSFF	0	0	0	0	0	0.12
FSFS	30	7	9	6	52	58.87
FSSF	0	1	0	0	1	0.78
FSSS	66	178	35	118	397	390.52
SFFF	0	0	0	0	0	0.00
SFFS	0	0	0	0	0	0.17
SFSF	0	0	0	0	0	0.00
SFSS	1	0	0	0	1	1.11
SSFF	0	0	0	0	0	0.01
SSFS	7	0	7	0	14	4.80
SSSF	0	0	0	0	0	0.06
SSSS	15	1	5	2	23	31.85
Total	124	192	61	127	504	503.99

[a] The order of loci from left to right is *αGpdh*, *Mdh*, *Adh*, and *Hex-C*.

up to four factors. χ^2 analysis was applied to the data, and the χ^2 value was partitioned into each component with one degree of freedom (*cf.* Mukai *et al.*, 1974). The results are presented in Table II. For the other populations, similar results were obtained. From these results, the following findings were obtained: when significant χ^2 values are obtained, polymorphic inversions (I or J) are always involved. Thus further analyses were conducted after excluding the polymorphic inversion-carrying chromosomes. In this case, no significant linkage disequilibrium was detected.

TABLE II
The χ^2 Value for Each Component of Linkage Disequilibria Involving the *αGpdh-1* (=A), *Mdh-1* (=B), *Adh* (=C), *In(2L)t* (=I) and *In(2R)NS* (=J) (IG population of *D. melanogaster*)

Component	$\chi^2_{df=1}$	Component	$\chi^2_{df=1}$
IA	54.29[b]	IBC	5.85[a]
IB	5.94[a]	IBJ	—
IC	62.13[b]	ICJ	0.99
IJ	2.34	ABC	0.55
AB	0.07	ABJ	—
AC	20.37[b] (2.72)†	ACJ	4.27[a]
AJ	0.00	BCJ	—
BC	2.65	IABC	0.94
BJ	—	IACD	—
CJ	0.51	IACJ	8.86[a]
IAB	0.71	IBCJ	—
IAC	16.67[b]	ABCJ	—
IAJ	0.08		

[a] Significant at the 5% level.
[b] Highly significant.
† The figure in parentheses indicates the $\chi^2_{df=1}$ value when inversion-free chromosomes alone were used. Thus, the 20.37 value is due to the difference in gene frequency between *In(2L)t*-carrying chromosomes and *In(2L)t*-free chromosomes.
— Stands for the component which cannot be tested because of low gene frequencies at the loci involved.

The significant linkage disequilibria between polymorphic inversions and isozyme genes in the IG population can be explained by the founder effect. As will be described in the following section, it has been found that chromosomes carrying *In(2L)t*, the frequency of which is about 60%, have originated from only a few chromosomes with very little genetic variation (perhaps one) (Mukai *et al.*, 1980). If there is little recombination between the inversion and isozyme genes located near the breakpoints of the inversion, the existence of linkage disequilibrium can be understood.

Thus, we could not find linkage disequilibria induced by epistasis. This result is consistent with the finding obtained for the Raleigh, N. C. population (Mukai and Voelker, 1977). Very similar results were obtained for other populations (Yamaguchi *et al.*, 1980).

In order to estimate the magnitude of linkage disequilibrium due to random genetic drift in the populations (the magnitude due to sampling of chromosomes for the test being excluded), the following analysis was made. The coefficient of linkage disequilibrium (D) is known to be gene frequency dependent. In order to partially avoid this difficulty, the correlation coefficient of gene frequencies (R) is used. The following relationship is known for the $\chi^2_{df=1}$ value due to linkage disequilibrium between two factors:

$$R^2 = \frac{\chi^2_{df=1}}{n} \qquad (1)$$

where n is the number of chromosomes employed. In the actual calculation, the value (R'^2) calculated by dividing the average of all $\chi^2_{df=1}$ values within each population and year by n was close to the corresponding value of $\hat{R}^2$ calculated from the $\chi^2_{df=1}$ values of the two-factor interactions. Thus, the R'^2 values were used instead of the corresponding R^2 values in order to obtain a smooth relationship between R^2 and n over populations and years.

Using the second chromosomes excluding those carrying polymorphic inversions, the total χ^2 value in each of the IG, IW, OG, and HS populations was partitioned into the χ^2 components with one degree of freedom. The same thing was done for each of the 1968, 1969, 1970, and 1974 populations of N. C. (*cf.* Mukai *et al.*, 1974; Mukai and Voelker, 1977). The values of 1,000 R'^2 were calculated for each population using the average of the $\chi^2_{df=1}$ values. In addition, pooled data on the 1968, 1969, and 1970 populations of Raleigh, N. C. were also analyzed. In these calculations, one extremely large value ($\chi^2_{df=1}=9.89$, BC of IW) was excluded, since it is clear that this abnormality was induced by a low gene frequency of Mdh^F ($\hat{q}_F=0.050\pm0.010$) and sampling. The results are presented graphically in Fig. 2. The estimate of the R'^2 value is dependent on the sample size (n). This relationship can clearly be seen in this figure. The R'^2 value in the original population can be approximately predicted by the limiting value of 1,000 R'^2 when the sample size (n) approaches infinity. For both the Japanese and Raleigh populations, the R'^2 values in the populations appear to be very small. The average map distance of the four loci investigated is 30.82 cM. Considering double cross-over and the lack of recombination in *Drosophila* males, the effective average recombination value of the above cases might be approximately 10%. Using the relationship $R^2\cong 1/4N_ec$ under the condition of $N_ec\gg 1$, where N_e is the effective size of the population and c is the recombination value (Hill and Robertson, 1968; Ohta and Kimura, 1969), the R^2 value due to random genetic drift may be estimated assuming selective neutrality of the polymorphic isozyme

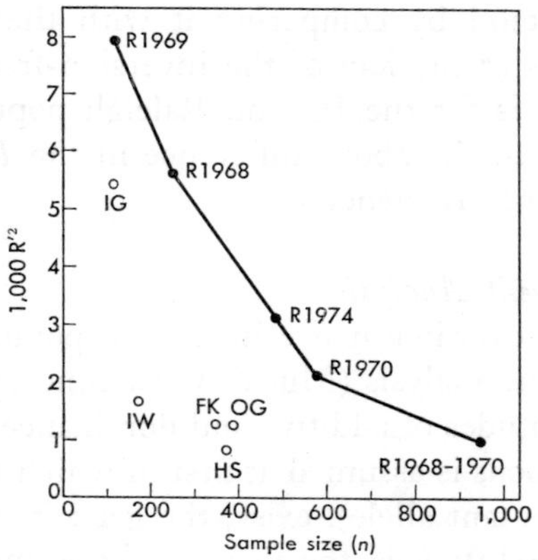

Fig. 2. Relationship between the estimates of 1,000 R'^2 and sample size in various populations investigated. Chromosomes carrying polymorphic inversions were excluded. IG, IW, OG, and HS mean the Ishigaki-Jima (1976), Ishigaki-Jima (1977), Ogasawara, and Hirosaki populations. FK stands for the Fukushima population (unpublished data). R 1969 stands for the Raleigh population (1969) and so on.

genes investigated. In fact, Mukai *et al.* (1974) showed approximate selective neutrality of these genes with respect to viability and fertility. The effective sizes of the populations (N_e) for lethal genes were estimated to be *ca.* 4,700, 6,000, and 2,600 for the IG, OG, and HS populations, respectively, and the effective size of the Raleigh population was estimated to be approximately 20,000 (Mukai and Yamaguchi, 1974; Yamaguchi *et al.*, 1980), If N_e of the island (IG and OG) and marginal (HS) populations is about 10,000, then R^2 becomes 0.00025 (or 1,000 R^2=0.25). Comparing this value with an approximate value that can be obtained from Fig. 2, the linkage disequilibrium in the original Japanese populations can be explained entirely by random genetic drift, and the largest fraction of χ^2's in the present analysis is due to sampling error. One of the most important findings in the present analysis is that the R'^2 values of the Japanese populations are generally smaller than those of the Raleigh populations in contrast to expectations on the basis of the effective sizes of the populations. This result may be partially explained by the difference in the effect of migration: in the Raleigh population, a number of immigrants must come in every generation from adjacent populations with slightly different gene frequencies. On the other hand, immigrants can seldom be expected in island or isolated populations. Thus, the meaning of the small χ^2 values due to linkage disequilibrium in the Japanese populations can be understood by comparing it with the Raleigh population. The gene frequencies at the loci of the inversion-free chromosomes used for the present analysis for the IG and Raleigh populations are approximately the same. Thus, the above difference in the R'^2 values is not due to a difference in gene frequencies.

2. *Variance Component Analysis*

The presence of overdominance in natural populations can be tested by variance component analysis (Mukai, Cardellino *et al.*, 1974). In order to calculate the magnitudes of additive and dominance variances for fitness, a model for a single locus is assumed at first in which two alleles, A ("normal" allele) and a (mutant allele), exist; the gene frequency of A is p and that of a is q ($p+q=1.0$; h is the degree of dominance of the mutant allele, and s is the selection coefficient against the mutant (aa) homozygote (Wrightian fitness model). With this model, the additive (σ_a^2) and dominance (σ_d^2) variances are given by:

$$\sigma_a^2 = 2pq[(p-q)h+q]^2s^2; \qquad (2)$$

$$\sigma_d^2 = p^2q^2(1-2h)^2s^2. \qquad (3)$$

Under the condition of $h \gg \sqrt{\mu/s}$, where μ is the mutation rate from A to a, the equilibrium gene frequency (q) becomes:

$$\hat{q} = \frac{\mu}{hs}. \qquad (4)$$

This means that $\hat{q}$ is a very small quantity as compared with 1. Then almost the whole genetic variance becomes an additive variance ($\sigma_a^2 \cong 2pqh^2s^2$) and the dominance variance becomes nearly zero (Mukai, Cardellino *et al.*, 1974).

In the case of overdominance ($h<0$), the equilibrium gene frequency is:

$$\hat{q} = \frac{h}{2h-1}. \qquad (5)$$

Then the additive variance becomes zero and the dominance variance turns out to be:

$$\sigma_d^2 = \left[\frac{(h-1)hs}{2h-1}\right]^2. \qquad (6)$$

Thus, if we estimate the additive and dominance variances from an equilibrium population and compare their magnitudes, it is possible to judge whether or not overdominance is common in the population.

It is difficult to estimate the genetic variance components for fitness. Therefore we have estimated the components of genetic variance for viability, an important components of fitness, by assuming that viability and fertility are positively correlated, or that viability is positively correlated with fitness. It is then necessary to verify this assumption.

It is possible to estimate the total homozygous load (T) due to newly arisen mutations and to partition it into the lethal load (L) and the detrimental load (D) (Greenberg and Crow, 1960). In the estimation of the homozygous viability of the $+_i$ chromosome by the *Cy* method (Wallace, 1956), a cross, $Cy/+_i \times Cy/+_i$, is made, where *Cy* is a dominant marker and included in *In(2LR)SM1*, which suppresses recombination in the second chromosome in the heterozygotes. In the offspring, phenotypically *Cy* and wild-type flies segregate at the rate of a:b (the expected ratio is 2:1). The *Cy* homozygote is lethal. The viability is expressed as $2a/(b+1)$. For the estimation of heterozygote viability, $Cy/+_i$ and $Cy/+_j$ are crossed. The viability is expressed in the same way as in homozygotes.

For the test of spontaneous mutations, a single chromosome is multiplied by the marked inversion technique and spontaneous mutations are accumulated at random at the minimum pressure of natural selection. In certain generations, the homozygous viability of each chromosome line can be tested by the *Cy* method. The homozygous load can be partitioned into its components as follows:

$$\left.\begin{aligned} T &= \ln A - \ln B \\ D &= \ln A - \ln C \\ L &= \ln C - \ln B \end{aligned}\right\} \qquad (7)$$

where A is the homozygous viability of the original chromosome, B is the average viability of all homozygote lines, and C stands for the average viability of homozygote lines excluding lethal-carrying lines. The $D:L$ ratio was estimated twice: 0.975 (Mukai and Yamazaki, 1968) and 0.954 (Mukai *et al.*, 1972). It should be noted that in the case of newly arisen mutations the gene frequency (q) must be very small, or $q = n\mu$ where n is the number of generations for which mutations were accumulated and μ is the spontaneous mutation rate per locus per generation at any locus.

In the case of an equilibrium population, the homozygous load can be estimated by the *Cy* method using chromosomes extracted at random.

However, we could not avoid using the average viability of random heterozygotes as A. Thus, the D value estimated should be adjusted by the random genetic load of viability. This adjustment was made assuming that mutant genes are deleterious to both viability and fertility to the same extent and also considering the effective size of the population. The adjusted D value is called D'. The $D':L$ ratio was estimated for four populations: 1.007 (N. C. population), 1.278 (Florida population), 0.976 (Wisconsin population), and 0.922 (Yamanashi population) (Mukai, 1975). These values are close to those of the newly arisen mutations described above. It is generally accepted that "recessive" lethal genes are heterozygously slightly deleterious to both the viability and fertility of their carriers. The averages of the products of selection coefficients of mutant genes (s) and their degrees of dominance (h) are about 0.015 ($s_l=1$, $h_l=0.015$) for lethal genes and 0.012 ($s_d=0.03$, $h_d=0.4$) for mildly deleterious genes and they are approximately the same. Thus, insofar as viability is concerned, lethal genes and mildly deleterious genes are approximately equally selected our. Thus, non-lethal deleterious genes should be deleterious to both viability and fertility in heterozygous conditions. Otherwise, the $D':L$ ratio would not be close to the $D:L$ ratio of newly arisen mutations. Thus, as a first approximation, we can assume that the effect of a mutant gene on viability is approximately proportional to that on fertility.

Two population, from IG, Okinawa and from Aomori, were analyzed. In the former population, 294 second chromosomes with the standard gene arrangement (abbreviated as N) and 308 second chromosomes carrying *In(2L)t* (abbreviated as I) were employed. These chromosomes were maintained individually as $Cy/+_i \times Cy/+_i$ strains after extracting by the *Cy-Pm* method. Using these chromosome lines, 43 seven $\times$ seven partial diallel crosses were made: 19 sets of $N \times N$, 20 sets of $I \times I$, and 4 sets of $N \times I$. In each set, 7 lines were used for rows ($Cy/+_i\, i=1$–7) and the other 7 lines were used for columns ($Cy/+_j\, j=8$–14). Thus, 602 lines, among which 294 were N and 308 were I, were employed. Reciprocal crosses were made with 2 simultaneous replications. This mating is represented diagramatically in Fig. 3. The viability of $+_i/+_j$ flies in the offspring of $Cy/+_i \times Cy/+_j$ was expressed as in the Cy method described above. Since there was no significant difference between the reciprocal crosses, they are considered to be replications. Thus, the analysis of variance was conducted for each set after each viability index was standardized by the mean viability of that set. A typical case is shown in Table III. In each set, the additive ($\sigma_A{}^2$) and dominance ($\sigma_D{}^2$) variances on a chromo-

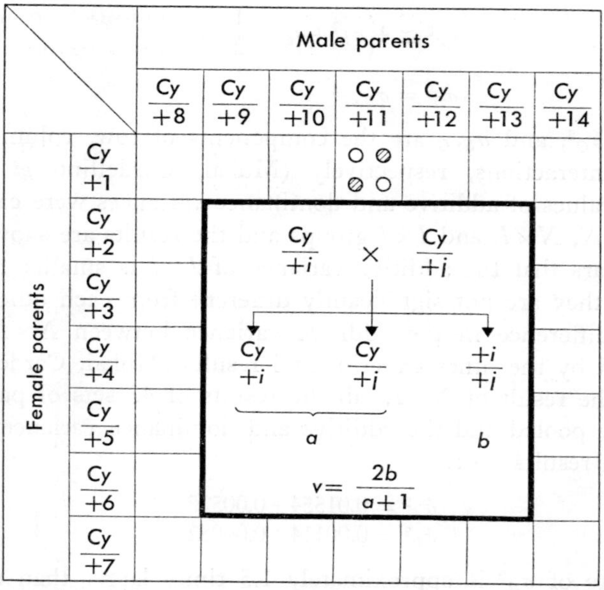

Fig. 3. Mating scheme of partial diallel cross (factorial experiment). ○● stand for reciprocal crosses.

TABLE III

A Typical Analysis of Variance for a 7×7 Partial Diallel Cross for the Second Chromosomes of *D. melanogaster* Sampled from the IG population (1978)

Source	DF	SS	MS	F
Females	6	0.264458	0.0441	6.34[a]
Males	6	0.579907	0.0967	13.90[b]
F×M	36	0.250271	0.0070	0.58
Error	147	1.774719	0.0120	
Total	195	2.869355		

[a] Significant at the 5% level.
[b] Significant at the 1% level.
$\hat{\sigma}_A{}^2=0.0207$, $\hat{\sigma}_D{}^2=-0.0013$.

some basis were estimated using the following formulas taking into account the fact that the *In(2LR)SM1* or *Cy* chromosomes do not always completely suppress the effects of deleterious genes located on the homologous chromosomes:

$$\sigma_A^2 \cong 4\left(\sigma_R^2+\sigma_C^2-\frac{1}{2}\sigma_{R\times C}^2\right) \tag{8}$$

$$\sigma_D^2 = \sigma_{R\times C}^2 \tag{9}$$

where σ_R^2, σ_C^2, and $\sigma_{R\times C}$ are the components of row, column, and row $\times$column interactions, respectively (Mukai, Cardellino *et al.*, 1974). The mean values of additive and dominance variances were calculated for each of $N\times N$, $N\times I$, and $I\times I$ groups and the results are shown in Table IV. It appears that the additive variance of $I\times I$ is smaller than that of $N\times N$, but they are not significantly different from each other. Since no significant difference in the additive variance between $N\times N$ and $I\times I$ is supported by the other experimental results (Mukai, Cardellino *et al.*, 1974) and the result of $N\times I$, all the results of 43 sets of partial diallel crosses were pooled and the additive and dominance variances were estimated. The results are:

$$\left.\begin{array}{l}\hat{\sigma}_A^2 = 0.01554\pm0.00523\\ \hat{\sigma}_D^2 = 0.00114\pm0.00082\end{array}\right\} \tag{10}$$

The estimate of σ_A^2 is approximately 1.5 times larger than that for the Raleigh, N. C. population, but the dominance variance is almost the same.

From the Aomori population, 140 second chromosomes were extracted and 10 seven$\times$seven partial diallel crosses were made in the same manner as above, without considering the existence of polymorphic inversions in the second chromosomes. The analysis was also made in the same way as above. The estimated additive and dominance variances are as follows:

$$\left.\begin{array}{l}\hat{\sigma}_A^2 = 0.00276\pm0.00091\\ \hat{\sigma}_D^2 = 0.00011\pm0.00014\end{array}\right\} \tag{11}$$

The above data were reanalyzed excluding the crosses containing inversions, but no significant difference was obtained (Kusakabe and Mukai, 1981).

TABLE IV
The Additive and Dominance Variances with Their 95% confidence Intervals of the Second Chromosomes of *D. melanogaster* Sampled from the IG population

	Additive variance	Dominance variance
Standard ($N\times N$)	0.01960±0.00999	0.00135±0.00145
Standard$\times$Inversion ($N\times I$)	0.02025±0.00912	0.00104±0.00120
Inversion ($I\times I$)	0.01075±0.00640	0.00070±0.00261
Pooled	0.01554±0.00523	0.00114±0.00082

From the above two analyses it is clear that dominance variances are much smaller than the estimates of the additive variance and are close to 0. This result agrees well with that from the Raleigh, N. C. population for which the additive and dominance variances were estimated to be $\hat{\sigma}_A^2=0.00960\pm0.00250$ and $\hat{\sigma}_D^2=0.00115\pm0.00046$ (Mukai, Cardellino *et al.*, 1974).

Thus, from the results of partial diallel crosses, it is also concluded that overdominance is rare in natural populations of *D. melanogaster*. Further analyses will be made in the following sections from the present data.

3. *Isolated Island Population*

Drosophila melanogaster populations typically have several cosmopolitan polymorphic inversions. Each of these is supposed to have occurred *uniquely* sometime in the past, and to have spread to various places in the world. Utilizing the history of one type of inversion in one population, it may be possible to estimate the average selection intensity for very weakly selected polymorphic loci.

A sufficient number of second chromosomes were extracted in April 1976 (Inoue *et al.*, 1979) and 1979 by the multiple inversion technique with each coming from a different individual of *D. melanogaster* sampled in the IG, Okinawa population. These chromosome lines were maintained at 18°C and balanced with *Cy* [*In(2LR)SM1*]. The homozygote and heterozygote viabilities of these chromosomes were estimated using the *Cy* method and the relative viability was expressed as described above.

For the determination of the karyotype of the chromosomes, *Cy*/+ males from each chromosome line were crossed with *cn bw* (cinnabar, brown eye color, white phenotype) female flies with the standard gene arrangement. The third instar larvae, whose genotypes were *cn bw*/++, were dissected and the salivary gland chromosomes were observed. Two cosmopolitan inversions, *In(2L)t* and *In(2R)NS*, were found in the population. Their frequencies were 0.63 and 0.59 for *In(2L)t* and 0.37 and 0.40 for *In(2R)NS*, in 1976 and 1979, respectively. There was no linkage disequilibrium between them (Yamaguchi *et al.*, 1980). The homozygotes were classified into two categories: *In(2L)t*-carrying and *In(2L)t*-free. The *In(2R)NS* and unique inversions, whose frequency was very low, were disregarded since there was no linkage disequilibrium between this inversion and *In(2L)t*, which suppresses almost all exchanges in the left arm of the second chromosome. The distribution pattern of standardized

TABLE V
Distributions of Homozygote Viabilities of *In(2L)t*-carrying and *In(2L)t*-free Second Chromosomes of *D. melanogaster*

Range of viability	*In(2L)t*-carrying	*In(2L)t*-free
0	15	65
0–0.05	93	81
(0–0.10)	(139)	(87)
0.05–0.15	59	11
0.15–0.25	11	8
0.25–0.35	18	10
0.35–0.45	13	7
0.45–0.55	13	12
0.55–0.65	32	23
0.65–0.75	58	41
0.75–0.85	73	53
0.85–0.95	74	63
0.95–1.05	54	40
1.05–1.15	34	15
1.15–1.25	6	4
1.25–1.35	7	1
Total	545	369

viabilities, pooled over the 2-year samples, are shown in Table V. The frequencies of lethal chromosomes, which are defined as chromosomes with viability indices of less than 0.1 following the conventional definition given by Greenberg and Crow (1960), were 0.255 ± 0.019 and 0.236 ± 0.022 for the *In(2L)t*-carrying and standard chromosomes, respectively. These two values were not significantly different from each other. The frequencies of completely lethal chromosomes differ in the two groups (Table V), but there is a spurious cause. The *In(2LR)SM1*, which includes *Cy*, is not as perfect a crossover suppressor with *In(2L)t* chromosomes as with standard chromosomes. Hence, the lethal gene may occasionally be transferred to the *Cy* chromosome, and *vice versa*, leading to a lethal chromosome being misclassified as incompletely lethal (Yamaguchi *et al.*, 1980). Thus, in the comparison of homozygous viabilities of *In(2L)t*-carrying and standard chromosomes, only chromosome lines whose homozygous viabilities are greater than 0.15 were employed. Their percentage frequencies are graphically presented in Fig. 4. The distribution patterns do not differ significantly ($\chi^2_{df=10}=6.11$, $0.80<P<0.90$). The mean viabilities were 0.7931 ± 0.0125 and 0.7846 ± 0.0138 for the *In(2L)t*-carrying and standard chromosomes, respectively.

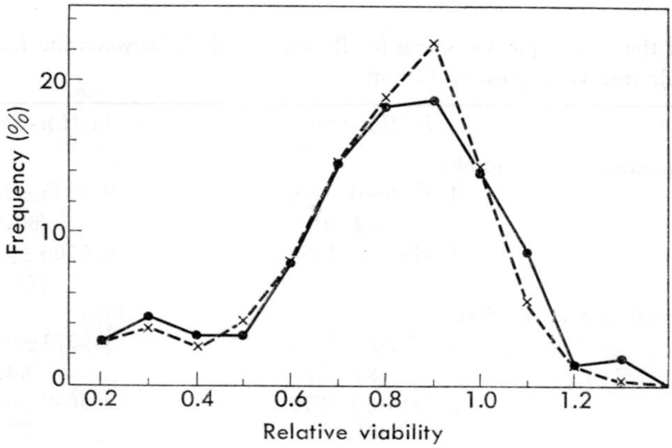

Fig. 4. Distribution patterns of homozygous viabilities of the *In*(2*L*)*t*-carrying (●) and the *In*(2*L*)*t*-free (×) chromosomes sampled from the IG population.

1) Estimation of the number of generations since the In(2*L*)*t-carrying chromosome entered the island population*

There is reason to suppose that a single or a few chromosomes with little genetic variation were introduced into the island population long after the standard chromosomes were established. There is no significant difference in the frequencies of lethal-carrying chromosomes between the inversion-carrying and standard chromosomes. In addition, the distribution patterns of homozygous viabilities of the *In*(2*L*)*t*-carrying and the standard chromosomes are almost completely the same (Fig. 4). Since about 100–200 generations are required for such mutants (including mildly deleterious genes) to nearly reach equilibrium (Mukai, 1979), it may be argued that more than 200 generations have passed since the *In*(2*L*)*t*-carrying chromosome was introduced into IG.

The genotypic variances of sternopleural and abdominal bristle numbers for the same experimental materials as above were estimated (Ichinose and Mukai, 1981). A summary of the results is presented in Table VI, which indicates that the genotypic variances for these bristle characters are smaller in the *In*(2*L*)*t*-carrying chromosomes than those for the standard chromosomes. Since some 2,000 generations are required for such genes to equilibriate (Mukai, 1979), this indicates that less than 2,000 generations have passed since the *In*(2*L*)*t*-carrying chromosomes entered the population.

The time since immigration can be narrowed down further. Nei and Li (1979) have shown that when d_0 and d_n are the differences between the

TABLE VI
Comparison of the Genotypic Variances for Bristle Numbers between the *In(2L)t*-carrying and the *In(2L)t*-free Chromosome Group

Group	*In(2L)t*-free	*In(2L)t*-carrying
(A) Sternopleural bristle number		
Female	1.3106±0.2036 (82.90)	0.9346±0.1477 (80.06)
Male	1.1785±0.1880 (78.61)	0.7706±0.1233 (78.19)
(B) Abdominal bristle number		
Female	4.9312±0.7701 (82.01)	3.9773±0.6260 (80.77)
Male	3.3514±0.5270 (81.03)	2.5640±0.4121 (77.45)
(C) Homozygous viability	0.04281±0.00472 (164.6)	0.04716±0.00475 (213.1)

For the sake of comparison, the genotypic variances for homozygote viabilities were added in (C). The numbers in parentheses stand for the approximate degrees of freedom.

frequency of an allele in the inversion-carrying and in the standard chromosomes at generations 0 and n, and c stands for the recombination value between the gene and the inversion,

$$d_n \cong d_0(1-c)^n \tag{12}$$

This formula is appropriate for estimating n since it does not depend on the inversion frequency remaining constant, which clearly has not been true, but it does depend on the constancy of the frequency of the allele in the standard chromosome, which is reasonable. The reasons for the increase in the inversion frequency are unknown, but are irrelevant for this discussion. There is an appropriate enzyme locus within the inversion, *αGpdh*, with two alleles: $\alpha Gpdh^F$ and $\alpha Gpdh^S$. The rate of recombination by double crossover between this locus and *In(2L)t* has been estimated to be 0.00022 (Mukai and Voelker, 1977). The frequencies of the $\alpha Gpdh^S$ allele averaged over 1976 and 1977 are 0.0187±0.0057 and 0.1647±0.0182 for the *In(2L)t*-carrying chromosomes and the standard chromosomes. Thus, using d_0=0.1647, d_n=0.1460, and c=0.5×0.00022, the value of n can be estimated by Eq. (12) to be 1,052. (The factor 0.5 in the value for c is because of the absence of exchange in *Drosophila* males.) The frequencies of F alleles at the *Adh* and *Mdh* loci (see Fig. 1), which both carry F and S alleles, were 0.3487±0.0234 and 0.0702±0.0126 in the standard chromosomes and 0.0374±0.0078 and 0.0221±0.0061 in the *In*(2*L*)*t*-carrying chromosomes, respectively. At both loci, the frequencies of allele *F* are very low in the inversion-carrying chromosomes.

The most reasonable interpretation is that the *In*(2*L*)*t* chromosome was monomorphic at the time it entered the population about 1,000 generations ago and that the small numbers of $\alpha Gpdh^S$, Mdh^F, and Adh^F alleles were introduced mainly by rare recombination events and by small input from even rarer mutations. Assuming these three to be typical of isozyme loci in general, we conclude that the left arm of the chromosome (the region governed by the inversion) is essentially monomorphic. The right arm has been randomized by recombination.

2) *Estimation of the average selection coefficient*

In order to compare the difference between homozygous and heterozygous viabilities in the inverted *versus* the standard chromosomes, the following parameters are defined:

- A=mean viability of heterozygotes between the standard chromosomes (abbreviated as N/N′).
- a=mean viability of homozygotes of the standard chromosomes (N/N).
- B=mean viability of heterozygotes between the *In*(2*L*)*t*-carrying chromosomes (I/I′).
- b=mean viability of homozygotes of the *In*(2*L*)*t*-carrying chromosomes (I/I).
- $L_{I(R)N}$=homozygous load relative to the average viability of the random mating population for the standard chromosomes.
- $L_{I(R)I}$=homozygous load relative to the average viability of the random mating population for the *In*(2*L*)*t*-carrying chromosomes.

Assuming that gene actions at different loci and the effects of various environmental factors are independent and that there is no genotype × environment interaction, the following relationships hold:

$$L_{I(R)N} = \ln A - \ln a \tag{13}$$

$$L_{I(R)I} = \ln B - \ln b. \tag{14}$$

Thus, the difference between $L_{I(R)N}$ and $L_{I(R)I}$ (D) is measurable and expressed as follows:

$$D = L_{I(R)N} - L_{I(R)I} = \ln A - \ln a - \ln B + \ln b = \ln \frac{Ab}{aB}. \tag{15}$$

It should be noted that in the present case D is concerned with loci, which in the standard chromosomes are polymorphic with near-neutral selection intensity, since lethal and mildly deleterious genes equally contribute to the homozygous loads of the *In(2L)t*-carrying and standard chromosome populations.

If there is overdominance at all polymorphic loci, the average selection coefficient can be estimated using the D value. If p_{ik} is the frequency of the k-th allele at the i-th locus and s_{ik} is the viability decrease in the homozygote relative to the heterozygote (assuming that all heterozygotes are equal),

$$A-a = \sum_{i,k} (p_{ik}s_{ik} - p_{ik}^2 s_{ik}) = \sum_{i,k} p_{ik}(1-p_{ik})s_{ik} \cong \sum_{i=1}^{n} \bar{H}_i \bar{s}_i \cong n\bar{H}\bar{s}, \tag{16}$$

where H_i and $\bar{H}$ are the mean frequency of heterozygotes at the i-th locus ($i=1-n$) and for all polymorphic loci. In this rough approximation it has been assumed that H and s are independent.

Since $B-b \cong 0$ because of the near-monomorphism of the inverted region of the *In(2L)t* chromosomes,

$$D \cong n\bar{H}\bar{s} \tag{17}$$

giving a basis for the estimation of $\bar{s}$.

From the present experimental data, a and b are estimated to be $\hat{a}=0.78463 \pm 0.01376$ and $\hat{b}=0.79314 \pm 0.01253$. These are the values standardized by the average heterozygote viability of the population including the two types of chromosomes, *In(2L)t*-carrying and standard. However, the effect of double recombination between *In(2LR)SM1* and *In(2L)t* in heterozygotes should be considered. It increases the apparent viability of *In(2L)t* homozygotes. When the double recombination value on both sides of the *Cy* genes is p, then the apparent increment of the viability index is between about $1.85p$ and $0.5p$ if the real viabilities are 0.1 and 1.0. The p value was estimated to be about 0.02 (Yamaguchi *et al.*, 1980). Thus, $\hat{b}$ may be overestimated by approximately 0.01. Correcting this bias, $\hat{b}=0.78314 \pm 0.01253$.

For the estimation of A and B, the average viabilities obtained from partial diallel crosses (or factorial experiments) described above were used: $\hat{A}=0.92250 \pm 0.00770$ and $\hat{B}=0.93421 \pm 0.006965$. As in the case of homozygotes, some correction for the mean viability should be made for heterozygotes which carry *In(2L)t* in homozygous conditions. In this case, the viability index increases by about $0.5p$ on the average due to recombination on both sides of the *Cy* genes in parental *In(2LR)SM1/In(2L)t* females, where p is the double recombination value. The corrected $\hat{B}$ is approximately 0.92921 ± 0.00697. Although the standards used for these estimates are different from those used for the homozygotes described above, this does not cause any difficulty since we assumed gene actions and environmental factors to be independent and the difference is cancelled in the calculation of D (Eq. (15)).

The sampling variance of D [$V(D)$] may be estimated as follows:

$$V(D) \cong \frac{V(A)}{A^2} + \frac{V(a)}{a^2} + \frac{V(B)}{B^2} + \frac{V(b)}{b^2} \tag{18}$$

If the proportion of polymorphic loci is 0.3 (Lewontin and Hubby, 1966), the n in Eq. (17) may be estimated to be approximately 350, since the number of cistrons in the second chromosome is about 2,200 (*cf.* Judd *et al.*, 1972) and since *In*(*2L*)*t* almost completely suppresses recombination in the left arm of the second chromosome. $\bar{H}$ may be conservatively assumed to be 0.2 in *D. melanogaster*.

To summarize,

$A = 0.92250 \pm 0.00770$; $\hat{a} = 0.78463 \pm 0.01376$;
$\hat{B} = 0.92921 \pm 0.00697$; $\hat{b} = 0.78314 \pm 0.01253$;
$\hat{D} = -0.00915 \pm 0.02625$; from Eqs. (15) and (18);
$n = 350$; $\bar{H} = 0.2$; $\bar{s} = -0.00013 \pm 0.000375$; from Eq. (17).

In these calculations, $\bar{H}$ and n are treated as constants. Of course, the estimate of $\bar{s}$ is not significantly different from 0. The 95% confidence interval of $\bar{s}$ becomes as follows:

$$-0.00087 < \bar{s} < 0.00060.$$

From this result, it may be concluded that if selection coefficients are constant, they must be extremely small or possibly 0. However, there are reservations concerning this conclusion. The effects of allele manifesting overdominance in fertility but that are neutral in viability were not detected in the present experiments and the number of loci with such alleles may be extremely small.

4. *Direct Estimation of Viability by the Cy Method*

The above conclusion stating that overdominance for viability does not exist can be tested directly by the *Cy* method. In the estimation of the homozygous load, viabilities of heterozygotes consisting of random combination of second chromosomes extracted from natural populations were estimated. We have already shown that linkage disequilibrium can hardly be seen between isozyme genes. Thus, the average viabilities of

FF, *FS*, and *SS* genotypes of random heterozygotes were calculated for the *Adh* and *αGpdh* loci. The results of two Japanese populations (HS and OG) were used. Simultaneously the result from the Raleigh, N.C. populations was employed (Yoshimaru and Mukai, 1979). For the *Adh* locus, *In(2L)t*-carrying heterozygotes were excluded, but they were not excluded for the *αGpdh* locus since it has been shown that the average viabilities of the *In(2L)t/In(2L)t'* heterozygotes and the standard chromosome heterozygotes are the same. The results of the HS and OG populations were pooled and all the results are shown in Table VII. Furthermore, the average viabilities of the *FF* and *SS* homozygotes were pooled for comparison with the average heterozygote viabilities and these results are also shown in Table VII. None of the values are significantly different from each other. If the results of the two populations and the two loci are pooled, the result becomes as follows:

Heterozygotes: 0.9961 ± 0.0024 ($n=1{,}014$);
Homozygotes: 1.0017 ± 0.0028 ($n=1{,}701$).

TABLE VII
Mean Relative Viabilities of Genotypes for Isozyme Genes in a Heterozygous Genetic Background

A) Alcohol dehydrogenase locus (2-50.1)[a]

Genotype	Population			
	U.S.A.		Japan	
	n	Viability	n	Viability
FF	44	0.9926±0.0136	166	0.9999±0.0111
FS	212	0.9921±0.0064	244	0.9922±0.0084
SS	226	1.0022±0.0071	99	1.0200±0.0155
Pooled homozygote	270	1.0006±0.0063	265	1.0074±0.0090

B) α-Glycerol-3-phosphate dehydrogenase locus (2-17.8)

Genotype	Population			
	U.S.A.		Japan	
	n	Viability	n	Viability
FF	33	1.0051±0.0141	583	1.0022±0.0062
FS	276	1.0022±0.0059	282	0.9965±0.0086
SS	522	0.9988±0.0046	28	0.9962±0.0193
Pooled homozygote	555	0.9992±0.0044	611	1.0019±0.0060

[a] *In(2L)t*-carrying individuals were excluded.
n, the number of heterozygote crosses.

These values are not significantly different from each other and no evidence of overdominance was obtained. The result supports the above finding in the isolated population.

The experimental results examining whether or not overdominance is common in natural populations can be summarized as follows: 1) Linkage disequilibrium due to epistasis (including overdominance with multiplicative gene action) between polymorphic isozyme genes is almost nonexistent. 2) Dominance variance for viability is extremely small in comparison with additive variance. 3) Viabilities measured by using the characteristics of polymorphic inversion in an isolated population did not show any evidence for selection at near-neutral polymorphic loci (the maximum selection coefficient is on the order of 10^{-4}). 4) The results of direct estimation of the relative viabilities of the two isozyme loci did not show any evidence for overdominance. These results almost completely reject the overdominance at polymorphic loci.

FREQUENCY-DEPENDENT SELECTION

The theory stating that genetic variability can be maintained in a population through frequency-dependent selection with minority superiority had already been applied to data analysis in 1946 (Wright and Dobzhansky, 1946), but it was Kojima and Yarbrough (1967) who claimed that this type of selection is the mechanism for the maintenance of protein polymorphisms. They showed apparent evidence for frequency-dependent selection with respect to viability at the *Est-6* locus (esterase-6: 3-36.8) (Kojima and Yarbrough, 1967) as well as at the *Adh* locus (alcohol dehydrogenase: 2-50.1) (Tobari and Kojima, 1967). A similar experimental result at the *Adh* locus has been reported by Clarke and his associate (Clarke, 1979). Supporting evidence for frequency-dependent selection have been published by several investigators (*e.g.*, Morgan, 1976; Snyder and Ayala, 1979), while negative results have also been obtained (Dolan and Robertson, 1975). We at first tested whether or not frequency-dependent selection is responsible for the maintenance of isozyme polymorphism using the *Adh* locus by direct competition between two types of larvae.

1. Competition Experiment with Various Frequencies of Two Types of Genotypes

Several hundred second chromosomes were extracted from the OG and the HS, Aomori populations. One hundred and twenty-four inversion-free chromosomes carrying the Adh^F alleles were selected randomly from the OG samples, and a cage population was established. Using 120 inversion-free chromosome lines having the Adh^S alleles, another cage population was initiated. These cage populations were called the OG(*F*) and OG(*S*) populations, respectively.

In the same way as above, an HS(*S*) (cage) population was initiated using 56 inversion-free Adh^S-carrying chromosomes and an HS(*F*) (cage) population was established using 156 inversion-free Adh^F-carrying chromosomes. Additionally, two more cage populations were set up using the chromosomes from the HS population: HSD(*S*) and HSD(*F*) populations with 4 inversion-free Adh^S-carrying and 4 inversion-free Adh^F-carrying chromosomes, respectively.

Thus, linkage disequilibrium is not expected in the OG(*S*), OG(*F*), HS(*S*), and HS(*F*) populations, while it can be expected due to the small sample size in the HSD(*S*) and HSD(*F*) populations. Incidentally it has been shown that in the original OG and HS populations there was no linkage disequilibrium (Yamaguchi *et al.*, 1980).

Flies were allowed to lay eggs for 6–8 hr on five cups with a new culture medium in each cage. Then the cups were maintained continuously at 25°C. Twenty-seven to 30 hr after the beginning of oviposition, first instar larvae were picked out from the cups and transferred into experimental vials (3 cm × 10 cm). Larvae from both cages were placed in a vial at each of the following five levels of allele frequencies to compete with each other: (250 Adh^F/Adh^F *vs.* 50 Adh^S/Adh^S), (200 *vs.* 100), (150 *vs.* 150), (100 *vs.* 200), and (50 *vs.* 250). Five vials, one at each frequency, were prepared at a time. The adult flies emerging from these vials were collected during 8–16 days after the transfer, and their genotypes were determined by agar-gel electrophoresis. A cornmeal-molasses medium was always used. The formula is agar-powder: 6 g, molasses: 28 m*l*; sugar: 14 g; Ebios (dry yeast): 20 g; cornmeal: 110 g; propionic acid: 4 m*l*, and water: 1,000 m*l*.

Four experiments were conducted. In each experiment, the procedure described above was repeated 4–10 times. The first experiment was conducted with about 9 m*l* (8–10 m*l*) of the culture medium in a 2.5 cm × 10 cm vial, but the remaining experiments were made with 1.5 m*l* (1–2 m*l*). The average survival rates were about 55–75%. The relative survival rate of the Adh^S/Adh^S genotype compared to Adh^F/Adh^F were calculated using the electrophoresis data, and they were standardized relative to the values of the 150 *SS*/150 *FF* classes. All the results are summarized in

TABLE VIII
The Results of Experiments Testing for Frequency-dependent Selection

Experiment	Number of replications	Average survival rate	Ratio of *SS* to *FF* larvae				
			50/250	100/200	150/150	200/100	250/50
1[a]	7	0.6989 ±0.0092	1.0178	0.9875	1	1.0342	0.9903
2[b]	4	0.7445 ±0.0104	1.0710	1.0818	1	1.1891	1.0396
3[b]	10	0.6024 ±0.0145	0.9605	0.9729	1	1.0087	0.9286
4[b]	5	0.5472 ±0.0134	1.0809	1.0317	1	1.0504	1.0480
Average			1.0161 ±0.0101	1.0049 ±0.0078	1	1.0513 ±0.0122	0.9853 ±0.0099

[a] 9 m*l* of food was supplied.
[b] 1.5 m*l* of food was supplied.
The materials in Exps. 1 and 2 were from the OG population, and those in Exps. 3 and 4 were from the HS population. Experiment 4 was conducted by using only *4SS* and *4FF* lines.

Table VIII. From this table, no evidence of frequency-dependent selection can be detected. It should be noted that 1) linkage disequilibrium did not induce frequency-dependent selection; and 2) the magnitude of competition for food was not related to the induction of frequency-dependent selection.

2. *Cy Method*

Data for the estimation of random heterozygote viabilities for the second chromosomes (Mukai and Yamaguchi, 1974) were used to test for frequency-dependent selection. The *Cy* method described above was employed to estimate viability.

Crosses were divided into the following three types on the basis of the genotypes involved in the matings with respect to the *Adh* locus (Table IX).

Since the *Cy* chromosomes always carry Adh^F alleles, the frequencies of the Adh^F alleles in the experimental vials are different among the three types, where we assume that the Adh^F alleles in the *Cy* chromosomes and wild chromosomes are identical. Assuming that the relative viabilities of *FF*, *FS*, and *SS* at the *Adh* locus are given by $1+a-bp$, 1, and $1-a+bp$ (p is the frequency of the Adh^F allele in a vial, and a and b are appropriate constants), the average viability indices [$2\times$(wild type)/(curly$+1$)] of the three types of random heterozygotes are expected to be

TABLE IX
The Genotypes of Parents and Offspring in the *Cy* Method

	Parents	Offspring	Offspring genotypic ratio *FF FS SS*	Frequency of *F*
(I)	$\frac{Cy(F)}{+(F)} \times \frac{Cy(F)}{+'(F)}$	$\frac{Cy(F)}{+(F)}, \frac{Cy(F)}{+'(F)}, \frac{+(F)}{+'(F)}$	3, 0, 0	1.0
(II)	$\frac{Cy(F)}{+(F)} \times \frac{Cy(F)}{+(S)}$	$\frac{Cy(F)}{+(F)}, \frac{Cy(F)}{+(S)}, \frac{+(F)}{+(S)}$	1 : 2, 0	$\frac{2}{3}$
(III)	$\frac{Cy(F)}{+(S)} \times \frac{Cy(F)}{+'(S)}$	$\frac{Cy(F)}{+(S)}, \frac{Cy(F)}{+'(S)}, \frac{+(S)}{+'(S)}$	0, 2 : 1	$\frac{1}{3}$

$$\text{(I)}\quad \frac{1}{c} \qquad \text{(II)}\quad \frac{2}{c\left(2+a-\frac{2}{3}b\right)} \qquad \text{(III)}\quad \frac{1-a+\frac{1}{3}b}{c} \tag{19}$$

where c is the heterozygous viability of the *Cy* chromosomes relative to non-curly, excluding the *Adh* locus. This model is not completely general, but may be appropriate as a first approximation. In this model the three values are equal if both a and b are zero, that is, if neither frequency-dependent selection nor intrinsic difference in viabilities among the three genotypes exists.

The viabilities of random second chromosome heterozygotes from the Raleigh, N. C. population were estimated. From the original data, the average viabilities of three types of crosses were estimated and no difference was detected among them (Mukai *et al.*, 1974). Because Adh^S was almost completely linked with the polymorphic inversion [*In*(*2L*)*t*], there might have been some bias in the previous calculations. That is, the average viability of the crosses of $Cy/+(S) \times Cy/+(S)$ was thought to be underestimated, because random heterozygotes consisting of two different *In*(*2L*)*t*-carrying chromosomes were less viable than the others. Thus, we selected only inversion-free random heterozygotes from the original data and calculated the average viability for each of the three types of crosses. Using the random heterozygotes (each with four simultaneous replications), the average relative viabilities of the above three types of crosses were calculated. In addition, the random heterozygotes from the OG population and the HS population were also analyzed, and the results were pooled. All the results are given in Table VII(A). The results of $Cy(F)/+(F)$ (♀♀) $\times Cy(F)/+'(S)$ (♂♂) and its reciprocal cross were pooled. From Table VII(A) it can clearly be seen that there is no significant difference among the mean viabilities of the three types of crosses. An analysis of variance was performed for their viabilities, but no significant difference was found.

The above method of analysis was applied to the *αGpdh* locus of the Raleigh, N. C. population and the two Japanese populations. The results can be seen in Table VII(B). The results of the analysis of variance show no difference in viabilities among three genotypes.

Thus, no evidence for frequency-dependent selection was obtained with respect to viability.

3. *Variance Component Analysis*

From the results of the variance component analysis described above, the existence of frequency-dependent selection for polymorphic isozyme genes at a majority of the loci can be rejected. The following single-locus model is usually used for the study of frequency-dependent selection. There are two alleles, A_1 and A_2, whose frequencies are assumed to be p and q, respectively. Then the fitness (or viability) of the three genotypes A_1A_1, A_1A_2, and A_2A_2, and their frequencies may be expressed as follows:

	A_1A_1	A_1A_2	A_2A_2
Frequency	p_0^2	$2p_0q_0$	q_0^2
Fitness (or viability)	$1-m(p-p_0)$	1	$1+m(p-p_0)$

where p_0 and q_0 are the equilibrium gene frequencies of A_1 and A_2, respectively. Using the *Cy* method for the estimation of viability, we assume that the allele which is located in the *Cy* chromosome is A_1. The gene frequency of A_1 (p) in the test vial is 2/3 in the case where A_1A_2 is being tested and 1/3 in the case where A_2A_2 being tested. The reason is that the former cross is $Cy(A_1)/A_1 \times Cy(A_1)/A_2$ and the latter is $Cy(A_1)/A_2 \times Cy(A_1)/A_2$ and the offspring are $Cy(A_1)/A_1$, $Cy(A_1)/A_2$, and A_1A_2 for the former and $Cy(A_1)/A_2$, $Cy(A_1)/A_2$, and A_2A_2 for the latter (see the chart on p. 85).

Under the above conditions, the additive (σ_a^2) and dominance (σ_d^2) variances can be calculated from Eqs. (2) and (3). The viability index ($2b/(a+1)$, see Fig. 3) was considered. The results are as follows (*cf.* Mukai and Nagano, 1981; Tachida *et al.*, 1981):

$$\sigma_a^2 = \frac{m^2p_0^3(1-p_0)}{18} \tag{20}$$

$$\sigma_d^2 = \frac{m^2p_0^2(1-p_0)^2}{9} \tag{21}$$

On a chromosome basis, the ratio (R) of dominance variance (σ_D^2) to the additive variance (σ_A^2) is:

$$R = \frac{\sigma_D^2}{\sigma_A^2} = \frac{2\sum m^2 p_0^2(1-p_0)^2}{\sum m^2 p_0^3(1-p_0)}. \tag{22}$$

It may be reasonable to assume that m and p_0 are not correlated. Then, R becomes as follows:

$$R = \frac{2\sum p_0^2(1-p_0)^2}{\sum p_0^3(1-p_0)}. \tag{22'}$$

At the risk of oversimplification, we may assume that the gene frequencies at all loci are approximately equal. Then, R becomes: $2(1-\bar{p}_0)/\bar{p}_0$, where $\bar{p}_0$ is the average equilibrium gene frequency. From the present experimental data, $\hat{\sigma}_R^2+\hat{\sigma}_C^2=4.46\times10^{-3}$ (the IG population) and 5.38×10^{-3} (the Florida population) and $\hat{\sigma}^2_{R\times C}=1.12\times10^{-3}$ (the IG population) and 6.34×10^{-4} (the Florida population) (Mukai, 1977; Tachida *et al.*, 1981). Using these values, $\bar{p}_0$ can be estimated to be 0.89 for the IG population and 0.94 for the Florida population.

It is impossible to determine from the present data whether frequency-dependent selection is operating at a majority of loci or not, since we do not know the gene frequencies of mildly deleterious genes, but at least it can be concluded that this type of selection is not operating at the loci controlling protein polymorphisms. The reason for this is that the gene frequencies at these loci are intermediate (about 0.3 or 0.7 (Lewontin and Hubby, 1966)). Thus, the generality of frequency-dependent selection can be rejected.

4. *Isolated Population*

The difference in the homozygous load between the chromosome groups with the standard gene arrangement and that with *In(2L)t* in the isolated population described above can be used to test the generality of frequency-dependent selection. In the above model of frequency-dependent selection, the genetic load can be calculated as follows: the random load relative to the A_1A_1 genotype (L_R) using the *Cy* method is

$$L_R = q_0 m\left(\frac{2}{3}p_0-\frac{1}{3}\right). \tag{23}$$

The homozygous load relative to the A_1A_1 genotype (L_I) is:

$$L_I = q_0 m\left(p_0-\frac{1}{3}\right). \tag{24}$$

Thus, the homozygous load relative to the average heterozygote of the population becomes as follows:

$$L_{I(R)} = L_I - L_R = \frac{1}{3} p_0 q_0 m. \tag{25}$$

Assuming that there are n loci in the chromosome where frequency-dependent selection with minority superiority is operating and that m and the equilibrium gene frequency are not correlated, the $L_{I(R)}$ on a chromosome basis may be expressed as follows:

$$L_{I(R)} = \frac{1}{3} n \cdot \bar{m} \cdot \overline{p_0 q_0} = \frac{1}{6} n \cdot \bar{m} \cdot \bar{H} \tag{26}$$

where H is the average heterozygosity of the loci where frequency-dependent selection is operating.

As described above, the $L_{R(I)}$ for the polymorphic near-neutral loci which are located in the left arm of the second chromosomes turns out to be -0.00915 ± 0.02625 using the characteristics of *In(2L)t* and its history on a small isolated island. Assuming the average heterozygosity of the polymorphic loci is 0.2 and $n=350$, $\bar{m}$ can be estimated to be -0.000784 ± 0.002250, which is an extremely small value and the 95% confidence interval includes 0. In conclusion, it may be said that frequency-dependent selection at polymorphic isozyme loci is rare, and even if this type of selection is operating at loci where polymorphisms have been maintained, its effect must be on the order of 10^{-4} or less (probably 0), and it is questionable whether or not such weak selection can actually maintain protein polymorphism.

In summary, the results obtained from four experiments: competition experiment, an experiment using the *Cy* method, variance component analysis for viability, and an experiment on the characteristics of a polymorphic inversion on an isolated island have clarified that perceptible frequency-dependent selection is not the principal mechanism for the maintenance of protein polymorphisms in natural populations. This conclusion is also supported by the observation that there is no linkage disequilibrium in population (Mukai *et al.*, 1974; Yamaguchi *et al.*, 1980), since, if frequency-dependent selections with minority superiority and multiplicative gene action are the general rule, strong linkage disequilibrium should be formed in a random mating population (Yamazaki, 1974).

DIVERSIFYING SELECTION

Recently, considerable attention has been paid to the maintenance of genetic variability in random mating populations due to genotype × environment interaction or diversifying selection. There are many types of diversifying selection. For example, the following type can easily be considered as a candidate in natural populations. If there are many niches and the fitnesses of the genotypes are ordered differently in different niches but random mating takes place over all niches, then the gene frequency becomes intermediate with a reasonable amount of dominance and even when there is no dominance under certain conditions (Levene, 1953; Gillespie and Langley, 1974), and a high level of heterozygosity can be maintained in the population. In the present section, we are first going to discuss diversifying selection with respect to mutant genes as a whole and, secondly, we intend to specify that diversifying selection most probably operates on mildly deleterious genes or viability polygenes which are supposed to be different from structural genes. That is to say, diversifying selection does not operate in the maintenance of protein polymorphisms.

1. *Homeostasis of Heterozygotes*

One of the necessary condition for the maintenance of genetic variability in a random mating population is the greater stability of fitness or viability of heterozygotes as compared with corresponding homozygotes. We have extracted many second chromosomes and estimated the homozygous and heterozygous viabilities of these chromosomes by the *Cy* method (*cf.* Wallace, 1956). The estimation of viability was conducted using several simultaneous replications as before. Using these viability indices, it is possible to estimate environmental variances for viability in homozygotes and heterozygotes. Using the *Cy* method, two types of eggs, *Cy* and wild-type, are laid. Some proportion of the *Cy* (or wild-type) eggs will become adult *Cy* (or wild-type) flies. The proportion of *Cy* (or wild-type) eggs that survive to become adults will follow a binomial distribution. Thus, the viability index is the ratio of two terms from two independent binomial distributions (C. C. Cockerham, personal communication). Following this approach, the approximate sampling variance of the above viability index (V_s) was calculated. The result is:

$$V_s \cong \frac{4n}{M^2m^3}[Mm(M-2n)+Mn(M-m)+(M-2n)(M-m)] \tag{27}$$

where M is the number of eggs for *Cy* flies in a vial, the number of eggs for wild-type flies is $M/2$, and m and n are the observed numbers of *Cy* and wild-type flies, respectively (Mukai *et al.*, 1981). Unfortunately, it is impossible to determine M, but under the present experimental conditions, it may be between 500 and 800. The effect of the difference between 500 and 800 on V_s is found to be trivial and in the present work M is assumed to be 700.

The environmental variance (V_{EN}) of the viability index for each homozygous line or cross was calculated as follows:

$$V_{EN} = \hat{V}_0 - \hat{\bar{V}}_s \tag{28}$$

where V_0 is the variance of the viabilities among simultaneously replicated cultures within each line or cross (error variance) and $\hat{\bar{V}}_s$ is the average of the expected sampling variances of individual replicated cultures within a line or cross. For heterozygotes and homozygotes whose viability indices are larger than 0.5, the environmental variances were calculated. In this way, the possible effect of major genic semi-lethals may be excluded.

The viabilities of the second chromosomes of the Raleigh, N. C. population were analyzed. The average environmental variances of homozygotes and heterozygotes together with the numbers of lines and crosses (n) (Mukai *et al.*, 1981) are shown as follows:

Homozygotes	0.0141 ± 0.0012	$n=328$;
Heterozygotes	0.00671 ± 0.00058	$n=674$.

From these results, it can be concluded that the environmental variance of heterozygotes is significantly smaller than that of homozygotes.

A similar analysis was done for the second chromosome group which originated from a single stem chromosome and mutations were accumulated in this group at a minimum pressure of natural selection for 32–60 generations. The amount of genetic variation was approximately one-half of that in the above natural population. However, the results were different from those obtained above. There was no significant difference between homozygotes and heterozygotes. The figures obtained are as follows:

Homozygotes	0.00730 ± 0.00055	$n=130$;
Heterozygotes	0.00899 ± 0.00065	$n=170$.

These estimates cannot be compared at face value with those of natural chromosomes since the culture conditions, especially the sizes of the vials, were different.

It is clear from the experimental results that the stability of heterozygotes (developmental homeostasis) is not due to the heterozygosity *per se* and that it is a product of natural selection. It is highly likely that a small proportion of mutations are heterozygously homeostatic and their frequencies increase due to natural selection. Indeed, Gillespie and Langley (1974) have shown that, when the environmental variance for homozygotes is large and the fitness of the heterozygote is the average of the corresponding homozygotes polymorphisms can be maintained under certain conditions. Thus, it has been shown that a small fraction of newly arisen mutations may satisfy the necessary condition to be maintained in a population by diversifying selection.

2. *Variance Component Analysis—Large Additive Variances*

As described above, extremely small dominance variances for viability were estimated for several *D. melanogaster* populations. This finding directly rejects the universality of overdominance and frequency-dependent selection. Under mutation-selection balance, the additive variance (σ_A^2) for viability becomes approximately $\sigma_A^2 \cong 2\Sigma\hat{q}h^2s^2$, where h is the degree of dominance, s is the selection coefficient with respect to viability against the mutant homozygote, and $\hat{q}$ is the equilibrium gene frequency which can be expressed as μ/chs. chs is the selection coefficient against heterozygotes with respect to fitness as a whole and h and s are concerned with viability as defined above. Thus, the additive variance for viability can be expressed as $\sigma_A^2 \cong 2(\Sigma\mu_l s_l h_l + \Sigma\mu_d s_d h_d)/c$ where subscripts l and d stand for the lethal and detrimental genes, respectively. From the data $\Sigma\mu_d = 0.14$, $\Sigma\mu_l = 0.005$ per second chromosome, and $\bar{s}_d \cong 0.03$, $\bar{h}_d \cong 0.4$, $\bar{s}_l = 1$, and $\bar{h}_l \cong 0.02$ (Mukai, 1964; Mukai and Yamazaki, 1968; Mukai and Yamaguchi 1974; Mukai *et al.*, 1972). Then, the additive variance becomes 0.0018 for the second chromosomes under the condition of $c=2$. The observed additive and dominance variances of viability for various populations of *D. melanogaster* are graphically shown in Fig. 5 together with the homozygous lethal loads. It is clear from this figure that there are much larger additive variances than the value expected on the basis of the balance between mutation and selection. Furthermore, it should be pointed out that there is a south to north cline with respect to the magnitude of additive variance and in the north population (Aomori) the estimate of σ_A^2 is very close to the expected value under the mutation-selection balance. Thus, it is speculated with high probability that diversifying selection is occurring in natural populations, especially in southern populations. If we take a sample from such a population and estimate the components

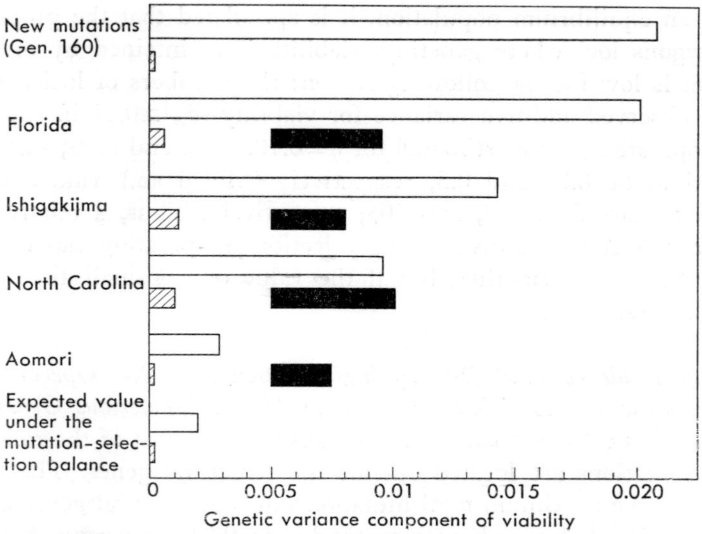

Fig. 5. Additive variances of relative viabilities of the second chromosomes sampled from various populations of *D. melanogaster*. □ additive variance; ▨ dominance variance; ■ lethal load.

of genetic variance in a constant environment such as our laboratory, then the additive variance will become large since gene frequencies are large and there is no absolute overdominance in a constant environment, and also there is no frequency-dependent selection as shown above. In the south there are a variety of fruits. Consequently, there may be a great variety of yeasts for *D. melanogaster* to eat and the variability of niche for this type of fly can be expected to be high. This may be one of the bases for diversifying selection.

It should be pointed out here that in one of the northern populations (the Aomori population), the observed additive variance is very close to the expected value. It is consistent that the average environmental variance of heterozygotes is similar to that of homozygotes in contrast to the cases of southern populations (Kusakabe and Mukai, 1981).

As an alternative to the above explanation, it may be stated that the south to north cline is due to the cline of mutation rates, that is, the mutation rate is generally higher in the south than in the north. This may be partially true, but this alternative hypothesis will be rejected if we pay attention to the slight incline of homozygous lethal loads from north to south.

In an equilibrium population, it is speculated that the proportion of heterozygous loci where genetic variability is maintained by diversifying selection is low for the following reason: the numbers of loci responsible for the observed additive variance for viability ($\hat{\sigma}_A^2 \cong 0.01$ in the Raleigh, N.C. population) were estimated for $\hat{q}=0.10$, 0.20, and 0.30; s and h were assumed to be 0.03 and 0.5, respectively (Mukai and Yamazaki, 1968). The results are 246, 138, and 105, respectively. Thus, a relatively small number of loci where diversifying selection is operating can explain the observed genetic variability, but if the value of s is small, the number of loci increases.

3. Some Evidence That Diversifying Selection Does Not Operate on Polymorphic Structural Loci but Operates on Mildly Deleterious Genes

Mukai and Cockerham (1977) suggested that most of the mildly deleterious mutations are located outside the structural genes. The reason is as follows: the maximum total mutation rate of structural genes estimated by the starch-gel electrophoresis method is *ca.* 0.04 per second chromosome per generation (based on recent findings the actual value may be 0.01 per second chromosome per generation), considering the detection rate of missense mutations by starch-gel electrophoresis (Ramshaw *et al.*, 1979; Fuerst and Ferrell, 1980), while the minimum rate of mildly deleterious mutations is about 0.14–0.17 per second chromosome per generation. Thus, most of the mildly deleterious mutations should be located outside the structural loci. This is the basis on which we drew the conclusion described above.

1) Comparison of environmental variances of viabilities between heterozygotes of standard chromosomes and those carrying In(2L)t in a homozygous condition

In the section on overdominance it was shown that approximately 1,000 generations may have passed since a few *In(2L)t*-carrying chromosomes with little genetic variation (possibly a single chromosome) were introduced into the IG population. After that event, the frequency of the chromosomes carrying this inversion has increased to about 60%. This level of frequency has been maintained for at least 7 years (Watanabe and Watanabe, 1977; Yamaguchi *et al.*, 1980). That is reason to speculate that the frequencies of lethal and deleterious genes have reached equilibria, but those of near-neutral and neutral genes including polymorphic isozyme genes have not and are still nearly monomorphic in the *In(2L)t*-carrying chromosomes. Thus, if we compare the stabilities of viability between heterozygotes of the standard chromosomes (N/N') and those carrying *In(2L)t* in the homozygous condition (I/I'), we can isolate the heterozygous effects of polymorphic near-neutral genes on the stability of individuals, and most of the genes controlling protein polymorphisms are included in this category.

Using the chromosomes extracted from the IG population, the environmental variances were estimated for about 1,400 crosses in both N/N' and I/I' using Eqs. (27) and (28). In each cross, two simultaneous replications were made. The results are as follows:

Standard (N/N') 0.0106 ± 0.0011 ($n=1{,}340$);
In(2L)t (I/I') 0.0117 ± 0.0045 ($n=1{,}414$)

where n stands for the number of crosses. It is evident that there is no significant difference in environmental variance between N/N' and I/I' crosses.

There may be approximately 1,100 structural genes in the left arm of the second chromosome (Judd *et al.*, 1972), and about one-third of the loci are polymorphic in the standard chromosome group. The average heterozygosity at these polymorphic loci may be at least 0.2. Thus, a typical N/N' individual carries about 75 ($1{,}100 \times 1/3 \times 0.2 = 73.3$) heterozygous structural loci, which do not serve to increase the stability of viability of the individuals. This finding may extend to the majority of heterozygous structural loci, and it is concluded that diversifying selection with respect to viability which we observed may not act at the structural loci.

2) *Additive variance and the average frequency of genes controlling protein polymorphisms*

If we compare the Florida and the Raleigh, N. C. populations with respect to the additive variance and the average of the frequencies of genes controlling protein polymorphisms, some additional information supporting the above conclusion is obtained: the additive variance of the Florida population was estimated to be 0.0202 ± 0.0038 and the dominance variance was 0.00063 ± 0.00057 (see Fig. 5). The additive variance of the latter population is approximately one-half the additive variance of the former (see Fig. 5). The gene frequencies at several enzyme loci were reported for both the populations (Johnson and Schaffer, 1973). These loci are the 4 loci described above and the α-amylase locus (*Amy*, map position 2-77.3) in the second chromosome and the esterase-6 (*Est-6* map position 3-36.8), esterase-C (*Est-C*, map position 3-49), octanol dehydrogenase (*Odh*, 3-49.2), and acid phosphatase (*Acph*, map position 3-101.1) loci in the third chromosome. The frequencies of the most common alleles

TABLE X
Frequencies[a] of the Most Common Alleles at Seven Loci in the North Carolina and Florida Populations of *D. melanogaster*

Isozyme	Florida	North Carolina
Mdh	0.992	0.963
Adh	0.857	0.692
αGpdh	0.882	0.825
Est-C	0.854	0.916
Est-6	0.635	0.606
Acph	0.947	0.960
Odh	0.826	0.915
Average	0.856	0.840

[a] Data from Johnson and Schaffer (1973).
Mdh, malate dehydrogenase; *Adh*, alcohol dehydrogenase; *αGpdh*, α-glycerol-3-phosphate dehydrogenase; *Est-C*, esterase-C; *Est-6*, esterase-6; *Acph*, acid phosphatase; *Odh*, octanol dehydrogenase.

at seven loci are presented in Table X. The averages of the most common allele frequencies are 0.840 for the N. C. population and 0.856 for the Florida population (Johnson and Schaffer, 1973). Thus, it can be expected that the average heterozygosities of protein loci of the two populations are not so different even if the difference in the frequencies of polymorphic inversions between the two populations is considered. Nevertheless, the additive variance of the Florida population is twice as large as that of the N. C. population. Therefore, it is most probable that the difference in additive variance between the two populations is due to non-structural mildly deleterious genes or viability polygenes.

A comparison similar to the above was made between the IG and Aomori populations. As described above, the additive variances are estimated to be 0.01554 ± 0.00523 for the IG population and 0.00276 ± 0.00091 for the Aomori population. The former is 5.6 times larger than the latter. The average heterozygosities at the loci controlling isozymes or proteins are currently being estimated (Kusakabe and Mukai, unpublished). Thus far ten loci have been examined. These loci are the *αGpdh*, *Adh*, *Hex-C*, *Amy*, *Est-6*, *Pgm* (phosphoglucomutase, map position 3-43.4), *Odh*, *Pt-3*, *Pt-4* (larval protein-3 and -4), and *6Pgdh* (6-phosphogluconate dehydrogenase, map position 1-0.64). The frequencies of the most common alleles and others, the numbers of detected alleles, and the expected heterozygosities (H) are given in Table XI. The expected heterozygosity at each locus was calculated as follows:

TABLE XI
Average Heterozygosity of Ten Loci in the IG and Aomori Populations of *D. melanogaster*

Locus	Population	Number of chromosomes sampled	Frequency of the most common allele	Number of alleles detected	Expected hetero-zygosity
αGpdh	I	435	0.920	3	0.148
	A	239	0.855	2	0.273
Adh	I	410	0.768	2	0.356
	A	239	0.573	2	0.489
Hex-C	I	435	1.000	1	0.000
	A	240	0.883	2	0.207
Amy	I	436	0.927	6	0.139
	A	240	0.929	5	0.135
Est-6	I	436	0.672	3	0.448
	A	240	0.821	3	0.295
Pgm	I	415	0.990	2	0.020
	A	240	0.888	3	0.201
Odh	I	386	0.824	2	0.290
	A	235	0.974	2	0.051
6Pgdh	I	360	0.978	2	0.043
	A	189	0.582	2	0.487
Pt-3	I	94	0.745	3	0.404
	A	98	0.786	3	0.358
Pt-4	I	158	0.867	3	0.232
	A	112	0.857	2	0.245
Average	I			2.7	0.208±0.069
	A			2.6	0.274±0.045

I, Ishigaki-Jima; A, Aomori. *Pgm*, phosphoglucomutase; *Pt-3*, laval protein-3; *Pt-4*, laval protein-4; *6Pgdh*, 6-phosphogluconate dehydrogenase.

$$H = 1 - \sum x_i^2 \tag{29}$$

where x_i is the frequency of the i-th allele at an arbitrary locus. The average heterozygosity ($\bar{H}$) was estimated by taking the mean of H values over the estimated loci. These results are also given in Table XI. It can be clearly seen that the average heterozygosities of the IG and Aomori populations are approximately equal. This tendency may be extended to all polymorphic loci. This is one of the characteristics of populations that can be expected from the neutral theory of protein polymorphisms (Kimura and Ohta, 1971).

From the above experimental results, it is reasonable to speculate that the difference in additive variance between the above two populations,

0.01278 or about 65% of that of the IG population, is not due to structural genes but may be due to mildly deleterious genes. Furthermore, since the additive variance of the Aomori population is very close to the value expected on the basis of mutation-selection balance (*ca.* 0.0018), it is highly probable that protein polymorphisms scarcely contribute to the genetic variance of viability.

In conclusion, it is most likely that diversifying selection does not operate for the maintenance of protein polymorphisms.

DISCUSSION AND CONCLUSION

No evidence for balancing selection was obtained for the polymorphic structural genes in the present set of experiments. It is important to clarify why some investigators have obtained experimental results indicating that some forms of balancing selection are responsible for the maintenance of protein polymorphisms. Overdominance has been enthusiastically supported by Milkman (1967), Wills *et al.* (1970), and Wills (1973) together with truncation selection. However, if their claims are correct, strong linkage disequilibrium should be formed. Furthermore, under the conditions of linkage disequilibrium, dominance variance should become inflated. The present experimental results have shown that neither linkage disequilibrium among polymorphic isozyme genes nor appreciable amounts of dominace variance for viability exists. Indeed, the 95% confidence interval of the average selection coefficient of homozygotes for near-neutral and neutral genes ($\bar{s}$) was estimated to be $-0.00087<\bar{s}<0.00060$. This finding essentially rejects the universality of overdominance.

No supporting evidence for frequency-dependent selection was obtained in the present experiments. We have examined whether or not frequency-dependent selection is manifested under the conditions of linkage disequilibrium of the locus in question in relation to neighboring loci, but no evidence for this was obtained. It is necessary to find out how Kojima, Clarke, and their associates obtained experimental evidence for frequency-dependent selection.

Diversifying selection has not been extensively tested. Experimental results of Powell (1971), Powell and Wistrand (1978), and McDonald and Ayala (1974) appear to support the contention that diversifying selection is responsible for the maintenance of protein polymorphisms. The present experimental results agree with the findings of Powell, Ayala, and others that diversifying selection operates for the maintenance of genetic variability in natural populations but differ from their results in that it is suggested that diversifying selection operates on genes located outside structural genes in contrast to the polymorphic structural genes. It is most likely that they have included the effects of non-structural genes that are linked to polymorphic structural genes. It is necessary to clarify this point, and a large-scale experiment concerning this subject is going on in our laboratory.

In conclusion, it can be said that no evidence was obtained that balancing selection maintains protein polymorphisms in random mating populations of *D. melanogaster*.

SUMMARY

Three typical mechanisms of balancing selection, *i.e.*, overdominance, frequency-dependent selection and diversifying selection were examined to see whether or not they are responsible for the maintenance of protein or isozyme polymorphisms in populations of *D. melanogaster*. The results of extensive tests for linkage disequilibrium, estimation of genetic variance components for viability, estimation of relative viabilities of homozygotes and heterozygotes for polymorphic genes, and the environmental variance for viability in the second chromosomes provide no evidence for the three types of balancing selection described above.

REFERENCES

Clarke, B. (1979). The evolution of genetic diversity. *Proc. Roy. Soc. Lond. B.* **205**, 453–474.

Dolan, R. and Robertson, A. (1975). The effect of conditioning the medium in *Drosophila*, in relation to frequency-dependent selection. *Heredity* **35**, 311–316.

Franklin, I. and Lewontin, R. C. (1970). Is the gene the unit of selection? *Genetics* **65**, 707–734.

Fuerst, P. A. and Ferrell, R. E. (1980). The stepwise mutation model: an experimental evaluation utilizing hemoglobin variants. *Genetics* **94**, 185–201.

Gillespie, J. H. and Langley, C. H. (1974). A general model to account for enzyme variation in natural populations. *Genetics* **76**, 837–848.

Greenberg, R. and Crow, J. F. (1960). A comparison of the effect of lethal and detrimental chromosomes from *Drosophila* populations. *Genetics* **45**, 1154–1168.

Harris, H. (1966). Enzyme polymorphisms in man. *Proc. Roy. Soc. Ser. B.* **164**, 298–310.

Hill, W. G. and Robertson, A. (1968). Linkage disequilibrium in finite populations. *Theor. Appl. Genet.* **38**, 226–231.

Ichinose, M. and Mukai, T. (1981). Studies on quantitative characters of *Drosophila me-*

lanogaster: Difference in genetic variance components of bristle characters between the standard and *In*(2*L*)*t*-carrying chromosomes (in preparation).

Inoue, T., Iwamoto, C., and Mukai, T. (1979). Genetic variability of viability in a southern population of *Drosophila melanogaster*. *Japan. J. Genet.* **54**, 453.

Johnson, F. M. and Schaffer, H. E. (1973). Isozyme variability in species of the genus *Drosophila*. VII. Genotype-environment relationships in populations of *D. melanogaster* from the Eastern United States. *Biochem. Genet.* **10**, 149–163.

Judd, B. H., Shen, M. W., and Kaufman, T. C. (1972). The anatomy and function of a segment of the X chromosome of *Drosophila melanogaster*. *Genetics* **71**, 139–156.

Kimura, M. (1968). Evolutionary rate at the molecular level. *Nature* **217**, 624–626.

Kimura, M. and Ohta, T. (1971). Protein polymorphism as a phase of molecular evolution. *Nature* **229**, 467–469.

Kojima, K. and Yarbrough, K. M. (1967). Frequency-dependent selection at the esterase-6 locus in *Drosophila melanogaster*. *Proc. Natl. Acad. Sci. U.S.* **57**, 645–649.

Kusakabe, S. and Mukai, T. (1981). Evidence of diversifying selection for mildly deleterious genes and of neutrality of protein polymorphisms (in preparation).

Levene, H. (1953). Genetic equilibrium when more than one ecological niche is available. *Am. Nat.* **97**, 331–333.

Lewontin, R. C. and Hubby, J. L. (1966). A molecular approach to study of genic heterozygosity in natural populations. II. Amounts of variation and degree of heterozygosity in natural populations of *Drosophila pseudoobscura*. *Genetics* **54**, 595–609.

McDonald, J. F. and Ayala, F. J. (1974). Genetic response to environmental heterogeneity. *Nature* **250**, 572–574.

Milkman, R. (1967). Heterosis as a major cause of heterozygosity in nature. *Genetics* **55**, 493–495.

Morgan, P. (1976). Frequency-dependent selection at two enzyme loci in *Drosophila melanogaster*. *Nature* **263**, 765–766.

Mukai, T. (1964). The genetic structure of natural populations of *Drosophila melanogaster*. I. Spontaneous mutation rate of polygenes controlling viability. *Genetics* **50**, 1–19.

Mukai, T. (1975). Dynamics of mutant genes in populations. In *Basis of Human Genetics* (ed. M. Kimura), pp. 183–212. Iwanami-Shoten, Tokyo (in Japanese).

Mukai, T. (1977). Lack of experimental evidence supporting selection for the maintenance of isozyme polymorphisms in *Drosophila melanogaster*. In *Proceedings of the Second Taniguchi International Symposium on Biophysics, Molecular Evolution, and Polymorphism* (ed. M. Kimura), pp. 103–126. National Institute of Genetics, Mishima.

Mukai, T. (1979). Polygenic mutations. In *Quantitative Genetic Variation* (eds. J. N. Thompson, Jr. and J. M. Thoday), pp. 177–196. Academic Press, New York.

Mukai, T. and Cockerham, C. C. (1977). Spontaneous mutation rates at enzyme loci in *Drosophila melanogaster*. *Proc. Natl. Acad. Sci. U.S.* **74**, 2514–2517.

Mukai, T. and Nagano, S. (1981). The genetic structure of natural populations of *Drosophila melanogaster*. XVI. Experimental results suggesting diversifying selection for viability polygenes. *Genetics* (to be submitted).

Mukai, T. and Voelker, R. A. (1977). The genetic structure of natural populations of *Drosophila melanogaster*. XIII. Further studies on linkage disequilibrium. *Genetics* **86**, 175–185.

Mukai, T. and Yamaguchi, O. (1974). The genetic structure of natural populations of

Drosophila melanogaster. XI. Genetic variability in a local population. *Genetics* **76**, 339–366.

Mukai, T. and Yamazaki, T. (1968). The genetic structure of natural populations of *Drosophila melanogaster*. V. Coupling-repulsion effect of spontaneous mutant polygenes controlling viability. *Genetics* **59**, 513–535.

Mukai, T., Chigusa, S. I., and Kusakabe, S. (1981). The genetic structure of natural populations of *Drosophila melanogaster*. XV. Nature of developmental homeostasis for viability. *Genetics* (accepted).

Mukai, T., Tachida, H., and Ichinose, M. (1980). Selection for viability at loci controlling protein polymorphisms in *Drosophila melanogaster* is very weak at most. *Proc. Natl. Acad. Sci. U.S.* **77**, 4857–4860.

Mukai, T., Watanabe, T. K., and Yamaguchi, O. (1974). The genetic structure of natural populations of *Drosophila melanogaster*. XII. Linkage disequilibrium in a large local population. *Genetics* **77**, 771–793.

Mukai, T., Cardellino, R. A., Watanabe, T. K., and Crow, J. F. (1974). The genetic variance for viability and its components in a local population of *Drosophila melanogaster*. *Genetics* **78**, 1195–1208.

Mukai, T., Chigusa, S. I., Mettler, L. E., and Crow, J. F. (1972). Mutation rate and dominance of genes affecting viability in *Drosophila melanogaster*. *Genetics* **72**, 335–355.

Nei, M. and Li, W.-H. (1979). Non-random association between electromorphs and inversion chromosomes in infinite populations. *Genet. Res.* **35**, 65–83.

Ohta, T. and Kimura, M. (1969). Linkage disequilibrium due to random genetic drift. *Genet. Res.* **13**, 47–55.

Powell, J. R. (1971). Genetic polymorphisms in varied environments. *Science* **174**, 1035–1036.

Powell, J. R. and Wistrand, H. (1978). The effect of heterogeneous environments and a competitor on genetic variation in *Drosophila*. *Am. Nat.* **112**, 935–947.

Prakash, S., Lewontin, R. C., and Hubby, J. L. (1969). A molecular approach to the study of genic heterozygosity in natural populations. IV. Patterns of genic variation in central, marginal and isolated populations of *Drosophila pseudoobscura*. *Genetics* **61**, 841–858.

Slatkin, M. (1972). On treating the chromosome as the unit of selection. *Genetics* **72**, 157–168.

Schaffer, H. E. and Johnson, F. M. (1974). Isozyme allelic frequencies related to selection and gene-flow hypothesis. *Genetics* **77**, 163–168.

Snyder, T. P. and Ayala, F. J. (1979). Frequency-dependent selection at the PGM-1 locus of *Drosophila pseudoobscura*. *Genetics* **92**, 995–1003.

Stone, W. S., Wheeler, M. R. Johnson, F. M., and Kojima, K. (1968). Genetic variation in natural island populations of members of the *Drosophila nasuta* and *Drosophila ananassae* subgroups. *Proc. Natl. Acad. Sci. U.S.* **59**, 102–109.

Ramshaw, J. A. M., Coyne, J. A., and Lewontin, R. C. (1979). The sensitivity of gel electrophoresis as a detector of genetic variation. *Genetics* **93**, 1039–1049.

Tachida, H., Matsuda, M., Kusakabe, S., and Mukai, T. (1981). Variance component analysis for viability in an isolated population of *Drosophila melanogaster*. *Genet. Res.* (to be submitted).

Tobari, Y. N. and Kojima, K. I. (1967). Selective modes associated with inversion karyotypes in *Drosophila ananassae*. I. Frequency-dependent selection. *Genetics* **57**, 179–188.

Yamaguchi, O., Ichinose, M., Matsuda, M., and Mukai, T. (1980). Linkage disequilibrium in isolated populations of *Drosophila melanogaster*. *Genetics* **96**, 507–522.

Yamazaki, T. (1974). Organization of linked genes under frequency-dependent selection of minority advantage. *Japan. J. Genet.* **49**, 33–36.

Yamazaki, T. (1977). The effects of overdominance on linkage in a multilocus system. *Genetics* **86**, 227–236.

Yamazaki, T. and Maruyama, T. (1972). Evidence for the neutral hypothesis of protein polymorphism. *Science* **178**, 56–58.

Yoshimaru, H. and Mukai, T. (1979). Lack of experimental evidence for frequency-dependent selection at the alcohol dehydrogenase locus in *Drosophila melanogaster*. *Proc. Natl. Acad. Sci. U.S.* **76**, 876–878.

Vigue, C. L. and Johnson, F. M. (1973). Isozyme variability in species of the genus *Drosophila*. VI. Frequency-property-environment relationships of allelic alcohol dehydrogenases in *D. melanogaster*. *Biochem. Genet.* **9**, 213–227.

Wallace, B. (1956). Studies on irradiated populations of *Drosophila melanogaster*. *J. Genet.* **54**, 280–293.

Watanabe, T. K. and Watanabe, T. (1977). Enzyme and chromosome polymorphisms in Japanese natural population of *Drosophila melanogaster*. *Genetics* **85**, 319–329.

Wills, C. (1973). In defense of naive pan-selectionism. *Am. Nat.* **107**, 23–34.

Wills, C., Crenshaw, J., and Vitale, J. (1970). A computer model allowing maintenance of large amounts of genetic variability in Mendelian populations. I. Assumptions and results for large population. *Genetics* **64**, 107–123.

Wright, S. and Dobzhansky, Th. (1946). Genetics of natural populations. XII. Experimental production of some of the changes caused by natural selection in certain populations of *Drosophila pseudoobscura*. *Genetics* **31**, 125–156.

4

Molecular Evolution, Protein Polymorphism and the Neutraι Theory (1982) (ed. M. Kimura), pp. 123–134.
Japan Scientific Societies Press, Tokyo / Springer-Verlag, Berlin

Genic Variability in Natural Populations of the Haploid Plant, *Conocephalum conicum*

TSUNEYUKI YAMAZAKI

Laboratory of Population Genetics, Faculty of Science, Kyushu University, Fukuoka 812, Japan

Since the discovery of an unexpected amount of genic variation in natural populations of *Drosophila pseudoobscura* (Lewontin and Hubby, 1966) and man (Harris, 1966), innumerable studies have been published concerning the amount of enzyme or protein polymorphisms in various diploid organisms. There is much debate as to how these genic polymorphisms are maintained in natural populations and their role is in the adaptive evolution of organisms.

At present there are two main hypotheses to explain the above data. One is the neutral hypothesis of enzyme polymorphisms proposed by Kimura in 1968. The alternative is the balancing selection hypothesis in which the genic polymorphisms observed in natural populations are interpreted to be adaptive and are maintained by some form of balancing selection (*e.g.*, Prakash *et al.*, 1969; Ayala *et al.*, 1974). One of the most famous models for the balancing selection is overdominance. Frequency dependent selection by minority advantage (*e.g.*, Ehrman and Petit, 1968; Yamazaki, 1974), and diversifying selection in which the fitness of individuals changes temporally or spatially (Levene, 1953; Gillespie, 1973; Takahata and Kimura, 1979), are also among the models studied very intensively. In general, genetic variability is maintained in a population quite efficiently by the heterozygous advantage over homozygotes. Heterozygous advantage in the strict sense is not always required for the maintenance of genetic variability in a population under the model of frequency dependent selection or diversifying selection. It is, however, true that diploidy itself plays an important role in the maintenance of the polymorphisms even under these two selection models. For example, the geometric mean of the fitness of heterozygotes needs to be greater than those of homozygotes for the genetic variability to be maintained in temporally variable environments (Gillespie, 1973). The conditions for the maintenance of genetic polymorphisms are somewhat different, depending on the models considered. If there is a certain amount of dominance under any balancing selection model, in other words, if the fitness of the heterozygote is greater than the average of both homozygotes, genic polymorphisms are rather easily maintained in a population. Even if mutations are not maintained in an equilibrium population by balancing selection, the rate of elimination of variability from populations is drastically reduced by a certain amount of recessivity. In fact, most experimental evidence shows that the degree of dominance in the heterozygote is somewhere between 0.03–0.4 depending on the nature of mutations (Mukai and Yamaguchi, 1974).

On the other hand, overdominance or dominance does not exist in haploid organisms, and mutations show their effects immediately after their occurrence since no shelter is available in the haploid condition.

We, therefore, generally expect the number of genic polymorphisms of haploid organisms to be much less than those of diploid organisms under most selection models. Only under the model of selective neutrality of genetic variability do we expect the equality of polymorphisms between haploid and diploid organisms. Several experimental studies have been reported so far on the genic polymorphisms of haploid organisms (Milkman, 1973; Spieth, 1975; Krazakowa and Szweykowski, 1979). In this report I examine 11 enzyme loci of the haploid moss, *Conocephalum conicum*, collected near Fukuoka city in the southern part of Japan, and compare the amount of heterozygosity with that of diploid organisms.

ELECTROPHORETIC ANALYSIS OF SAMPLES OF *C. CONICUM*

The haploid moss, *C. conicum*, used in the present studies was collected from seven different populations near Fukuoka in the southern part of

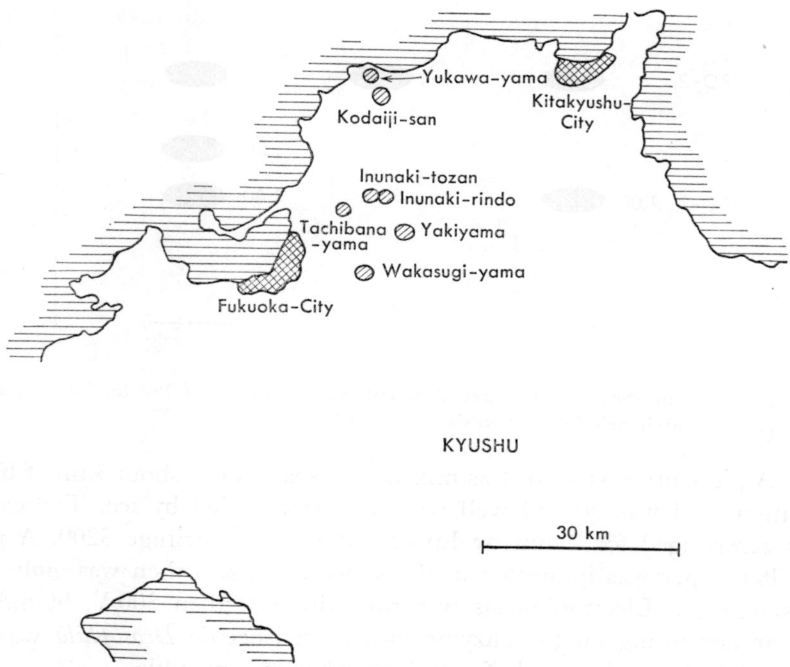

Fig. 1. Distributions of seven natural populations of *C. conicum*, showing geographical locations sampled.

Japan from 1978 to 1980. A population here is defined as mosses growing along a valley or a stream. The collection sites within a population were separated by at most about 2 km along a small mountain stream in the case of the Yakiyama population. In the six other populations all the collections were done at sites less than 1 km apart. The geographical locations and distances of the above seven populations are shown in Fig. 1.

The moss usually grows in small patches of less than 1 m² but occasionally forms carpets of more than 10 m in length. I collected a carpet of about 100 cm² from each site and called it " a sample." Only one sample was collected from each continuous carpet of moss to minimize the possibility of picking the same clone more than once. For electrophoresis one shoot was selected from a sample.

Out of 20 enzyme systems tried for starch gel electrophoresis and staining, only six enzyme stainings worked well so that their genic variability could be examined.

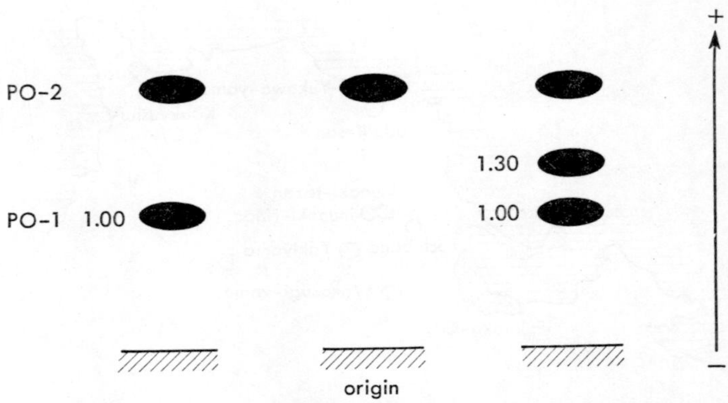

Fig. 2. Diagram showing the three different band patterns of PO loci (PO-1, PO-2) observed by starch gel-electrophoresis of *C. conicum*.

A piece of moss shoot was mixed with sea sand in about 3 ml of buffer solution, and was ground well with a mortar cooled by ice. The extract was centrifuged for 2 min by Eppendorf micro centrifuge 3200. A piece of filter paper was immersed in the supernatant and then was embedded in starch gel. Electrophoresis was run with less than 300 V, 90 mA for 1–2 hr depending on the enzyme locus. An isogenic *Drosophila* was run as a control together with *C. conicum* whenever possible.

1) *Peroxidase* (*PO*)

Gel buffer (pH 8.05): NaOH 2.1 g, H_3BO_3 18.0 g, added to H_2O 1,000 ml. Bridge buffer (pH 8.55): NaOH 10.2 g, H_3BO_3 55.5 g, added to H_2O 3,000 ml. Staining: O-dianisidine (pinch), β-naphthol (pinch), acetone 20 ml, Tris buffer (Tris 12.1 g acetic acid 30 ml, added to H_2O 1,000 ml) 20 ml, H_2O 70 ml, 3% H_2O_2 1.0 ml. Electrophoresis was run for 1 hr under 240 V, 90 mA. The following band patterns appeared as shown in Fig. 2; in PO-1 three band patterns 1.00, 1.00/1.30, and null were observed. The exact cause of the appearance of the 1.00/1.30 band in a haploid organism is not known; gene duplication might be one explanation. I treated it as one allele. At the PO-2 locus only one band pattern was observed.

2) *Esterase* (*EST*)

Gel buffer: 0.1 M Tris-borate EDTA buffer (Tris 39.3 g, boric acid 3.06 g, EDTA 2.22 g, added to H_2O 4,000 ml). Electrode buffer: the same as gel buffer. Staining: phosphate stock A (NaH_2PO_4 27.8 g in H_2O 1,000 ml) 50 ml, phosphate stock B (Na_2HPO_4 53.6 g in H_2O 1,000 ml) 10 m*l*, H_2O 40 ml, N-propanol 10 ml, α-naphthyl acetate 3.0 ml, fast garnet GBC

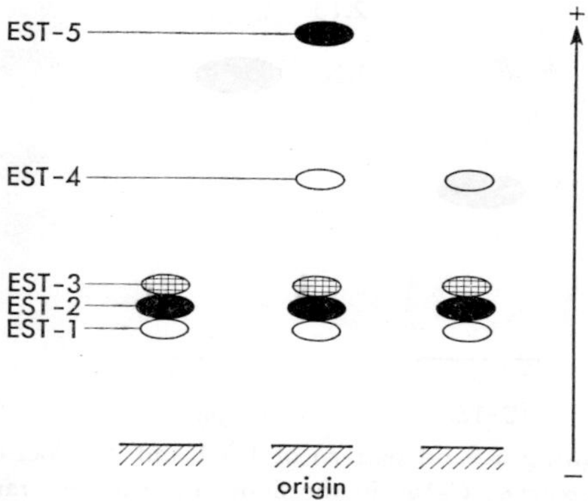

Fig. 3. Diagram showing the three different band patterns of EST loci (EST-1, EST-2, EST-3, EST-4, and EST-5) observed by starch gel-electrophoresis of *C. conicum*.

(pinch). Electrophoresis was run for 1 hr 45 min at 90 mA, 280 V. The following band patterns appeared as shown in Fig. 3.

3) *Malate dehydrogenase* (*MDH*)

Electrode buffer: Tris-citrate EDTA buffer (Tris 217.98 g, citric acid 108.5 g, EDTA 7.2 g, H_2O 18,000 ml). Gel buffer (pH 7.1): electrode buffer 1,000 ml, EDTA 1.5 g, H_2O 5,000 ml. Staining: 0.1 M Tris HCl buffer (pH 8.9) 50 ml, DPN* 10 mg, NBT* 10 mg, PMS* 5 mg, malic acid (pinch). Electrophoresis was run for 1 hr 30 min at 300 V, 90 mA.

4) *Tetrazolium oxidase* (*TO*)

This enzyme was detected as a byproduct of the other enzyme staining. It appears as a white spot in MDH or isocitrate dehydrogenase (IDH) stained gel.

5) *Isocitrate dehydrogenase* (*IDH*)

Buffer system: the same as in MDH. Staining: 0.1 M Tris HCl buffer (pH 8.0) 50 ml, TPN 10 mg, NBT 10 mg, PMS 5 mg, 0.1 M $MgCl_2$ 1 ml, sodium isocitrate (pinch). Electrophoresis was run for 1 hr 30 min at 300 V, 90 mA. The band patterns are shown in Fig. 4.

6) *Glutamate oxaloacetate transaminase* (*GOT*)

Gel buffer: 0.1 M Tris borate EDTA buffer (pH 8.9). Electrode buffer: the same as gel buffer. Staining: GOT substrate solution [α-ketoglutaric

---

* DPN, diphosphopyridine nucleotide; NBT, nitro blue tetrazolium; PMS, phenazine methosulfate; TPN, triphosphopyridine nucleotide.

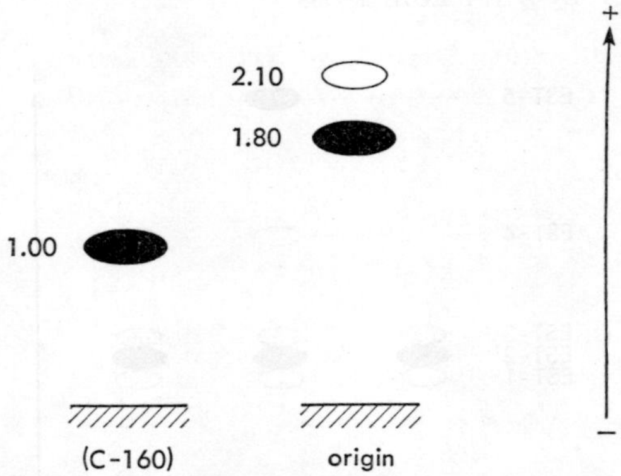

Fig. 4. Diagram showing a band pattern of IDH and TO loci observed by starch gel-electrophoresis of *C. conicum*. C-160 in the figure indicates a strain of *Drosophila melanogaster* run together as a control. The white and black circle indicate TO and IDH, respectively.

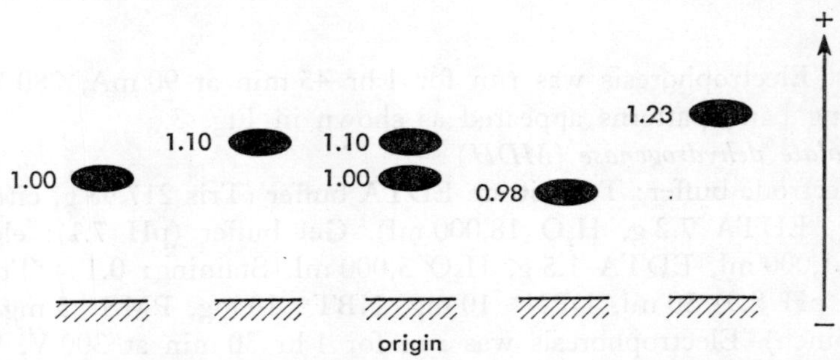

Fig. 5. Diagram showing five different band patterns of GOT locus observed by starch gel-electrophoresis of *C. conicum*.

acid (free acid) 1.461 g, L-aspartic acid (potassium salt) 5.324 g, EDTA 2.0 g, $NaHPO_4$ 56.8 g, H_2O 2,000 ml] (pH 7.4) 50 ml, H_2O 50 ml, fast blue BB salt (pinch). Staining was done for 5–10 min at 37°C. Electrophoresis was run for 1 hr 30 min at 200 V, 70 mA. The band patterns are shown in Fig. 5.

POPULATION VARIABILITY AND HETEROZYGOSITY

The seven natural populations sampled in this study, Yakiyama, Inunaki-rindo, Inunaki-tozan, Tachibanayama, Kodaijisan, Yukawayama, and Wakasugiyama, are all located less than 30 km from Fukuoka city.

We are almost certain that each local population tested is a Mendelian population or comprises an identical biological entity derived from the same clone, since all samples were taken from the same mountain stream. Sampling distance within a local population may be short enough (maximum 3 km) to ensure fertilization within this distance since there is fairly abundant water in the streams, especially after a rain. Water is necessary for completing fertilization in the moss since the male gamete of moss in general is sperm, not pollen as in higher seed plants. The sperm produced on the antheridium of the male individual, must swim through water to the archegonium of the female plant. If this species propagates only by asexual means, all the individuals in a local population must be genetically identical except for those derived from rare spontaneous mutations. If the latter means is common we do not have to examine many samples from a local population since all the individuals should have the same genotype. At first I therefore sampled only a few individuals from each local population to test how much differentiation was present among different local populations. All the populations were found to have the same majority alleles (see Table I). The results, however, differed from what I expected. Even within a local population abundant genic variability was found. Out of 11 enzyme loci 4, MDH, PO-2, EST-1, and EST-2, were monomorphic. The other seven loci, IDH, TO, PO-1, GOT, EST-3, EST-4, and EST-5, were polymorphic though two EST loci (EST-4, EST-5) had only a null allele, in addition to one bandmorph. IDH, TO, and GOT were truly polymorphic in the sense that they had more than two bandmorphs. The average heterozygosity of the Yakiyama population which was most extensively studied was 0.174; the average of all the samples was 0.167. The amount of heterozygosity here was obtained as follows: Het.$=1-\sum x_i^2$, where x_i is the frequency of the ith allele at a locus. Genic variability was also found in other local populations though the sample sizes were too small to estimate gene frequencies. These were, therefore, calculated only for the Yakiyama population (see the top row of Table I). The presence of a large amount of polymorphisms within a population strongly indicates that this haploid plant exercises sexual reproduction much more commonly than is usually supposed. The fact that both male and female plants are often seen in mixture within a distance of a few meters in nature substantiates the prevalence of sexual reproduction in this species.

From these data alone we cannot determine whether genetic differentiation exists between different local populations. We, at present, do not know if " local population " is a biologically meaningful unit, a deme. At any rate the present data are sufficient, however, to show that the

TABLE I
Allozyme Frequencies in *C. conicum*

Populations	IDH 1.00	IDH 1.01	IDH 0.99	MDH 1.00	MDH 0.92	TO 1.00	TO 1.01	TO 0.99	TO 0.93	TO 0.89	PO-1 1.00	PO-1 1.00/1.30	PO-1 Null
Yakiyama	88 (0.967)[a]	2 (0.022) [0.064][b]	1 (0.011)	88 (1.00) [0]	0 (0)	88 (0.967)	1 (0.011)	1 (0.011) [0.065]	1 (0)	1 (0.011)	73 (0.849)	**2 (0.023) [0.262]**	11 (0.128)
Inunaki-rindo	3	0	0	3	0	3	0	0	0	0	2	**0**	0
Inunaki-tozan	3	0	0	2	1	3	0	0	0	0	4	**0**	0
Tachibanayama	2	0	0	2	0	2	0	0	0	0	1	**0**	1
Kodaijisan	2	0	0	2	0	2	0	0	0	0	4	**0**	0
Yukawayama	3	0	0	3	0	3	0	0	0	0	3	**0**	0
Wakasugiyama	1	0	0	1	0	0	0	0	1	0	1	**0**	0
Total	102	2	1	101	1	101	1	1	1	1	88	**2**	12
Frequency	0.971	0.019	0.010	0.990	0.010	0.962	0.010	0.010	0.010	0.010	0.863	**0.020**	0.118
Heterozygosity		0.057		0.020				0.074				**0.241**	

[a] The number in parenthesis indicates the allele frequency of the Yakiyama population.
[b] The number in brackets indicates the heterozygosity of the Yakiyama population.

TABLE I (continued)

Populations	PO-2	GOT					EST-1	EST-2	EST-3		EST-4		EST-5	
	1.00	1.00	0.98	1.00/ 1.10	1.10	1.23	1.00	1.00	1.00	1.01	1.00	Null	1.00	Null
Yakiyama	69 (1.00) [0]	43 (0.518)	1 (0.012)	1 (0.012) [0.569]	33 (0.398)	5 (0.060)	78 (0) [0]	78 (0) [0]	71 (0.922) [0.144]	6 (0.078)	62 (0.795) [0.326]	16 (0.205)	47 (0.603) [0.479]	31 (0.397)
Inunaki-rindo	2	2	0	0	1	0	3	3	3	0	3	0	0	3
Inunaki-tozan	4	3	0	0	1	0	4	4	4	0	4	0	3	1
Tachibanayama	2	1	0	0	1	0	2	2	2	0	2	0	1	1
Kodaijisan	4	4	0	0	0	0	4	4	4	0	4	0	3	1
Yukawayama	3	1	0	0	2	0	3	3	3	0	3	0	1	2
Wakasugiyama	1	—	—	—	—	—	—	—	—	—	—	—	—	—
Total	85	54	1	1	38	5	94	94	87	6	78	16	55	39
Frequency	1.00	0.545	0.010	0.010	0.384	0.051	1.00	1.00	0.935	0.065	0.830	0.172	0.585	0.415
Heterozygosity	0			0.553			0	0	0.122		0.282		0.486	

$\overline{\text{Hetero.}}=0.174$, $\overline{\text{Hetero.}}=0.167$.
(Yakiyama) (All populations)

amount of enzyme polymorphisms of haploid moss, *C. conicum*, are not different from those of most diploid organisms. We have examined several *Drosophila* populations in Japan for the last few years (Yamazaki, in preparation). The average heterozygosity of 16 enzyme loci was 0.170. The above two figures were obtained using exactly the same technique by the same investigator. Average heterozygosities of the two different organisms (haploid and diploid) were almost identical. There is, however, one difference in results between the above two organisms. Null alleles are not as commonly observed in *Drosophila* as in *Conocephalum*. Three out of eleven loci examined had null alleles. In the present discussion a null allele was treated as one allele. Of course, the interpretation of the data is not simple because the amount of heterozygosity depends not only on selection patterns but also on several other factors, such as population size, population structures, and historical events. The equality of heterozygosity, however, is rather difficult to explain by balancing selection as mentioned previously.

The assumption of the neutrality of allozyme polymorphisms is the simplest and easiest interpretation of data. The present data seem to be consistent with those of other haploid organisms such as *Escherichia coli* (Milkman, 1973) and *Neurospora intermedia* (Spieth, 1975), though there are significant differences between the above two organisms and the moss studied here in the way they face natural selection. The former often endure unfavorable environments as dormant spores while the latter species maintains its ordinary life style as a haploid plant under most environmental changes. In addition, *Neurospora* in nature often endures and survives in a diploid-like heterokaryon state. In bacteria even the exchange of genetic materials between different taxa is possible through transformation or transduction. Considering these differences, the data of the truly haploid higher plant, *C. conicum*, are meaningful for elucidating the role of enzyme polymorphisms in evolution.

Recently the PO locus of another haploid moss, *Plagiochila asplenioides*, was also found to be polymorphic (Krzakowa and Szweykowski, 1979). All these data seem to be consistent and indicate that there are no differences in the amount of genetic variability between haploid and diploid organisms. These results are, at least, consistent with the expectation by the neutral hypothesis of protein polymorphisms, though drawing conclusions from these data alone seems premature since there are many other factors which influence the amount of genetic variability in natural populations (see Yamazaki, 1977).

SUMMARY

The amount of genetic variability expected in natural populations is different for diploid and haploid organisms under most selection models: diploid organisms are expected to be more polymorphic. I examined several natural Japanese populations of haploid moss, *C. conicum*, by starch gel electrophoresis. Seven enzyme loci out of 11 examined turned out to be polymorphic. The average " heterozygosity " of the 11 loci was 0.167. This value is almost identical to those obtained in diploid organisms of various sources. Moreover, abundant genic variability was found within a local population, indicating the prevalence of sexual reproduction in natural populations of this species. The equality of genic variability between diploid and haploid organisms is most easily explained by the selective neutrality of allozyme polymorphisms, though other possibilities are not excluded from the present data.

REFERENCES

Ayala, F. J., Tracey, M. L., Barr, L. G., McDonald, J. F., and Pérez-Salas, S. (1974). Genetic variation in natural populations of five *Drosophila* species and the hypothesis of the selective neutrality of protein polymorphisms. *Genetics* **77**, 343–384.

Ehrman, L. and Petit, C. (1968). Genotype frequency and mating success in the *Willistoni* species group of *Drosophila*. *Evolution* **22**, 649–658.

Gillespie, J. H. (1973). Polymorphism in random environments. *Theor. Popul. Biol.* **4**, 193–195.

Harris, H. (1966). Enzyme polymorphisms in man. *Proc. Roy. Soc. Ser. B.* **164**, 298–310.

Kimura, M. (1968). Evolutionary rate at the molecular level. *Nature* **217**, 624–626.

Krzakowa, M. and Szweykowski, J. (1979). Isozyme polymorphism in natural populations of a Liverwort, *Plagiochila asplenioides*. *Genetics* **93**, 711–719.

Levene, H. (1953). Genetic equilibrium when more than one ecological niche is available. *Am. Nat.* **87**, 131–133.

Lewontin, R. C. and Hubby, J. L. (1966). A molecular approach to the study of genic heterozygosity in natural populations. II. Amount of variation and degree of heterozygosity in natural populations of *Drosophila pseudoobscura*. *Genetics* **54**, 595–609.

Milkman, R. (1973). Electrophoretic variation in *Escherichia coli* from natural sources. *Science* **182**, 1024–1026.

Mukai, T. and Yamaguchi, O. (1974). The genetic structure of natural populations of *Drosophila melanogaster*. XI. Genetic variability in a local population. *Genetics* **76**, 339–366.

Prakash, S., Lewontin, R. C., and Hubby, J. L. (1969). A molecular approach to the study of genic heterozygosity in natural populations. IV. Patterns of genic variation in central, marginal and isolated populations of *Drosophila pseudoobscura*. *Genetics* **61**, 841–858.

Spieth, P. T. (1975). Population genetics of allozyme variation in *Neurospora intermedia*. *Genetics* **80**, 785–805.

Takahata, N. and Kimura, M. (1979). Genetic variability maintained in a finite population under mutation and auto-correlated random fluctuation of selection intensity. *Proc. Natl. Acad. Sci. U.S.* **76**, 5813–5817.

Yamazaki, T. (1974). Organization of linked genes under frequency-dependent selection of minority advantage. *Japan. J. Genet.* **49**, 33–36.

Yamazaki, T. (1977). Enzyme polymorphism and functional difference: Mean, variance, and distribution of heterozygosity. In *Proceedings of the Second Taniguchi International Symposium on Biophysics, Molecular Evolution, and Polymorphism* (ed. M. Kimura), pp. 127–147. National Institute of Genetics, Mishima.

5

Molecular Evolution, Protein Polymorphism and the Neutral Theory (1982) (ea. M. Kimura), pp. 137–148.
Japan Scientific Societies Press, Tokyo / Springer-Verlag, Berlin

Means and Variances of Heterozygosity and Protein Function

TAKASHI GOJOBORI

Center for Demographic and Population Genetics, The University of Texas at Houston, Houston, Texas 77025, U.S.A.

In the evolutionary process negative selection due to the functional constraint of proteins seems to be more prevalent than positive selection which increases the fitness of an individual. It is known that although the rate of amino acid substitution for a given protein is roughly constant over species (Zuckerkandl and Pauling, 1962), the rate varies considerably from protein to protein. This variation can best be explained by different functional constraints of proteins, as pointed out by Margoliash and Smith (1965), Zuckerkandl and Pauling (1965), King and Jukes (1969), Dickerson (1971), and Kimura and Ohta (1974).

The functional constraint of proteins is also expected to affect the level of genetic variability in populations. In the last decade many authors have attempted to establish the relationship between protein function and genetic variability in populations. Gillespie and Kojima (1968), Kojima *et al.* (1970), and Gillespie and Langley (1974) suggested on the basis of their work with *Drosophila* and some other organisms that enzymes requiring a single physiological substrate (Group I) tend to have a lower variability than enzymes with multiple substrates (Group II). Johnson

(1974) classified enzymes into three groups, *i.e.*, variable substrate enzymes, regulatory enzymes, and nonregulatory enzymes. Examining the available data on protein polymorphism, he concluded that the first group has the highest average heterozygosity whereas the third group has the lowest. Although these studies have been criticized on the grounds of unsatisfactory classification and some other reasons (*e.g.*, Selander, 1976), there seems to be a general tendency, at least in *Drosophila*, that substrate-nonspecific enzymes have a greater average variability than substrate-specific enzymes (Yamazaki, 1977). Especially, Yamazaki (1977) examined the difference between the observed variance of heterozygosity in *Drosophila* and its expected value for neutral alleles, although he did not take into account the variation of parameters involved. He pointed out that substrate-specific enzymes have lower variances than the expected values. Except Yamazaki's study, however, these studies were mainly concerned with the possibility of the existence of balancing selection in heterogeneous environments rather than the functional constraint of proteins.

If a protein has a strong functional constraint, many mutational changes in the protein are expected to be damaging to the protein function. Therefore, we would expect that, with other things being equal, a protein with stronger functional constraint will have more deleterious mutations than a protein with less constraint. Recently Ohta (1976, 1977), Li (1977, 1978), and Kimura (1979) studied the population dynamics of slightly deleterious mutations in finite populations and showed that the existence of deleterious mutations reduces the average heterozygosity substantially compared with the case where all mutations are neutral. Furthermore, Li (1978) and Kimura (1981) independently showed that the variance of heterozygosity relative to the mean is reduced appreciably when a large proportion of mutations are slightly deleterious, although the degree of reduction is milder for Kimura's model than Li's model for given parameters. This suggests that if sufficient data are available, the presence of functional constraints can be detected by studying the relationship between the mean and variance of heterozygosity. I have therefore studied this relationship for various proteins, surveying gene frequency data from many different organisms in the literature. The purpose of this paper is to present the results of this study. The relationship between the variance of heterozygosity and protein function will also be discussed.

THEORETICAL BACKGROUND

In the present study we shall investigate the empirical relationship between the mean and variance of heterozygosity in relation to the theoretical one expected under the assumption of neutral mutations. If the effective population size (N) is the same for all populations and all populations are independent of each other, the theoretical mean (H) and variance (V) of heterozygosity over the populations for neutral alleles are given by

$$H=\frac{M}{M+1}, \tag{1}$$

$$V=\frac{2M}{(M+1)^2(M+2)(M+3)}, \tag{2}$$

where $M=4Nu$ and u is the mutation rate for the locus under consideration (Kimura and Crow, 1964; Stewart, 1976). In practice, N varies from population to population, so that Eqs. (1) and (2) are not directly applicable. If the distribution of N is known, then the variation of N can be incorporated into the formulas (*cf.* Nei *et al.*, 1976). Unfortunately, the distribution of N is unknown. However, if the variance of $M=4Nu$ is small compared to the mean, the mean and variance of heterozygosity are approximately given by

$$\bar{H}=\frac{\bar{M}}{\bar{M}+1}-\frac{\bar{M}^2c}{(\bar{M}+1)^3}, \quad \text{(Nei, 1975)} \tag{3}$$

$$\bar{V}=\frac{2\bar{M}}{(\bar{M}+1)^2(\bar{M}+2)(\bar{M}+3)} +\bar{M}^2c\left\{\frac{1}{(\bar{M}+1)^4}+\frac{4(3\bar{M}^5+22\bar{M}^4+52\bar{M}^3+27\bar{M}^2-51\bar{M}-51)}{(\bar{M}+1)^4(\bar{M}+2)^3(\bar{M}+3)^3}\right\} \tag{4}$$

where $c=V(M)/\bar{M}^2$, and $\bar{M}$ and $V(M)$ are the mean and variance of M.

The relationship between $\bar{H}$ and $\bar{V}$ is plotted in Fig. 1 for various values of c. The curve for $c=0$ represents the case where N is the same for all populations. For the range of mean heterozygosities considered in Fig. 1, the variance for a given mean heterozygosity increases with increasing c. However, the curves for $c\geq 1$ should be taken with caution, because Eqs. (3) and (4) do not hold well for $c\geq 1$.

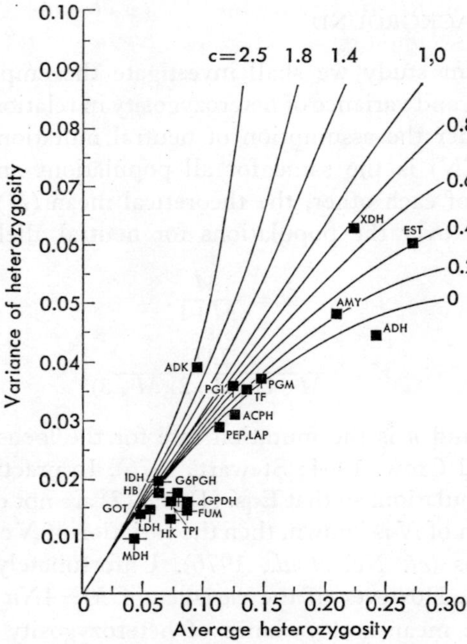

Fig. 1. Relationships between the mean and variance of heterozygosity for various values of *c*. In the data points, a downward bar denotes substrate-specificity, a left-hand bar denotes that the enzyme belongs to a single pathway and a right-hand bar denotes that the enzyme belongs to main pathways. It is clear that most of the enzymes whose variance of heterozygosity is lower than the expected line with no variance in a parameters have many bars.

DATA ANALYSIS

There are two problems in applying the above theory to actual data. First, it is often difficult to tell whether or not a locus studied in a species is the same as the locus with the same name in another, since there are often several loci for the same enzyme. However, since we are interested in the effect of the functional constraint of proteins, it does not matter very much whether we use exactly the same locus or not. In practice, the number of loci for the same enzyme varies from species to species. We regarded these loci as independent loci and computed the heterozygosity for each locus separately as though they were from different species.

The second problem is to determine the sample unit for computing the mean and variance of heterozygosity for a given enzyme or protein. There are three approaches. The first is to take the average of the heterozygosities of all local populations within a species as a sample unit, and then to calculate the variance of heterozygosity over species for a given protein. This approach has drawbacks in that it will tend to underestimate the variance owing to the first average operation and that the number of samples will be greatly reduced, because each species has only one sample. The second is to take the heterozygosity of an arbitrary local population as the representative value of a species. This approach also has a drawback in that the number of samples will be greatly reduced. The third is to take the heterozygosity of each local population as a sample unit. Although this approach will tend to underestimate the variance if the local populations of a species are correlated with each other, it has an advantage in that the number of sample units is large compared with the other methods. Furthermore, if we analyze many kinds of species together, the tendency to underestimate the variance would decrease. In the following analysis, therefore, we shall use the third approach. The data in the present analysis are taken from 14 *Drosophila* species, 14 *Anolis* species, and 31 other species.

TABLE I
Mean and Variance of Heterozygosities of Each Functional Group

Protein	Number of samples	Mean	Variance
ADH	207	0.241	0.0446
αGPDH	209	0.085	0.0159
FUM	50	0.072	0.0126
G6PDH	79	0.074	0.0165
IDH	294	0.061	0.0188
LDH	311	0.047	0.0137
MDH	347	0.042	0.0091
PGM	371	0.147	0.0372
XDH	86	0.223	0.0624
ACPH	131	0.124	0.0308
ADK	113	0.093	0.0390
GOT	304	0.054	0.0143
HK	121	0.086	0.0142
Amy	13	0.208	0.0482
Est	752	0.271	0.0600
Pep. Lap	223	0.112	0.0288
PGI	145	0.124	0.0356
TPI	48	0.076	0.0167
Hb	34	0.062	0.0175
Tf	63	0.134	0.0352

For data sources, readers may refer to Gojobori (1979). We have studied 20 different proteins, which are given in Table I.

The results of the analysis are presented in Table I and Fig. 1. It is seen from Fig. 1 that the proteins with high average heterozygosity generally have a larger variance of heterozygosity than those with low heterozygosity. Interestingly, 12 of the 20 proteins have a variance smaller than the value expected from the neutral model with no variation in M (*i.e.*, $c=0$), and many of them have a low average heterozygosity. Theoretically, it is known that when the sample size is small the estimate of variance is biased downward. Fuerst *et al.* (1977) examined the effect of sampling errors on the relationship between the average and variance of heterozygosity for samples of $2n=100$ genes at 20 loci (or species) by means of computer simulation. According to their Fig. 1, the estimates of variances are significantly lower than the theoretical curve for $c=0$ when the average heterozygosity is about 0.03 or smaller. In the present case, however, since the average heterozygosities are higher than 0.04 in all proteins and the sample size (59 species) used here is larger than 20 it is unlikely that this is due to sampling errors.

DISCUSSION

The existence of the enzymes of which the variances are lower than the expected values of the neutral case suggests that these enzymes have a stronger functional constraint than the others. I have therefore examined the functions of these enzymes. For this purpose I have classified enzymes according to the number of metabolic pathways a particular enzyme is involved in and whether it belongs to a main pathway or subpathway. According to Ishimoto *et al.* (1971), metabolic pathways can be divided into general and important pathways for various organisms, that is, main pathways, and other pathways, or subpathways. We expect that the enzymes involved in main pathways have a stronger functional constraint than those involved in subpathways. The reason is that main pathways are supposed to have more rigid homeostasis than other pathways because of their indispensability for the organism's survival. Therefore, even if a small change occurs in the enzyme molecule for a main pathway, there is a high probability that the molecular function will be impaired. A similar situation might occur for enzymes involved in a single pathway as well as for substrate-specific enzymes. The function of each protein group studied is given in Table II. Column 2 of this table indicates the substrate-

TABLE II
The Relative Importance of Functional Groups with Respect to Metabolic Pathways

Protein	Substrate specificity[a]	Main or sub-pathway[b]	Number of pathways[b]	Comment (functions of pathways)
ADH	n	Sub	3	Glycolysis, vitamin A biosynthesis, and tryptophan biosynthesis
αGPDH	s	Sub	1	Very important in flying muscles of insects
FUM	s	Main	1	TCA cycle
G6PDH	s	Main	1	Pentosephosphate cycle
IDH	s	Main	1	TCA cycle
LDH	s	Main and sub	2	Glycolysis, tryptophan biosynthesis
MDH	s	Main	1	TCA cycle
PGM	s	Main and sub	2	Glycolysis, sucrose, and starch
XDH	n	Main and sub	7	Various
ACPH	n	Main and sub	8	Various
ADK	n		Many	$ATP+AMP \rightleftarrows 2ADP$
GOT	s	Main and sub	3	Amino acid biosynthesis
HK	s	Main	2	Glycolysis
Amy	n			
Est	n		Many	Various
Pep. Lap	n		Many	Various
PGI	s	Main and sub	3	Similar to PGM
TPI	s	Main	1	Glycolysis
Hb				Carrier protein
Tf				Carrier protein

[a] Substrate-specific enzyme (s) and nonspecific enzyme (n).
[b] Metabolic map (Ishimoto *et al.*, 1971) is referred to for this classification.

specificity of each protein, and s and n represent "specific" and "non-specific," respectively. Column 3 shows whether each enzyme is involved in main metabolic pathways or subpathways. Column 4 gives the number of pathways in which each enzyme is involved, and Column 5 gives information about metabolism. The relationship between this metabolic function of the enzymes and the variance of heterozygosity is also presented in Fig. 1. It is clear from this figure that many of the 12 enzymes of which the variances are lower than the expected values for the neutral mutations with $c=0$ are involved in main pathways or single pathways, and many of them are substrate-specific enzymes. On the other hand, only one of the proteins that have higher variances than the values expected from the neutral model with $c=0$ is involved in main pathways or single pathways, and only two of them show substrate-specificity.

Another possibility is that smaller molecules are more likely to be subject to stronger functional constraints than larger molecules because a change in the former would have a higher probability of impairing the function of the molecules (*cf.* Koehn and Eanes, 1978). Figure 1 shows that the 20 proteins used can be divided into three groups in terms of the mean and variance of heterozygosity. The first group has a high average heterozygosity and includes XDH, Est, Amy, and ADH.* The second group has a medium level of heterozygosity and includes ADK, PGM, PGI, Tf, ACPH, and Pep. Lap.* The third group has a low heterozygosity and comprises IDH, Hb, GOT, LDH, MDH, G6PDH, αGPDH, TPI, FUM, and HK*. If the molecular size is important in determining the mean and variance of heterozygosity through functional constraint, these three groups are expected to have different molecular weights. I have therefore computed the average subunit molecular weight for each group, using Darnall and Klotz's (1975) data. In this computation esterases were excluded, since their molecular weight varies considerably from gene to gene. The average subunit molecular weights obtained for the high-, medium-, and low-heterozygosity groups were 77,200, 52,700, and 41,600, respectively. This confirms Koehn and Eanes' (1977) and Nei *et al.*'s (1978) earlier results about the relationship between molecular weight and heterozygosity and is compatible with the hypothesis that smaller molecules have a stronger functional constraint than large molecules.

Recently Kimura (1981) conducted a computer simulation to study the theoretical relationship between the average and the variance of heterozygosity, taking into account slightly deleterious mutation as well as the population size effect. He introduced the concept of effectively neutral mutation (Kimura, 1979). Assuming that the selective disadvandage (denoted by s') of the mutant alleles follows a gamma distribution,

$$f(s') = Ce^{-\alpha s'}s'^{\beta-1} \qquad \text{(Kimura, 1979)} \qquad (5)$$

where $C=\alpha^{\beta}/\Gamma(\beta)$ is a constant and $\Gamma(\cdot)$ represents a gamma function, he called mutations with s' less than $1/2N$ effectively neutral mutations. When β is equal to 0.5 and $4N\bar{s}'$ is less than 500 ($\bar{s}'$; the average of s'), the relationship between the average and the variance of heterozygosity

* XDH, xantine dehydrogenase; Est, esterase; Amy, amylase; ADH, alcohol dehydrogenase; ADK, adenylate kinase; PGM, phosphoglucomutase; PGI, phosphoglucoisomerase; Tf, transferrin; ACPH, acid phosphatase; Pep. Lap, peptidase; IDH, isocirtate dehydrogenase; Hb, hemoglobin; LDH, lactate dehydrogenase; MDH, malate dehydrogenase; G6PDH, glucose-6-phosphate dehydrogenase; αGPDH, α-glycerophosphate dehydrogenase; TPI, triosephosphate isomerase; FUM, fumarase; HK, hexokinase.

in his simulation was very similar to the relationship observed in Fig. 1 for the enzymes whose variances are lower than the expected values with $c=0$. Therefore our data can be explained by Kimura's model, although whether or not the values given to the parameters are realistic is not known at present.

So far we have argued that the variance of heterozygosity lower than the neutral expectation is due to the functional constraint of proteins. However, this conclusion is tentative, since in this paper we used the equilibrium theory of population genetics. In practice, it is quite possible that the allele frequency distribution has not reached the equilibrium status. It is known that the last glaciation (Würm-Wisconsin) ended only about 10,000 years ago, and thus in the majority of the present species the bottleneck effect is expected to remain. If this is the case, the variance of heterozygosity is expected to be smaller than the neutral expectation (Li and Nei, 1975). This is so for those loci with low average heterozygosity more particularly than for loci with high average heterozygosity, since the former group of loci would possibly have a lower mutation rate.

I have therefore examined the effect of bottlenecks on the variance of heterozygosity by using Li and Nei's (1975) formulas for a special case. In this examination I assumed that the average heterozygosity in the initial population was 0.0392 and allelic frequencies were 0.98 and 0.02 because of the severe bottleneck effect but after population expansion $4Nu$ has become 1 with $u=10^{-7}$. Under this assumption the mean and variance of heterozygosity increase with time, as shown in Table III. It is clear that the variance (V_t) in transient states is considerably smaller than the variance (V) which is obtainable from the mean (H_t) under the

TABLE III
Means (H_t) and Variance (V_t) of Heterozygosity for Neutral Alleles in the t-th Generation after Population Expansion

Generation (t)	H_t	V_t	V
10^3	0.0394	1.45×10^{-5}	0.0122
10^4	0.0410	1.47×10^{-4}	0.0127
10^5	0.0573	0.0016	0.0171
5×10^5	0.1227	0.0105	0.0320
10^6	0.1911	0.0223	0.0427
5×10^6	0.4376	0.0468	0.0469

These values were computed by Li and Nei's (1975) formulas. V is the variance that is obtainable from H_t under the assumption of equilibrium.

assumption of equilibrium. Particularly when H_t is about 0.12, V_t is about one-third of V. Therefore, our data can also be explained by the bottleneck effect.

In this connection it should be noted that the c value in Fig. 1 is likely to be substantially large. If it is as large as 1, all the observed variances except for one (ADK) are smaller than the equilibrium values. Therefore, it is possible that the bottleneck effect remains for almost all loci examined.

SUMMARY

The observed relationships between the mean and variance of heterozygosity for protein loci were compared with the theoretical relationships for neutral mutations, taking into account the variation in the population parameter $4Nu$. The 20 protein loci examined can be classified into two classes, *i.e.*, loci of which the variance of heterozygosity is lower than the expected value and others. Most of the former loci produce enzymes which possibly have a strong functional constraint because of their involvement in the main metabolic pathways or single pathways or because of the requirement of specific substrates. Thus, the variance lower than the neutral expectation can be explained by the slightly deleterious mutations. As an another possibility, the bottleneck effect was also discussed.

Acknowledgments

I would like to express my gratitude to Drs. M. Nei, W.-H. Li, and N. Takahata for their valuable suggestions and critical comments on this manuscript. This paper is based in part on my Ph.D. thesis at Kyushu University, Japan. I would like to thank my thesis advisor, Dr. H. Matsuda, for his help and encouragement during this study. I am also indebted to Dr. M. Kimura who suggested that I summarize this study. I wish especially to thank Dr. M. Kimura for his useful comments and encouragement. This work was partially supported by NIH Grant GM 20293 and NSF Grant DEB 76-06069.

REFERENCES

Darnall, D. W. and Klotz, I. M. (1975). Subunit constitution of protein: A table. *Arch. Biochem. Biophys.* **166**, 651–682.

Dickerson, R. E. (1971). The structure of cytochrome *c* and the rates of molecular evolution. *J. Mol. Evol.* **1**, 26–45.

Fuerst, P. A., Chakraborty, R., and Nei, M. (1977). Statistical studies on protein polymorphism in natural populations. I. Distribution of single locus heterozygosity. *Genetics* **86**, 455–483.

Gillespie, J. H. and Langley, C. H. (1974). A general model to account for enzyme variation in natural populations. *Genetics* **76**, 837–884.

Gillespie, J. H. and Kojima, K. (1968). The degree of polymorphisms in enzymes involved in energy production compared to that in non-specific enzymes in two *Drosophila ananassae* populations. *Genetics* **61**, 582–585.

Gojobori, T. (1979). Studies of protein polymorphism by data analyses with respect to loci. Ph. D. Thesis in Kyushu University, Fukuoka.

Ishimoto, M., Minakami, S., Oshima, T., and Wada, H. (1971). *Metabolic Maps*, 3rd ed. Kyoritsu Publishing Co., Ltd., Tokyo.

Johnson, G. B. (1974). Enzyme polymorphism and metabolism. *Science* **184**, 28–37.

Kimura, M. (1979). Model of effectively neutral mutations in which selective constraint is incorporated. *Proc. Natl. Acad. Sci. U.S.* **76**, 3440–3444.

Kimura, M. (1981). *Shudan Idengaku Gairon*, Revised Version, Chapter 10. Baifukan, Tokyo (in press) (in Japanese).

Kimura, M. and Crow, J. F. (1964). The number of alleles that can be maintained in a finite population. *Genetics* **49**, 725–738.

Kimura, M. and Ohta, T. (1974). On some principles governing molecular evolution. *Proc. Natl. Acad. Sci. U.S.* **71**, 2848–2852.

King, J. L. and Jukes, T. H. (1969). Non-Darwinian evolution. *Science* **164**, 788–798.

Koehn, R. K. and Eanes, W. F. (1977). Subunit size and genetic variation of enzymes in natural populations of *Drosophila*. *Theor. Popul. Biol.* **11**, 330–341.

Koehn, R. K. and Eanes, W. F. (1978). Molecular structure and protein variation within and among populations. In *Evolutionary Biology* (eds. M. K. Hecht *et al.*), Vol. 11, pp. 39–100. Plenum Press, New York.

Kojima, K., Gillespie, J. H., and Tobari, Y. N. (1970). A profile of *Drosophila* species enzymes assayed by electrophoresis. I. Number of alleles, heterozygosities, and linkage disequilibrium in glucose-metabolizing systems and some other enzymes. *Biochem. Genet.* **4**, 627–637.

Li, W.-H. (1977). Maintenance of genetic variability under mutation and selection pressures in a finite population. *Proc. Natl. Acad. Sci. U.S.* **74**, 2509–2513.

Li, W.-H. (1978). Maintenance of genetic variability under the joint effect of mutation, selection, and random drift. *Genetics* **90**, 349–382.

Li, W.-H. and Nei, M. (1975). Drift variances of heterozygosity and genetic distance in transient states. *Genet. Res.* **25**, 229–248.

Margoliash, E. and Smith, E. L. (1965). Structural and functional aspects of cytochrome *c* in relation to evolution. In *Evolving Genes and Proteins* (eds. V. Bryson and H. J. Vogel), pp. 221–242. Academic Press, New York.

Nei, M. (1975). *Molecular Population Genetics and Evolution*. North-Holland, Amsterdam and New York.

Nei, M., Chakraborty, R., and Fuerst, P. A. (1976). Infinite allele model with varying mutation rate. *Proc. Natl. Acad. Sci. U.S.* **73**, 4164–4168.

Nei, M., Fuerst, P. A., and Chakraborty, R. (1978). Subunit molecular weight and genetic variability of proteins in natural populations. *Proc. Natl. Acad. Sci. U.S.* **75**, 3359–3362.

Ohta, T. (1976). Role of very slightly deleterious mutations in molecular evolution and polymorphism. *Theor. Popul. Biol.* **10**, 254–275.

Ohta, T. (1977). Extension to the neutral mutation random drift hypothesis. In *Proceedings of the Second Taniguchi International Symposium on Biophysics, Molecular Evolution, and Polymorphism* (ed. M. Kimura), pp. 148–167. National Institute of Genetics, Mishima.

Selander, R. K. (1976). Genic variation in natural populations. In *Molecular Evolution* (ed. F. J. Ayala), pp. 21–45. Sinauer Associates, Inc., Sunderland, Mass.

Stewart, F. M. (1976). Variability in the amount of heterozygosity maintained by neutral mutations. *Theor. Popul. Biol.* **1**, 188–201.

Yamazaki, T. (1977). Enzyme polymorphism and functional difference: Mean, variance, and distribution of heterozygosity. In *Proceedings of the Second Taniguchi International Symposium on Biophysics, Molecular Evolution, and Polymorphism* (ed. M. Kimura), pp. 127–147. National Institute of Genetics, Mishima.

Zuckerkandl, E. and Pauling, L. (1962). Molecular disease, evolution, and genic heterogeneity. In *Horizons in Biochemistry* (eds. M. Kasha and B. Pullman), pp. 189–225. Academic Press, New York.

Zuckerkandl, E. and Pauling, L. (1965). Evolutionary divergence and convergence in proteins. In *Evolving Genes and Proteins* (eds. V. Bryson and H. J. Vogel), pp. 97–166. Academic Press, New York.

6

Molecular Evolution, Protein Polymorphism and the Neutral Theory (1982) (ed. M. Kimura), pp. 151–166.
Japan Scientific Societies Press, Tokyo / Springer-Verlag, Berlin

Stochastic Integrals and Their Application to Population Genetics

TAKEO MARUYAMA

National Institute of Genetics, Mishima 411, Japan

Stochastic models have a rather long history in population genetics. The model of finite populations has been studied extensively by S. Wright, M. Kimura, and others. These theoretical works have influenced experimental population genetics, but their impact on the topical problem of molecular evolution and protein polymorphism can be symbolized by the birth of the neutral hypothesis advanced by Kimura (1968), King and Jukes (1969), and Kimura and Ohta (1971). Most of the early stochastic models dealt with one locus at which two alleles are present, where the one-dimensional diffusion theory can be successfully applied. In these early studies, much effort was spent in searching for analytic solutions of problems. Particularly when the subject can be formulated by the one-dimensional theory, the problem can almost always be solved analytically (see Wright, 1969; Kimura, 1964; Crow and Kimura, 1970; Maruyama, 1977; Ewens, 1979). Although they may appear to be mathematically simple, the stochastic models and their analyses have played important roles and have been indispensable to population genetics and molecular evolutionary theory.

On the other hand, except for some special cases which turn out to be important biologically, analytic solutions to multiple allele problems and linked locus problems have been very difficult to obtain. The field of experimental population genetics now strongly demands solutions to such stochastic models so that further biological insight can be obtained. In fact, many problems that now appear to be biologically realistic seem to have hopelessly difficult analytic solutions. Therefore it is natural that a number of numerical methods have been developed. There are basically two different methods. One is to describe the process in terms of a classical diffusion equation and solve it numerically. The other is to carry out simulations using models. Since a diffusion equation or an equation derived from it gives the expectation of a quantity and the numerical solutions are more stable, it is usually more desirable to take this approach rather than a simulation. However, perhaps with a few exceptions, the numerical integration of a diffusion equation is limited to processes with no more than two or three space variables.

Recently we developed a new numerical method which enables us to handle various stochastic models in a very general form (Maruyama, Takahata, and Kimura unpublished; Maruyama, 1980 and 1981; Maruyama and Nei, 1981). Because of the availability of modern high-speed computers, the method has proven to be useful, permitting us to parameterize problems and seek concrete answers. Here I will outline the method and present some of the findings on various problems which are biologically important.

MATHEMATICAL METHOD

Let x_i be random variables representing gene frequencies or gamete frequencies. In most of the cases dealt with in population genetics the random variables form a stochastic process which can be regarded as a diffusion process. Then the set of random variables x_i satisfies a system of stochastic differential equations,

$$dx_i = \sum_j a_{ij}(x_1, x_2, \cdots)dB_j + b_i(x_1, x_2, \cdots)dt \tag{1}$$

where a_{ij}'s and b_i's are functions of x_i's, and B_j's are independent Brownian motions. The coefficients a_{ij} are elements of the non-negative definite square root of the covariance matrix whose i-th row, j-th column element is given by

$$\sigma_{ij} = \lim_{\Delta t \to 0} \frac{1}{\Delta t} E\{\Delta x_i \Delta x_j\} \tag{2}$$

where $E\{\ \}$ indicates the expectation. In practice, it is usually easy to determine the covariance matrix $[\sigma_{ij}]$. The $[a_{ij}]$ is the square root of the matrix $[\sigma_{ij}]$. The other coefficient b's represent the infinitesimal mean change in x_i. That is

$$b_i = \lim_{\Delta t \to 0} \frac{1}{\Delta t} E\{\Delta x_i\}.$$

The differential of the Brownian motion represents the random change which occurs in an infinitely small time interval.

Since Eq. (1) requires operations involving infinitesimally small time intervals, it is not possible to simulate it with a computer. As with differential equations of deterministic processes, Eq. (1) can be approximated by the difference equations,

$$\Delta x_i = \sum_j a_{ij}(x_1, x_2, \cdots)\Delta W_j + b_i(x_1, x_2, \cdots)\Delta t \tag{3}$$

where ΔW_j indicates a white noise with variance equal to the time increment Δt. Let t_k be discrete time incremented by Δt for $k=1, 2, \cdots$, and $x_i(t_k)$ be the random variable x_i at time t_k. Then we can rewrite Eq. (3) as

$$x_i(t_{k+1}) = x_i(t_k) + \sum_j a_{ij}(x_1(t_k), x_2(t_k), \cdots)\Delta W_j + b_i(x_1(t_k), x_2(t_k), \cdots)\Delta t. \tag{4}$$

This equation certainly gives the values of x_i at these discrete times t_k but does not give it between these specified times. Now let

$$x_i(t) = x_i(t_k) \tag{5}$$

for $t_{k+1} > t \geq t_k$. With this definition, the path constructed by Eq. (3) or equivalently by Eq. (4) is defined for all values of time, and each path forms a step function having discontinuity at t_k's. If we are dealing with differential equations of a deterministic process, it is fairly obvious that solutions given by a difference quation converge to those of the original equations as Δt becomes small. However, with a differential equation involving differentials in a random processes it is not simple and in fact requires highly technical mathematics to show the convergence of $x_i(t)$ given in Eqs. (4) and (5) to the solution of the original Eq. (1) (Skorokhod, 1965; McShane, 1974). Equation (1) is a symbolic representation of the behavior of sample paths, and actually, depending on the way Eq. (1) is integrated many different processes can be constructed which no longer satisfy the original Eq. (1). Therefore, it is important to note that among the schemes for integrating Eq. (1), the one given by Eqs. (4) and (5) simulates sample paths which converge to those specified by Eq. (1) (Wong and Zakai, 1965; McShane, 1974; Arnold, 1974).

In the case of differential equations for deterministic processes, there are many different improved methods for numerical analysis, such as the Runge-Kutta approximation which enables us to obtain a faster convergence than methods equivalent to Eqs. (4) and (5). However, with differential equations describing sample paths of random processes, a method yielding an improved convergence rate can often lead to a different process. The reason for this is that the second-order moment $\Sigma\{\Delta W_j\}^2$ does not vanish as the time increment decreases to zero, while a corresponding quantity in differential equations for deterministic processes can be made arbitrarily small by letting Δt be small (see Rumelin, 1980).

In some models of population genetics, we assume that the possible number of variables, which are usually either the gene frequencies or the gamete frequencies, is fixed. In many recent models dealing with molecular evolution, the number of variables is not constant, but changes over time. This is because the number of possible allelic states of a gene or gamete has been shown to be extremely large, but those represented in a population at any one time is rather small. Furthermore, in reality once a particular variable appears and then disappears from the population it will never return to the same population (Kimura and Crow, 1964).

Therefore it is necessary to incorporate newly arising variables into the system of Eq. (1), while the variables which disappear must be removed from the system. It is easy to eliminate a vanishing variable. We rename the variables by omitting the one which we want to remove. To add a new variable we do the following: during the time interval of $\Delta t = t_{k+1} - t_k$, the new variable

$$x_{n+1}(t_{k+1}) = \varepsilon \tag{6}$$

is introduced with a probability which can be specified by the model. For instance, in the case of a gene mutation at one locus in a population of N diploid organisms with mutation rate v, the probability is $2Nv\Delta t/\varepsilon$, provided that this is much less than unity. However, $2Nv\Delta t/\varepsilon$, the mean number of new mutants to be introduced into a population during Δt, is not necessarily small. Therefore, in general we introduce m new mutants all having equal frequency ε: $x_{n+1}(t_{k+1}) = \varepsilon, \cdots, x_{n+m}(t_{k+1}) = \varepsilon$, where m is a Poisson random number with a mean of $2Nv\Delta t/\varepsilon$. If $m=0$ no new mutants will be introduced, if $m=1$ one new mutant will be introduced, and so forth.

It is very important to make $\varepsilon > \Delta t$, particularly when the value of $2Nv$ is smaller than unity. If $\varepsilon < \Delta t$ with $2Nv < 1$, the simulation can lead to a process of a quite different nature at the boundary $x=0$. This property requiring that $\varepsilon > \Delta t$ for small $2Nv$ appears to be unique to this numerical method. The same requirement applies to cases with a fixed number of variables. The mutation from the j-th allele to the i-th allele cannot be performed by adding $2Nv_{ij}x_j(t)\Delta t$ to $x_i(t)$ and subtracting the same quantity from $x_j(t)$ when $2Nv_{ij}$ is small. Instead, it must be carried out by adding $m\varepsilon$ to $x_i(t)$ and subtracting the same from $x_j(t)$, where m is a Poisson random number with a mean equal to $2Nv_{ij}x_j(t)\Delta t/\varepsilon$ and $\varepsilon > \Delta t$.

There is more than one form of a stochastic differential equation representing the same process (Watanabe, 1971). The equation given by Eq. (1) is a standard one that is often used in the mathematical literature (for instance, Itô and McKean, 1965). However any matrix $A=[a_{ij}]$ having the property $AA^*=[\sigma_{ij}]$ can represent a diffusion coefficient matrix. Here A^* represents the transpose of A and σ_{ij} is the coefficients given in Eq. (2).

Stochastic models studied in population genetics usually involve diffusion coefficients resulting from multinomial sampling of gametes. In that case, $\sigma_{ij}=x_i(\delta_{ij}-x_j)$ where $\delta_{ij}=1$ if $i=j$ and otherwise equals 0. With this system of diffusion coefficients, several different representations are possible, besides the one given in Eq. (1) in which the square root of $[\sigma_{ij}]$ is used. The formulations of Pederson (1973), Itoh (1979), and Kimura (1980) can be used in cases of multinomial sampling. Itoh's presentation is simple and is given below. Itoh (1979) has shown that when $\sigma_{ij}=x_i(\delta_{ij}-x_j)$ the diffusion can be expressed as follows

$$dx_i(t) = \sum_{j=0,\ j\neq i}^{n} \sigma(i,j)\sqrt{x_i(t)x_j(t)}\,dB_{ij} \qquad (7)$$

for $i=1, 2, \cdots, n$, where $\sigma(i, j)=1$ if $i<j$, $\sigma(i, j)=-1$ if $i>j$, and B_{ij} for $i<j$ is an independent Brownian motion, whereas $B_{ji}=B_{ij}$. While Itoh's formula shown by Eq. (7) is simple in form, it tends to involve a larger number of independent Brownian motions B_{ij}, for the number grows proportional to the square of the number, n, of the random variables. There are processes in population genetics which cannot be represented by Eq. (7), and therefore require the square root of the $[\sigma_{ij}]$ matrix (Takahata, 1981).

The deterministic change in $x_i(t)$ is governed by $b_i(x_1, x_2, \cdots)\Delta t$ in Eq. (3). For instance, if an extant allele A_i mutates to new alleles at rate v,

$$b_1(x_1, x_2, \cdots)\Delta t = -vx_i(t)\Delta t. \tag{8}$$

Under natural selection,

$$b_i(x_1, x_2, \cdots)\Delta t = \frac{1}{2\overline{W}}\frac{\partial \overline{W}}{\partial x_i}\Delta t \tag{9}$$

in which $\overline{W}=\sum_i\sum_j(1+s_{ij})x_i(t)x_j(t)$ with $1+s_{ij}$ being the relative fitness of the genotype A_iA_j.

ACCURACY

For the following two reasons, the accuracy of the approximations represented by Eqs. (3)–(6) needs to be checked. One reason is that the difference Eq. (3) together with Eq. (4) converges rather slowly to a continuous limit of a diffusion process. The other reason is that the nature of the process can be greatly altered by the input of Eq. (6) for new mutants, particularly when sample paths have a high probability of being near boundaries. In the test simulation given below, I used $\Delta t=0.001$ and $\varepsilon=5\Delta t$ throughout.

Since the allele frequency distribution and heterozygosity are probably the most important quantities in population genetics, I have tested the accuracy of the method in predicting these quantities. For the numerical simulations, I used the infinite allele model of Kimura and Crow (1964), in which every mutant is assumed to be new. Two schemes of selection have been used. One model assumes that all mutants are selectively neutral, and the other is an overdominance model in which every heterozygote has relative fitness $1+s$, while fitness is equal to unity for every homozygote (see p. 158 for the equation). The average level ($\bar{h}$) of heterozygosity is defined as the mean of $1-\Sigma x_i^2$, and the allele frequency distribution, $\Phi(x)$, is the average number of alleles having frequency x. For the neutral model, both of these quantities can be obtained exactly (Kimura and Crow, 1964; see also Wright, 1949). The average heterozygosity for a small value of $2Ns$ in the case of overdominant selection can be studied by the formulas of Wright (1949), Watterson (1977), or Li (1978). The theoretical average $\bar{h}$ values obtained by these methods are presented in Table I, and are compared with the simulation results. It is clear that the agreement between the theory and numerical simulation is satisfactory in all cases. Figure 1 shows the theoretical allele frequency distribution of the neutral model together with the results of simulations. The theoretical

TABLE I
Mean Heterozygosities for Testing the Accuracy of the Computer Simulation Used

$4Vv$	$2Ns$	Mean heterozygosity	
		Theoretical	Simulation
0.02	0	0.0196	0.0195
0.1	0	0.0909	0.0897
0.2	0	0.1667	0.1658
0.5	0	0.3333	0.3316
0.004	2	0.0080	0.0079
0.012	6	0.0970	0.0964
0.026	13	0.4548	0.4501
0.060	30	0.6265	0.6360

The theoretical values for neutral alleles were obtained by $\bar{h}=4Nv/(1+4Nv)$ and those for overdominant alleles were obtained by numerical integration of the formula given by Wright (1949).

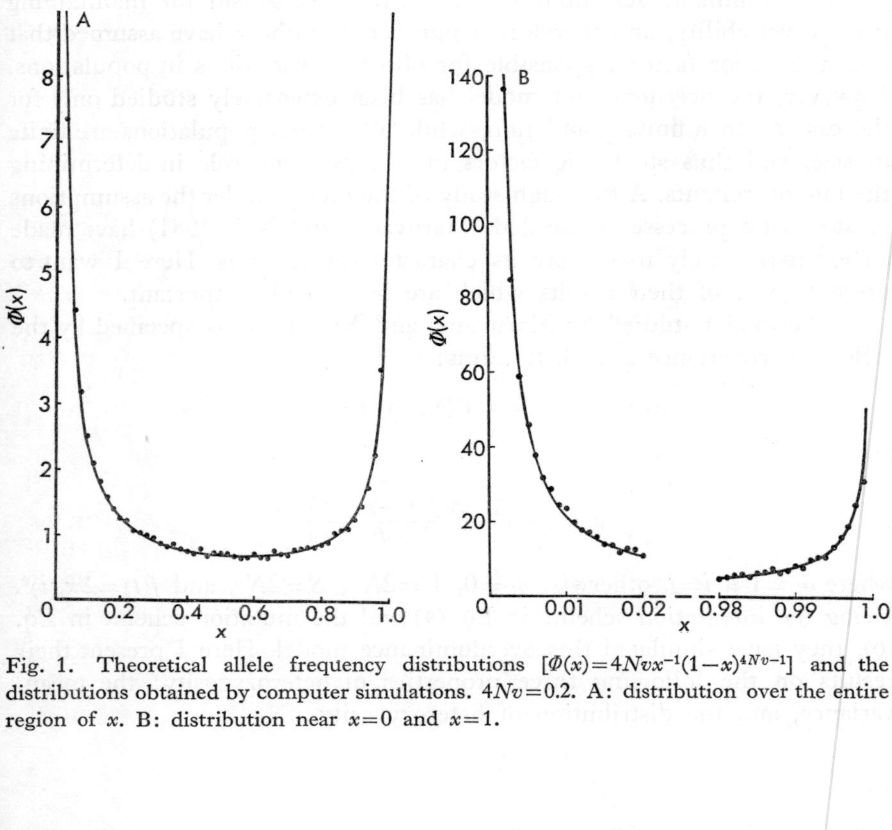

Fig. 1. Theoretical allele frequency distributions $[\Phi(x)=4Nvx^{-1}(1-x)^{4Nv-1}]$ and the distributions obtained by computer simulations. $4Nv=0.2$. A: distribution over the entire region of x. B: distribution near $x=0$ and $x=1$.

allele frequency distribution was obtained by Kimura and Crow's (1964) formula $\Phi(x)=4Nvx^{-1}(1-x)^{4Nv-1}$. Figure 1A shows the comparison between the theoretical distribution and the simulation results for the whole range of x partitioned into 50 intervals of equal length (0.02). Since in population genetics the distribution of rare alleles is also important, I have tested the distribution near $x=0$ and $x=1$. Figure 1B shows the comparison near $x=0$ and $x=1$ by partitioning [0, 0.02] and [0.98, 1.0] into 20 subintervals. The agreement between the theory and simulation is again satisfactory. Thus, this method of simulation appears to give quite accurate results when studying the population parameters we are interested in. It is important to note that even though the input parameter $\varepsilon=0.005$ was only five times larger than the smallest range of the frequency interval of [0, 0.001], the accuracy for the range turned out to be good (Fig. 1B).

OVERDOMINANCE MODEL

Overdominant selection is a powerful mechanism for maintaining genetic variability, and therefore a number of authors have assumed that it is the major factor responsible for observed variations in populations. However, the overdominant model has been extensively studied only for the case of an infinite population, while all natural populations are finite in size, and thus stochastic factors must play some role in determining the fate of mutants. A thorough study of the model under the assumptions of stochastic processes is needed. Maruyama and Nei (1981) have made an extensive study to explore its characteristic features. Here I want to present some of their results which are biologically important.

The model studied by Maruyama and Nei (1981) is specified by the following covariance and drift terms:

$$\sigma_{ij} = x_i(t)(\delta_{ij}-x_j(t))$$

and

$$b_i = \left\{-V+\frac{S(f(t)-x_i(t))}{1-sf(t)}\right\}$$

where $\delta_{ij}=1$ if $i=j$, otherwise $\delta_{ij}=0$, $V=2Nv$, $S=2Ns$, and $f(t)=\Sigma x_i(t)^2$. Using the integration scheme in Eq. (4) and the mutation scheme in Eq. (6), they have simulated this overdominance model. Here I present their results on the following three properties of heterozygosity: the mean, variance, and the distribution of heterozygosity.

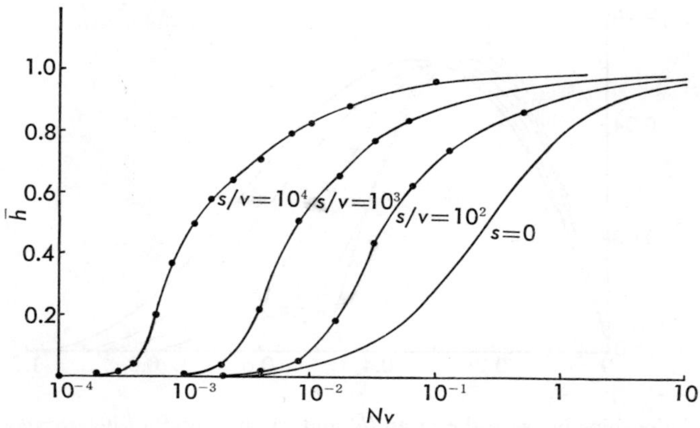

Fig. 2. Relationships between Nv and mean heterozygosity ($\bar{h}$) for overdominant alleles. When s and v remain constant, each curve represents the relationship between N and $\bar{h}$.

1. Mean Heterozygosity

The mean heterozygosity ($\bar{h}$) is shown as a function of Nv and Ns in Fig. 2. It is certain that the average heterozygosity for overdominant selection is always higher than that for neutral mutations when $4Nv$ remains the same. It is biologically very important to note that in the presence of overdominant selection $\bar{h}$ increases more rapidly, particularly when s/v is large. This implies that the average heterozygosity in the overdominance model is more sensitive to the variation in population sizes than that in the neutral model. A rapid change in $\bar{h}$ occurs in the region where $\bar{h}$ changes from about 0.05 to 0.6. This is indeed the range that includes the majority of observed cases in natural populations.

2. Variance of Heterozygosity

The relationship between the mean ($\bar{h}$) and variance ($\sigma_h{}^2$) of heterozygosity is important, and has been studied by Stewart (1976) for the neutral model. However, it has not been possible to obtain this relationship for other models. The present method certainly enables us to carry out calculations for any model, including the overdominance model, that can be formulated by Eq. (1). Figure 3 shows the relationship between $\bar{h}$ and $\sigma_h{}^2$. As the value of s/v becomes large, graphs of the relationships tend to a parabolic curve with a maximum at $\bar{h}=1/4$. Interestingly, the graphs assume their minima at $\bar{h}=1/2$, $2/3$, and $3/4$ and probably at all the other $\bar{h}=(n-1)/n$ for $n=5, 6, \cdots$.

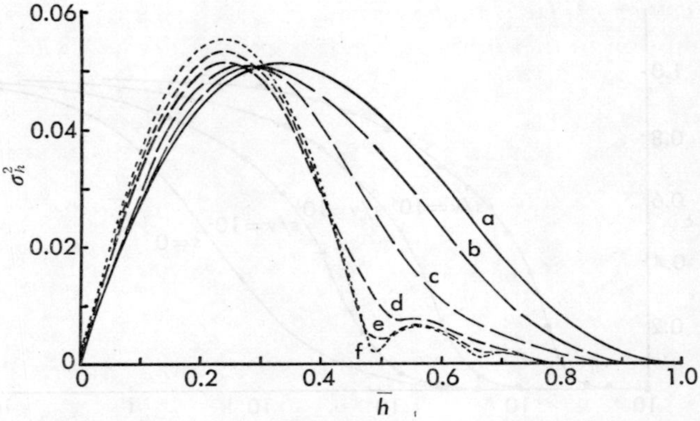

Fig. 3. Relationships between the mean ($\bar{h}$) and variance ($\sigma_h{}^2$) of heterozygosity. a, $s=0$. b, $s/v=10$. c, $s/v=100$. d, $s/v=1{,}000$. e, $s/v=10^4$. f, $s/v=10^5$.

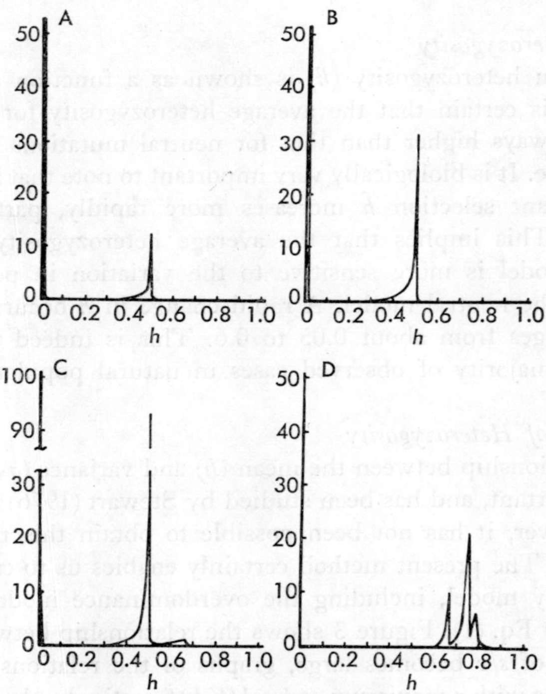

Fig. 4. Distributions of heterozygosity for various levels of $\bar{h}$. In the computation of these distributions it was assumed that $s/v=1{,}000$. A, $2Ns=10$, $\bar{h}=0.076$. B, $2Ns=12$, $\bar{h}=0.189$. C, $2Ns=20$, $\bar{h}=0.488$. D, $2Ns=80$, $\bar{h}=0.729$.

3. Distribution of Heterozygosity

In addition to the mean and the variance of hetrozygosity, the present method also enables us to obtain its distribution. This has been studied by Stewart (1976) who conjectured that the distribution for neutral mutations will have peaks at $h=0, 1/2, 2/3, \cdots$. In the overdominance model, these peaks are much more pronounced, and particularly when the value of s/v is large almost all the mass is localized at the peaks. Therefore, if the average heterozygosity is 0.5 or less, as observed in nature, a population stays mostly in the neighborhood of $h=0$ or $h=0.5$, but practically no mass is found at other intermediate values of h. For instance if $\bar{h}=0.20$, which is a typical value observed in nature, a population is at $h=0.5$ for 40% of the time and at $h=0$ for the rest. Typical cases of the distribution of heterozygosity are given in Fig. 4.

LINKAGE DISEQUILIBRIUM

During the past several years, linkage data, mostly for molecular polymorphisms, have become available in many species and there have been a number of attempts to use the linkage disequilibrium theory in testing rival hypotheses. Also some of the known linked loci in humans have been extensively studied and turn out to have a rather important bearing on human welfare, *e.g.*, the human lymphocyte antigen (HL-A) system discussed by Cavalli-Sforza and Bodmer (1971) and the endonuclease site linkages found by Kan and Dozy (1978). Since techniques for genotype identification and detection of linkage are advancing rapidly, we expect that more data suitable for linkage analysis will be produced in the near future and the linkage theory will become increasingly important.

The first mathematical theory of linkage for finite populations was advanced by Hill and Robertson (1968) and Ohta and Kimura (1969), but the theory advanced by these studies has limited application because it deals mainly with the mean values of D and $H_A H_B$ given below, while other quantities of distribution are also necessary. The present method is readily applicable and enables us to obtain almost any quantity, because the method provides sample paths of the process. I will demonstrate the method by obtaining some of the quantities associated with linkage problems, which was not possible previously.

We consider a population consisting of N diploid individuals ($2N$ gametes), and two linked loci where the recombination value between them is c. We assume that every mutant is new to the population, and the mutation rate is v at one locus and u at the other. Let A_i's be alleles at one locus and B_i's alleles at the other, and let $x_{ij}(t)$ be the frequency of gametes carrying A_i and B_j, where t denotes the time measured in units of $2N$ generations. Let $p_i(t)=\sum_j x_{ij}(t)$ and $q_j(t)=\sum_i x_{ij}(t)$. Then Eq. (1) for this process is

$$dx_{ij}(t) = \sum_{kl} a_{kl}{}^{ij} dB_{kl} + [-(U+V)x_{ij}(t) + C\{p_i(t)q_j(t) - x_{ij}(t)\}]dt \tag{10}$$

where $U=2Nu$, $V=2Nv$, $C=2Nc$, B_{kl}'s are independent Brownian motions, and $a_{kl}{}^{ij}$'s are elements of the non-negative definite square root of the covariance matrix

$$[x_{ij}(t)\{\delta_{ik}\delta_{jl} - x_{kl}(t)\}]$$

with $\delta_{ij}=1$ if $i=j$ and otherwise $\delta_{ij}=0$. Equation (10) can easily be changed into a system of difference equations, and then a term accounting for newly arisen mutants can be added.

The following three quantities were studied.

TABLE II
The Means and Standard Deviations of D of Eq. (7), H_AH_B of Eq. (8), and R of Eq. (9)

$2Nu$	$2Nc$	$\bar{D}$ $\times 100$	σ_D[a] $\times 100$	$\overline{H_AH_B}$ $\times 100$	σH_AH_B[a] $\times 100$	$\overline{D}/\overline{H_AH_B}$	$\bar{R}$	σ_R[a]
0.05	0	0.567	2.796	1.338	4.550	0.424	0.061	0.198
0.05	0.1	0.500	2.629	1.183	4.397	0.423	0.062	0.197
0.05	1	0.210	1.367	1.023	3.785	0.205	0.028	0.108
0.1	0	1.317	4.127	3.344	7.126	0.394	0.110	0.242
0.1	0.1	1.159	3.721	3.253	7.017	0.356	0.095	0.215
0.1	1	0.553	2.041	2.997	6.590	0.184	0.050	0.128
0.2	0	2.822	5.285	8.796	11.175	0.321	0.162	0.240
0.2	0.1	2.672	5.074	8.571	11.070	0.312	0.154	0.232
0.2	1	1.592	3.214	9.090	11.175	0.175	0.091	0.146
0.3	0	4.486	5.851	15.544	14.004	0.288	0.196	0.219
0.3	0.1	4.046	5.430	14.833	13.772	0.272	0.181	0.204
0.3	1	2.573	3.659	15.500	13.454	0.166	0.114	0.138
0.5	0	5.908	5.371	25.633	16.052	0.230	0.194	0.163
0.5	0.1	5.786	5.243	25.650	15.922	0.226	0.189	0.156
0.5	1.0	3.965	4.008	25.012	15.492	0.159	0.132	0.119
1.0	0	6.828	3.698	43.900	14.967	0.155	0.155	0.083
1.0	0.1	6.918	3.684	44.400	15.240	0.156	0.153	0.081
1.0	1.0	4.843	2.763	44.100	14.329	0.110	0.107	0.059

[a] Standard deviation.

$$D \equiv \sum_{ij} (p_i q_j - x_{ij})^2 \tag{11}$$

$$H_A H_B \equiv \{\sum_{i \neq j} p_i p_j\} \{\sum_{i \neq j} q_i q_j\} \tag{12}$$

and

$$R \equiv \frac{D}{H_A H_B} \tag{13}$$

where the time variable t is omitted. These quantities are biologically important, and are usually experimentally measurable. Hill (1975) has worked out formulas for the expectations of D and $H_A H_B$.

Assuming $V=U$, the means and variances of these quantities were calculated for a wide range of parameter values which seem to be biologically meaningful. The results are presented in Table II.

The present study reveals several important facts about the behavior of linked loci. First, the ratio of the two means, $E(D)/E(H_A H_B)$, is often much larger than the mean of the ratio $E(D/H_A H_B)$. Importantly, the experimentally observable quantity is usually the ratio R, rather than the

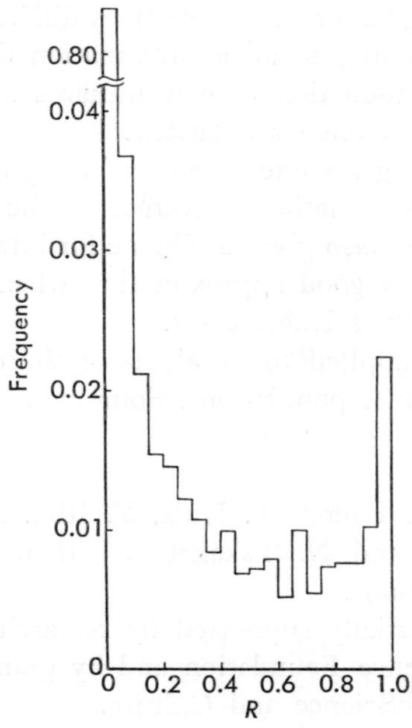

Fig. 5. Distribution of the linkage disequilibrium values (R). $2Nc=0.1$, $2Nv=2Nu=0.1$, and $\bar{R}$ (simulation)$=0.104$.

ratio of the means. Secondly, I want to mention that distributions of the linkage disequilibrium values, R, are often U-shaped, and have a large variance. One example of the distribution of R is given in Fig. 5. In order to cope with future problems in linkage analysis, much further study is necessary, and the method based on Itô's stochastic integrals as used in this paper will be a valuable tool.

SUMMARY

Based on stochastic integrals of the Itô type, a numerical method was developed for treating multi-dimensional stochastic processes of population genetics. This simulation method can reduce computational work substantially, and therefore appears to be a powerful tool for processes involving large dimensions.

Assuming that diffusion approximations are valid, we describe processes in a formal stochastic differential equation. Then, using a difference equation scheme of a special type, stochastic integrals representing sample paths of the processes can be calculated with the aid of a computer. Caution must be taken, however, because there are many different ways of integrating the equations, each leading to an approximation of a different process. The boundary and introduction of new mutants are also rather delicate and often can lead to erroneous solutions.

The present paper gives one scheme of integration under which the solution of a difference equation converges to the stochastic integral of the Itô type. This paper also gives a scheme for introducing new mutants which gives a reasonably good approximation when the method is tested against some analytically solvable cases.

The method was applied to problems of the overdominance model and of linked loci in finite populations. Some new findings are discussed.

Acknowledgments

I thank Drs. P. A. Fuerst, T. Hida, Y. Itoh, M. Kimura, I. Kubo, M. Nei, A. Shimizu, and N. Takahata for their valuable suggestions, comments, and discussions.

This work was partially supported by research grant DEB-8110220 from the National Science Foundation and by grant A-439017 from the Ministry of Education, Science and Culture.

REFERENCES

Arnold, L. (1973). *Stochastic Differential Equations: Theory and Applications*. John Wiley & Sons, New York.

Cavalli-Sforza, L. A. and Bodmer, W. F. (1971). *The Genetics of Human Populations*. Freeman, San Francisco.

Crow, J. F. and Kimura, M. (1970). *An Introduction to Population Genetics Theory*. Harper & Row, New York.

Ewens, W. J. (1979). *Mathematical Population Genetics*. Springer-Verlag, Berlin.

Hill, W. G. and Robertson, A. (1968). Linkage disequilibrium in finite populations. *Theor. Appl. Genet.* **38**, 226–231.

Hill, W. G. (1975). Linkage disequilibrium among multiple neutral alleles produced by mutation in finite population. *Theor. Popul. Biol.* **8**, 117–126.

Itô, K. and McKean, H. P. (1965). *Diffusion Processes and Their Sample Paths*. Springer-Verlag, Berlin.

Itoh, Y. (1979). Random collision process of oriented graph. *Inst. Stat. Math. Res. Mem.* **154**, 1–20.

Kan, Y. W. and Dozy, A. M. (1978). Polymorphism of DNA sequence adjacent to human β-globin structural gene: Relationship to sickle mutation. *Proc. Natl. Acad. Sci. U.S.* **75**, 5631–5635.

Kimura, M. (1964). Diffusion models in population genetics. *J. Appl. Probab.* **1**, 177–232.

Kimura, M. and Crow, J. F. (1964). The number of alleles that can be maintained in a finite population. *Genetics* **49**, 725–738.

Kimura, M. (1968). Evolutionary rate at the molecular level. *Nature* **217**, 624–626.

Kimura, M. and Ohta, T. (1971). Protein polymorphism as a phase of molecular evolution. *Nature* **229**, 467–469.

Kimura, M. (1980). Average time until fixation of a mutant allele in a finite population under mutation pressure: Studies by analytical, numerical and pseudo-sampling methods. *Proc. Natl. Acad. Sci. U.S.* **77**, 522–526.

King, J. L. and Jukes, T. H. (1969). Non-Darwinian evolution. *Science* **164**, 788–798.

Li, W.-H. (1978). Maintenance of genetic variability under joint effect of mutation, selection and random drift. *Genetics* **90**, 349–382.

Maruyama, T. (1977). Stochastic problems in population genetics. In *Lecture Notes in Biomathematics* (ed. S. Levin), Vol. 17. Springer-Verlag, Berlin.

Maruyama, T. (1980). On an overdominant model of population genetics. *Adv. Appl. Probab.* **12**, 274–275.

Maruyama, T. (1981). Stochastic problems in population genetics: Applications of Itô's stochastic integrals. In *Stochastic Nonlinear Systems in Physics, Chemistry, and Biology* (eds. L. Arnold and R. Lefever), pp. 154–161. Springer-Verlag, Berlin.

Maruyama, T. and Nei, M. (1981). Genetic variability maintained by mutation and overdominant selection in finite populations. *Genetics* **98**, 441–459.

McShane, E. J. (1974). *Stochastic Calculus and Stochastic Models*. Academic Press, New York.

Ohta, T. and Kimura, M. (1969). Linkage disequilibrium at steady state determined by random genetic drift and recurrent mutation. *Genetics* **63**, 229–238.

Pederson, D. G. (1973). Note: An approximate method of sampling a multinomial population. *Biometrics* **29**, 814–821.

Rumelin, W. (1980). Numerical treatment of stochastic differential equations. Report No. 12, Forschungsschwerpunkt Dynamische Systeme, Universität Bremen.

Skorokhod, A. V. (1965). *Studies in the Theory of Random Processes*. Addison-Wesley, Reading, Massachusetts.

Stewart, F. M. (1976). Variability in the amount of heterozygosity maintained by neutral mutations. *Theor. Popul. Biol.* **9**, 188–201.

Takahata, N. (1981). Genetic variability and rate of gene substitution in a finite population under mutation and fluctuating selection. *Genetics* **98**, 427–439.

Watanabe, S. (1971). On the stochastic differential equations for multidimensional diffusion process with boundary conditions. *J. Math. Kyoto Univ.* **11**, 169–180.

Watterson, G. A. (1977). Heterosis or neutrality? *Genetics* **85**, 789–814.

Wong, E. and Zakai, M. (1965). On the convergence of ordinary integrals to stochastic integrals. *Ann. Math. Stat.* **36**, 1560–1564.

Wright, S. (1949). Adaptation and selection. In *Genetics, Paleontology, and Evolution* (eds. G. L. Jepson, G. G. Simpson, and E. Mayr), pp. 365–389. Princeton Univ. Press, New Jersey.

Wright, S. (1969). *Evolution and Genetics of Populations*, Vol. 2. Univ. Chicago Press, Chicago.

7

Molecular Evolution, Protein Polymorphism and the Neutral Theory (1982) (ed. M. Kimura), pp. 169–190.
Japan Scientific Societies Press, Tokyo / Springer-Verlag, Berlin

The Disappearance of Duplicate Gene Expression

NAOYUKI TAKAHATA

National Institute of Genetics, Mishima 411, Japan

Duplicate genes produced by chromosomal structural changes such as unequal crossing-over and polyploidization are the raw material of evolution. Duplicate loci sheltered from natural selection have a high potential to diverge to form new genes through their functional diversification with the aid of successive mutations. *Ergo*, duplications are considered to have played a fundamental role in organismic evolution, particularly in the evolution of higher plants (Grant, 1971) and in lower vertebrates (Ohno, 1970).

Many different duplicate genes are now known because they encode homologous proteins in various species. Examining the extensive data available on human enzymes, Harris (1980) has concluded that 35% of the enzymes studied so far are coded by multiple loci. In particular, studies of human hemoglobin genes (Kan and Dozy, 1978; Weatherall and Clegg, 1979; Jeffreys, 1979; Little *et al.*, 1980) and immunoglobulin genes (Hood *et al.*, 1975; Tonegawa *et al.*, 1977) have provided us new insight about the structure, function, and regulation of the duplicates besides practical clinical problems. Duplicate genes often are expressed differentially in different tissues and different periods of development. Clearly, subsequent modification of the coding regions of duplicate genes as well as their associated regulatory regions have resulted in their different regulation of their expression in addition to the different protein structure encoded in the structural genes.

It is obvious that emergence of new functional genes is not the only fate of the duplicates. On the contrary, it is most likely that the majority of these redundant genes lose their function and become silent in the course of evolution. This is simply because most mutations lead to deterioration of duplicate genes. The recent finding of pseudo genes in two mouse strains (Nishioka *et al.*, 1980; Vanin *et al.*, 1980), which are not translated into functional peptides owing to the complete lack of introns (intervening sequences) and substitutions following frameshifts, suggests that the loss of duplicate gene expression is not a rare event in evolution. Actually, several α-like and β-like pseudo genes have been also found in other organisms including man (Lacy *et al.*, 1979; Hardison *et al.*, 1979; Fritsch *et al.*, 1980; Lauer *et al.*, 1980). This widespread occurrence of pseudo genes raises many questions about their evolution and function.

In polyploid organisms, there are also many studies that have reported the loss of duplicate gene expression in connection with the time when polyploidization occurred. Making use of isozyme techniques, interesting observations have been made in several groups of fishes (for example, see Ferris and Whitt, 1979 and references therein) and in *Xenopus* (Bisbee *et al.*, 1977) which have arisen *via* tetraploid ancestors in the relatively recent past. For example, catostomid fishes have lost 35–65% of duplicate locus expression after about 50 million years (Ferris and Whitte, 1977). Also, it is suggested that in roughly a third of the *Xenopus* null alleles have been fixed in about 20 million years (Zimmer and Wilson, unpublished). From these observations, one might suspect that the loss of duplicate gene expression is still in progress in these tetraploid organisms. Interestingly, however, none of the catostomid species exhibit a high level polymorphism for unfixed null alleles (Ferris and Whitt, 1977) although some authors have occasionally reported null alleles at some duplicate loci in other fish tetraploid species (for example, see Ohno, 1970; Engel *et al.*, 1973). These observations suggest the rate of formation and fixation of null alleles may be different for different taxa depending on population size, selection pressures, and the extent to which the transition from tetrasomy to disomy has occurred. At any rate, the observation of such substantial losses of duplicate gene expression in a variety of tetraploid species has stimulated theoretical works from the viewpoint of population genetics (Bailey *et al.*, 1978; Kimura and King, 1979; Takahata and Maruyama, 1979; Allendorf, 1979; Ferris *et al.*, 1979; Ferris and Whitt, 1979; Li, 1980).

In this paper, I give theoretical results concerning the fixation process of null alleles at duplicate loci. The results have been recently obtained by Takahata and Maruyama (1979) and Maruyama and Takahata (1981). We have focussed our attention on the mean and median times until fixation measured in unit of generation, heterozygosities and the trajectories of null allele frequencies on their way to fixation.

MODEL AND METHOD

Let us consider a panmictic population of finite size N and two loci, which have resulted from duplication. They can be either linked or unlinked. Although it is possible to treat both linked and unlinked cases in a single formulation, we shall deal with them separately, because the unlinked case is considerably simpler due to the independent segregation. Let us assume two alleles at each of the two loci, and denote by A and a the alleles at one locus, and by B and b the alleles at the other locus. We call A and B normal (functional) alleles, and a and b null (non-functional) alleles. These null alleles actually may consist of many different alleles at the nucleotide level, but we can refer to them collectively as a null allele, because we have assumed that they are functionally equivalent. We also assume recombination between duplicate genes at a rate c in the linked case, but ignore the effect of unequal crossing-overs. Unequal crossing-overs are considered to be an important mechanism responsible for coincidental evolution of duplicate genes (Smith, 1974, 1976; Hood *et al.*, 1975; Ohta, 1980), and therefore this assumption may be somewhat unrealistic in some cases as will be discussed later. Similar to other theoretical works on this subject (Bailey *et al.*, 1978; Kimura and King, 1979; Allendorf, 1979), with the exception of Nei and Roychoudhury's (1973) work, we assume that the alleles A and B unidirectionally mutate to a and b at a fixed rate v.

With the two alleles at each of the two loci, we have nine different genotypes. The relative fitnesses of these genotypes are given in Table I. Let us denote the frequencies of gametes ab, aB, Ab, and AB by x, y, z, and w, respectively and the gene frequencies of the null alleles a and b by $p(=x+y)$ and $q(=x+z)$. The time t is measured in unit of generation starting from the time of diploidization. In the case of allopolyploids

TABLE I
Genotypes and Relative Fitnesses ($s_{ij}=s_{ji}$)

	AA	*Aa*	*aa*
BB	$1-s_{11}$	$1-s_{12}$	$1-s_{13}$
Bb	$1-s_{21}$	$1-s_{22}$	$1-s_{23}$
bb	$1-s_{31}$	$1-s_{32}$	$1-s_{33}$

(for example, catostomids) disomic inheritance would be rapidly achieved, but in the case of autopolyploids (for example, salmonids) it may take many generations until the two loci are inherited independently.

To explore the problem we first use a numerical method for integrating a two-dimensional Kolmogorov backward equation in the case of unlinked loci. However, as this numerical method can be applied only for a few cases of our interest, we also use another powerful method to solve the problem. A new method enables us to study many problems in population genetics as well as in other scientic fields.

1. *Unlinked Loci*

A diffusion formulation of two locus two allele problems was given by Kimura (1964). Let $\phi(t; p, q; p', q')$ be the joint probability density that the frequencies of a and b are, respectively, p and q at $t=0$ and p' and q' at time t. Further, let $f(p', q')$ be any function of p' and q' and $u(t; p, q)$ be the expectation of $f(p', q')$ as a function of time and of the initial frequencies p and q, *i.e.*,

$$u(t;\, p, q) = \int_0^1\!\!\int_0^1 \phi(t;\, p, q;\, p', q')f(p', q')dp'dq'. \tag{1}$$

Then, $u(t; p, q)$ satisfies the Kolmogorov backward equation

$$\frac{\partial u(t;\, p, q)}{\partial t} = Au(t;\, p, q) \tag{2}$$

in which

$$A \equiv \frac{1}{2}p(1-p)\frac{\partial^2}{\partial p^2} + 2N\left\{v(1-p) + \frac{1}{2W}p(1-p)\frac{\partial W}{\partial p}\right\}\frac{\partial}{\partial p} + \frac{1}{2}q(1-q)\frac{\partial^2}{\partial q^2} + 2N\left\{v(1-q) + \frac{1}{2W}q(1-q)\frac{\partial W}{\partial q}\right\}\frac{\partial}{\partial q} \tag{3}$$

with

$$
\begin{aligned}
W = {} & 1-s_{11}(1-p)^2(1-q)^2-2s_{12}p(1-p)(1-q)^2-s_{13}p^2(1-q)^2 \\
& -2s_{21}(1-p)^2q(1-q)-4s_{22}p(1-p)q(1-q)-2s_{23}p^2q(1-q) \\
& -s_{31}(1-p)^2q^2-2s_{32}p(1-p)q^2-s_{33}p^2q^2
\end{aligned} \tag{4}
$$

where s_{ij}'s are selection coefficients given in Table I and the time is measured in units of $2N$ generations.

Equation (2) governs the fate of many different problems and a pertinent solution to each problem is obtained by imposing appropriate initial and boundary conditions. In particular, from Eq. (1) we obtain the initial condition such that $u(0;\, p,\, q)=f(p,\, q)$ since $\phi(0;\, p,\, q;\, p',\, q')=\delta(p'-p)\,\delta(q'-q)$, in which $\delta(\cdot)$ is the Dirac delta function. On the other hand, boundary conditions are determined depending on a function of $f(p',\, q')$. For instance, if we are interested in the proportion of heterozygote for at least one of the duplicate loci, then $f(p',\, q')=1-[p'^2+(1-p')^2]\,[q'^2+(1-q')^2]$, in which $[p'^2+(1-p')^2]\,[q'^2+(1-q')^2]$ is the proportion of homozygote at both loci. We will use the quantity as a measure of genetic variability in populations. In this case, the boundary conditions are given by $u(t;p,q)=0$ for all boundaries of the domain $(0\leq p\leq 1,\, 0\leq q\leq 1)$. To obtain the probability that null alleles are fixed at one of the duplicate loci by time t, we choose $u(t;p,q)$ to satisfy the conditions $u(t;p,q)=1$ for $p>1-\varepsilon$ and/or $q>1-\varepsilon$ and $u(t;p,q)=0$ for $p=0$ and/or $q=0$, in which ε is an arbitrary small number. On the other hand, the probability that fixation of null alleles has occurred at both loci by time t is given by the imposed conditions $u(t;p,q)=1$ for $p>1-\varepsilon$ and $q>1-\varepsilon$ and $u(t;p,q)=0$ at the remaining boundaries. In all these cases, however, the appropriate boundary conditions are automatically carried over from the initial conditions and therefore no additional restrictions are needed in calculating Eq. (2).

To carry out numerical calculations, we used a difference equation method (for example, see Maruyama, 1977) and covered the domain $(0\leq p\leq 1,\ 0\leq q\leq 1)$ by 20×20 square meshes each with side length 0.05 for most cases. The small number of ε mentioned above was, then, taken as 0.05. We made extensive calculations for various values of mutation rate v and selection coefficients s_{ij}. Selection modes employed here are; i) duplicate dominant functional genes, ii) two functional genes required, iii) partial dominance and epistasis, and iv) overdominance. Specified parameters for Ns_{ij} for each case are as follows: i) (a) $S_{33}=10$, (b) $S_{33}=20$, (c) $S_{33}=50$; ii) (d) $S_{23}=S_{33}=50$; iii) (e) $S_{12}=1$, $S_{13}=S_{22}=2$, $S_{23}=3$, $S_{33}=50$, (f) $S_{12}=2$, $S_{13}=S_{22}=4$, $S_{23}=6$, $S_{33}=50$, (g) $S_{12}=10$, $S_{13}=S_{22}=20$, $S_{23}=30$, $S_{33}=50$; iv) (h) $S_{12}=S_{23}=-4$, $S_{22}=-8$, $S_{33}=30$, in which all unspecified parameters for Ns_{ij} are 0, $S_{ij}=Ns_{ij}$ and $S_{ij}=S_{ji}$. Recently, in addition to the above cases Allendorf (1979) studied another selection mode assuming an intermediate selective optimum of the number of normal gene copies, namely $s_{11}=s$, $s_{12}=s_{21}=s/2(s>0)$ and $s_{33}=1$ in the present notations in Table I. In other words, the genotypes with two normal and two null alleles have the highest fitness.

A numerical method of the Kolmogorov equation used above has a wide range of applications for stochastic problems. In actual calculations using a computer, however, the application is limited greatly depending on the dimension of equations and also the maximum value of diffusion coefficients. Unfortunately, the interesting cases relevant to the present problem fall into this difficulty. In such cases, we have performed Monte Carlo experiments to supplement the numerical results. In the Monte Carlo experiments we directly simulated the stochastic differential equations themselves rather than the underlying model populations. The basic idea of the experiments was explored in Kimura (1980) and Maruyama and Takahata (1981). Here, we make use of the theory of stochastic differential equations of Ito (1944).

In order to simulate the stochastic differential equation corresponding to Eq. (2), we use a difference equation method. Appropriate difference equations to Eq. (2) are given by

$$\left.\begin{aligned} \Delta p(t) &= \sqrt{p(1-p)}B_1(\Delta t)+2N\left\{v(1-p)+\frac{1}{2W}p(1-p)\frac{\partial W}{\partial p}\right\}\Delta t \\ \Delta q(t) &= \sqrt{q(1-q)}B_2(\Delta t)+2N\left\{v(1-q)+\frac{1}{2W}q(1-q)\frac{\partial W}{\partial q}\right\}\Delta t \end{aligned}\right\} \quad (5)$$

where $B_i(\Delta t)$ are independent Brownian motions with mean 0 and variance Δt. In our calculation, we chose Δt so that none of the four terms on the right-hand sides of Eq. (5) became greater than 0.01*. We calculated the gene frequencies successively using Eq. (5) and the following approximations; $p(t+\Delta t)\simeq p(t)+\Delta p(t)$ and $q(t+\Delta t)\simeq q(t)+\Delta q(t)$. We can interpret $p(t)$ and $q(t)$ as a sample path (population) of the process governed by Eq. (2).

2. *Linked Loci*

When the duplicate loci are linked, a numerical calculation of the Kolmogorov equation used in unlinked loci becomes difficult because it

* However, to simulate the process correctly, a precaution is necessary in treating the behavior of rare alleles, *i.e.*, the first terms in the right hand side of Eq. (5) should be adjusted by using bimonial or Poisson random variables when p or q is small. This point was overlooked in Maruyama and Takahata (1981).

is necessary to deal with the three independent frequencies of four gamete types. Therefore, we used the second method of simulation described above, and examined the mean and median times of fixation and trajectories of null allele frequencies under various combinations of recombination value c, mutation rate v and population size N. However, we assumed the complete recessive lethality of null alleles throughout the whole studies of linked duplicate loci. Appropriate stochastic difference equations to this case are

$$\left.\begin{aligned} \Delta x(t) &= e_{11}B_1(\Delta t)+e_{12}B_2(\Delta t)+e_{13}B_3(\Delta t) \\ &\quad +2N\left[v(y+z)-\frac{\{x^2(1-x)+cD\}}{1-x^2}\right]\Delta t \\ \Delta y(t) &= e_{21}B_1(\Delta t)+e_{22}B_2(\Delta t)+e_{23}B_3(\Delta t) \\ &\quad +2N\left\{v(w-y)+\frac{x^2y+cD}{1-x^2}\right\}\Delta t \\ \Delta z(t) &= e_{31}B_1(\Delta t)+e_{32}B_2(\Delta t)+e_{23}B_3(\Delta t) \\ &\quad +2N\left\{v(w-z)+\frac{x^2z+cD}{1-x^2}\right\}\Delta t \\ \Delta w(t) &= -\{\Delta x(t)+\Delta y(t)+\Delta z(t)\} \end{aligned}\right\} \qquad (6)$$

where $D=xw-yz$ and $[e_{ij}]$ is a non-negative definite square root of the matrix

$$\begin{bmatrix} x(1-x), & -xy, & -xz \\ -xy, & y(1-y), & -yz \\ -xz, & -yz, & z(1-z) \end{bmatrix}.$$

Here, we note that there are many stochastically equivalent representation forms related to a Kolmogorov backward equation. Actually, we can obtain at least four different equations concerning a multinomial sampling process of gametes. Equation (6) is one of the four. Another equation is obtained by modifying Pederson's method (1973) of sampling a multinomial population, although he mentioned nothing about the relationship between his method and a stochastic difference equation. The other two different equations are given by Itoh (1979) and Kimura (1980) for the purpose of simulating a stochastic difference equation. Their formulae are seometimes more convenient in the actual calculation than Ito's.

Based on Eq. (6) we have carried out calculations to see the effect of linkage on the fate of null alleles arising at duplicate loci. In the view of linkage, the preceding section provides the case of $c=0.5$, and we intended to see how the fixation process of null alleles at duplicate loci changes as c becomes smaller and eventually reaches 0.

THEORETICAL STUDIES ON DUPLICATE GENES

Before discussing the full details of our results, I would like to review the history of studies on duplicate loci; mainly focussing attention upon theories. One of the earliest reports can be traced back about 60 years ago (Bridges, 1919).

By attempting to delimit the part played by recurrent mutation in evolution, Haldane (1933) first suggested disappearance of extra genes in polyploids and polysomics. His arguments were qualitative and somewhat speculative, but he perceived an important aspect of the fates of duplicate loci. He remarked: "While the duplication on polyploidy may be preserved because it protects against mutations in some thousands of loci at once, the single genes will tend to disappear one at a time because they only protect one locus apiece. The final results will be that most of the genes will return to the diploid condition, but their spatial relations will have been greatly altered." The quantitative study on this subject was begun by Fisher (1935). He considered the circumstances under which a lethal or otherwise deleterious gene would be able to establish itself while being "sheltered" by the duplicate locus. Taking the difference between the deleterious mutation rates in the two loci into account, he showed that the proportion of normal genes at the locus with the higher mutation rate decreases quite slowly, and claimed that the opportunities for the establishment of a deleterious mutant in place of the normal gene at one of duplicate loci seem somewhat slight. Although this is correct in a large population, the situation is quite different in a small population as will be seen later.

Nei (1969) examined possible consequences of gene duplication in evolution. In that, based on the data on the lethal mutation rate (Crow and Temin, 1964) and the theory of the neutral mutation hypothesis (Kimura, 1968), he concluded that the probability of fixation of lethal mutants in duplicate genes is quite high. It is therefore expected that higher organisms carry a considerable number of non-functional genes in their genome. He also studied the accumulation of non-functional genes on sheltered chromosomes such as the Y (or Z) chromosome (Nei, 1970) The problem of the inactivation of sheltered chromosomes (Muller, 1914; Haldane, 1933; Fisher, 1935; Frota-Pessoa and Aratangy, 1968) is analogous to that of the accumulation of non-functional genes at duplicate loci. Using the method of diffusion approximations (Wright, 1945; Kimura, 1964), he suggested that the non-functional genes could be accumulated on sheltered chromosomes within a reasonable period of evolutionary time, but which is highly dependent on effective population size, heterozygous disadvantage and recombination value. On the other hand, the probability that a lethal or deleterious gene is temporarly fixed at autosomal duplicate loci was studied by Nei and Roychoudhury (1973), assuming back-mutations and linkage. They showed that the probability is significantly high only if the effective population size is smaller than 10^4 though the value greatly depends on various other parameters involved.

Apart from the studies on the fate of duplicate gene mentioned above, Spofford (1969) investigated the process by which newly arisen duplication can be incorporated into a species gene pool. Although her mathematical treatment is somewhat restricitive and the assumption of single-locus heterosis may not be always realistic for many duplicate genes, it has been shown that single-locus heterosis is a sufficient mechanism for the incorporation of duplicate genes.

Experimental analyses of loss of duplicate gene expression after duplication have been extensively conducted during the past decade, using isozyme techniques. Many organisms and many duplicate enzyme loci have been studied to reveal the differential regulation of duplicate genes, the divergence of function of their gene products and the extent of loss of extra genes (for example, see Ohno *et al.*, 1968; Ohno, 1970; Allendorf and Utter, 1973; Markert *et al.*, 1975; Ferris and Whitt, 1977; Bailey and Lim, 1977; García-Olmedo *et al.*, 1978 and others). The accumulation of results based on these isozyme analyses has stimulated theoretical analyses of the transient states of duplicate genes under more realistic situations than given by Fisher (1935).

Ferris and Whitt (1977) determined the percentage of duplicate genes still expressed in different species lineages at specified time after polyploidization in a finite population of tetraploid fishes. Although their estimate of an effective population size for a typical catostomids species as 10^9 may have been too much generous, and they did not deal with the theory in depth, their conclusion that unfixed duplicate genes are at present being either maintained or quenched by selection seems to be reasonable. Later, I shall examine this more quantitatively, using the numerical and simulation results of Takahata and Maruyama (1979) and Maruyama and Takahata (1981). Bailey *et al.* (1978) performed Monte Carlo experiments on this subject incorporating mutation, random genetic drift and selection into the model, and they are the first, to my knowledge, who treated the problem under more realistic situations than done by Fisher (1935). However, their selection scheme was rather simple in that only double homozygous nulls are lethal and all other combinations are normal in fitness. Based on the simulation results, they claimed that the median time $T_{0.5}$ until fixation at one of duplicate loci is approximately given by $15N+v^{-3/4}$, and it is insensive to mutation rate v for large N. This rather striking nature of $T_{0.5}$ inconsistent with other studies may come from their less extensive simulations. To generate a more confident conclusion, it was necessary to cover a wide range of parameter values (see Table 2 in Takahata and Maruyama, 1979). Kimura and King (1979) derived the equation for the mean time until fixation under almost the same assumptions as Bailey *et al.*'s. They pointed out that the median time $T_{0.5}$ depends much on the mutation rate, and that it is a little smaller than the mean time. They also investigated the trajectories of null allele frequencies in infinite populations in which the model is the same as those of Fisher (1935) and Christiansen and Frydenberg (1977). Recently, under the selection model mentioned earlier, Allendorf (1979) demonstrated a rapid loss of duplicate gene expression. In this model, the rate is accelerated with increasing population size. If these most duplicate enzyme loci are responding in this fashion, the prolonged retention of duplicate genes which is actually observed (Ferris and Whitt, 1977) must be explained by other mechanisms such as the regulatory divergence of duplicate loci.

No apparent positive correlation between the amount of duplicate gene silencing and time of divergence since polyploization has first pointed out by Ferris and Whitt (1977) (see also Bailey *et al.*, 1978). Ferris *et al.* (1979) have studied this problem based on their mathematical model taking the role of speciation into account, and demonstrated that speciation events can be responsible for determining the rate of duplicate gene silencing. More recently, Li (1980) has studied the same problem under the special cases and method described in pp. 172–175. But, he has considered three different periods of time during which loss of duplicate gene expression can proceed at different rates. Based on this model, he has argued that the observed loss of duplicate gene expression in tetraploid fishes can be explained by adjusting the relative length of these periods in evolution. The mode of inheritance, the regulatory divergence of duplicate loci and speciation events since tetraploidization might have been important causes to determine the loss of duplicate gene expression.

DISCUSSION

Now I present the full details of our numerical and simulation results. Let us first examine the cases of unlinked loci. The results for the fixation probabilities obtained by the numerical method of Eq. (2) are given in Figs. 1 and 2. If the null alleles are selectively neutral, the fixation of null alleles occurs quite rapidly and sooner or later the normal genes will be lost at both loci. The figures show that more than 95% of duplicate genes become silent during about the first $20N$ and $3N$ generations for $Nv=0.1$ and 1.0, respectively. And also, the fixation probabilities at both loci (the broken curves) are quite high even in these early generations. Under this circumstance, the fixation probability for a single locus is calculated from the formula $f(1, t)$ (see p. 394, Crow and Kimura, 1970). As $f(1, t)$ is roughly equal to $1-(4Nv+1)e^{-vt}$ for large t, we can obtain the mean time required for 95% of duplicate loci to be lost

$$T_{0.95}=\frac{1}{Nv}\log\left\{\frac{1+4Nv}{\sqrt{0.05}}\right\}$$

in which we assumed independent two loci and $T_{0.95}$ is measured in units of N generations. The expected values of $T_{0.95}$ for $Nv=0.1$ and 1.0 are 18.3 and 3.1, respectively. The agreement between theoretical and numerical results is satisfactory.

On the other hand, it is seen from Figs. 3 and 4 that the highest level of heterozygosity is attained around the median time of fixation (indicated by arrows). It is important to note that the heterozygosity remains at a rather high level for a considerable length of time until the fixation at both loci is completed. This implies that if we make an observation of the process when the extinction of normal genes proceeds to half of duplicate loci, we should encounter a high level of polymorphism for null alleles at the other half. This neutrality assumption of null alleles is of course unrealistic, because otherwise null alleles would have been fixed before duplication. We will use this assumption soley as a reference point for other models.

Next, let us examine the cases of duplicate dominant functional genes which are most extensively studied so far. Compared with the the neutral cases discussed above for the purpose of reference, a retardation occurs in the fixation of null alleles at one of the duplicate loci. But even if a strong selection operates against the double homozygotes (*aabb*), the amount of retardation is rather slight unless Ns_{33} is extremely large. There-

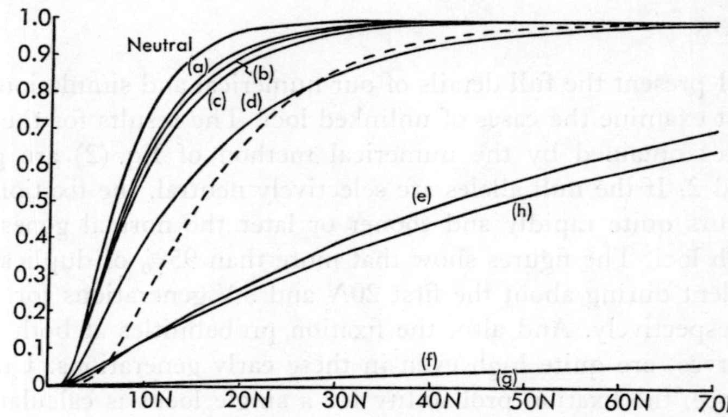

Fig. 1. Fixation probability of null allele at duplicate loci for $Nv=0.1$. (a) through (h) indicate the selection schemes. (a)–(c), duplicate dominant functional genes; (d), two functional genes required; (e)–(g), partial dominance and epstasis; (h), overdominance. For details, see text. The broken curve represents the probability that the fixation of null alleles occurs at both of duplicates under the neutrality assumption.

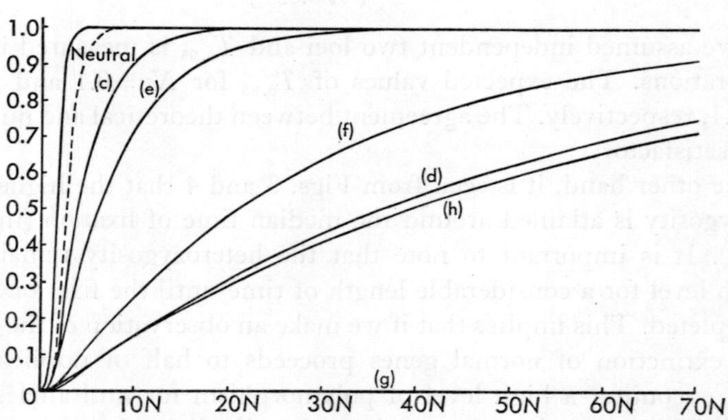

Fig. 2. Fixation probability of null allele at duplicate loci for $Nv=1.0$. The meanings of the curves are the same as in Fig. 1.

fore functional duplicate makes the recessive null mutants behave as if they were neutral. The so-called sheltering effect afforded by the other locus is most prominent in these cases. Although this type of completely recessive alleles might have played an important role in evolution, the data on several groups of fishes and *Xenopus* cannot be explained only by this selection model. There are three reasons for this. First, if the value of Nv is not extremely small and Ns_{33} is not very much larger than order

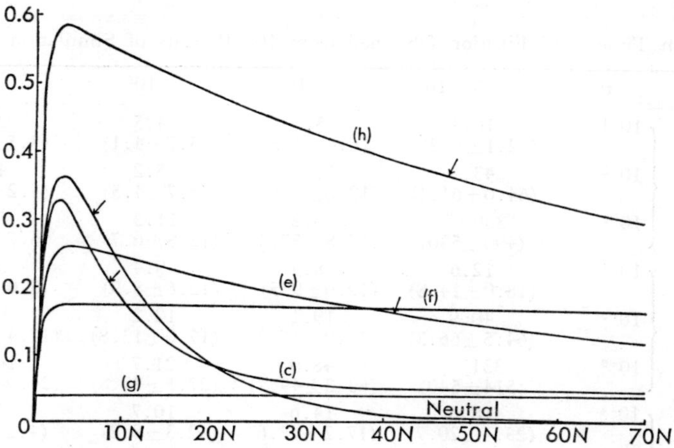

Fig. 3. Level of heterozygosity for the cases given in Fig. 1. The arrows indicate the median time of fixation seen from Fig. 1.

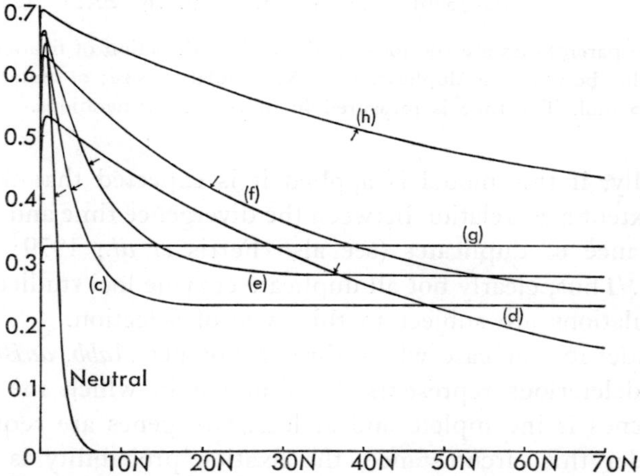

Fig. 4. Level of heterozygosity for the cases given in Fig. 2. The arrows indicate the median time of fixation seen from Fig. 2.

of 10^6, the divergence times since polyploidization estimated in several species are large enough for almost all loci to have lost either of the duplicates (see the bottom row in Table II). Second, if the loss is still in progress at duplicate loci, there should be a large amount of heterozygosity among related species. However, no such polymorphisms were observed for 20 some enzyme loci in 20 plus species of catostomids (Ferris and Whitt,

TABLE II
The Median Time until Fixation Obtained from 100 Repeats of Simulation

c	v	$N=10^3$	10^4	10^5	10^6
0	10^{-4}	10.4 (12.1±6.2)	5.1 (6.6±4.5)	4.5 (5.7±4.1)	2.9 (3.5±2.6)
	10^{-5}	43.5 (64.0±61.1)	11.6 (12.6±6.6)	5.2 (6.7±4.5)	4.1 (5.2±3.2)
	10^{-6}	330 (490±530)	36.2 (55.8±57.4)	11.3 (12.8±6.7)	5.8 (6.7±3.5)
10^{-3}	10^{-4}	12.6 (18.0±14.8)	8.9 (12.0±9.5)	9.4 (12.6±8.9)	3.8 (5.0±3.8)
	10^{-5}	40.9 (64.5±66.3)	19.1 (23.0±17.3)	12.2 (17.0±13.8)	9.6 (11.4±7.6)
	10^{-6}	331 (514±540)	48.9 (63.3±49.7)	21.7 (27.5±20.8)	16.0 (24.9±23.3)
0.5	10^{-4}	18.6 (23.4±20.7)	14.6 (17.2±12.6)	10.7 (12.3±8.2)	3.2 (4.2±3.0)
	10^{-5}	41.1 (57.7±51.2)	22.1 (28.1±22.2)	15.7 (18.4±13.6)	9.0 (11.6±8.4)
	10^{-6}	367.1 (517±550)	50.3 (73.4±70.3)	23.3 (28.6±20.5)	16.6 (22.7±19.7)

Numbers in the parentheses are the mean and standard deviation of fixation time. c, recombination value between the duplicate loci; N, population size; v, mutation rate from normal allele to null. The time is measured in units of N generations.

1977). Thirdly, if this model is applied it is expected that there should be to some extent a correlation between the divergence time and the amount of disappearance of duplicates (see also Ferris *et al.*, 1979; Ferris and Whitt, 1979). Thus, clearly not all duplicate enzyme loci studied in various natural populations are subject to this type of selection.

The model for the case where three genotypes *Aabb*, *aaBb*, and *aabb* are equally deleterious represents the situation in which the dominance of normal genes is incomplete and at least two genes are required to be normal. Under this circumstance, the fixation probability is greatly influenced by the value of Nv. If Nv is small, it is close to the neutral and duplicate dominant functional gene models. This is because the fixation can proceed at one of duplicates without producing the genotypes *Aabb* and *aaBb* to a large extent. However, when Nv becomes larger, many individuals with these genotypes are created and the selection against them acts efficiently. Thus, a considerable retardation of fixation time occurs in such a circumstance.

In the cases of partial dominance and epistasis, a quite different feature has been revealed. The fixation probability and the level of heterozygosity depend considerably on the strength of selection acting against heterozygotes. When the selection parameter (Ns) for heterozygotes is several times larger than Nv, the extinction of normal alleles at one of duplicates does occur, but it is greatly retarded (see the curves (e)–(g) in Figs. 1 and 2). Furthermore, if the values of Ns are increased 10 times greater than Nv, the fixation of null alleles ceases or takes an indefinitely long time. Note that this transition occurs in a rather narrow range of Ns values. This is intuitively expected from the well known population genetics concept that mutation pressure cannot overcome selection barrier. To retard the fixation time, it is necessary for selection against null alleles to operate on genotypes *aa*-- and --*bb* and/or all possible genotypes leading to the creation of these genotypes. This model seems to be compatible with the observations if we assume the selection parameter Ns varied among loci in related species. If this is the case, it is possible that some enzymes under weak selection have competely lost the expression at either of the two duplicate loci in some populations, while other enzymes under strong selection, have been still retained in their function at both duplicates. As a result this interpretation would explain the observation that approximately 50% of the duplicates on the average happen to have become silent and the frequency of null alleles is kept low at the present time. The "weakness" or "strongness" of selection may simply be a function of how divergent the regulation is (Whitt, personal communication). Although there is no direct evidence for it, this model has a room being able to incorporate the extent to which the regulatory divergence is different among duplicate enzyme loci.

A retention of functional duplicates can also be caused by overdominance (see the curves (h) in Figs. 1 and 2). In that case, however, the level of null heterozygosity should remain high over a considerable length of time (Figs. 3 and 4). Therefore, if the retention is due to this mechanism, we have good opportunities to observe heterozygotes at unfixed loci. The data, however, does not support this model.

Next, let us examine the cases of linked duplicate loci. The selection scheme used here is a special case in which the genotype *aabb* is lethal whereas the others are normal. Under this model, we studied the effects of recombination value c, mutation rate v and population size N on the time until fixation and the trajectories of null allele frequencies. The results are given in Table II and Figs. 5 and 6.

Table II shows that the mean fixation time is always slightly larger than the median time as in the cases of unliked loci. In addition, note

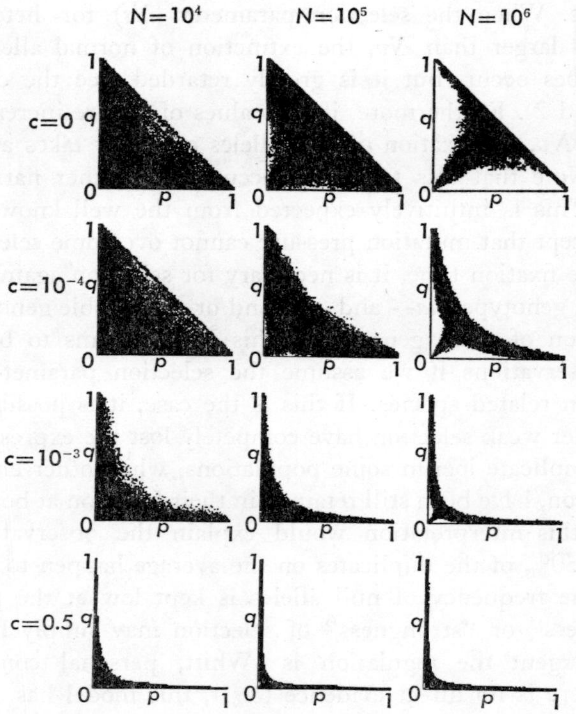

Fig. 5. Permissible regions, that null allelic frequencies can travel among 100 independent trials in simulation, are marked by black. Fixation of null alleles starting at their frequencies $p=q=0$ proceeds at one of duplicate loci, in which we fixed the mutation rate v as 10^{-6} and assumed the complete recessive lethality of null alleles.

that the values of the standard deviation of fixation time are roughly the same as the value of median time (see also Table 2 in Takahata and Maruyama, 1979). These findings indicate that the distribution of the fixation time is greatly skewed toward 0 and has a long tail extending to the opposite (see also Fig. 1 in Li, 1980). The pattern is in fact very much like that of the neutral case. It is clearly shown from the table that disappearance of duplicate genes is accelerated with increasing mutation rate, and that the more closely linked duplicate genes, the more rapidly does the extinction occur. In other words, the stronger sheltering effect is afforded by the other more tightly linked duplicate locus (Nei and Roychoudhury, 1973). In particular, this is conspicuous in large populations. More pre-

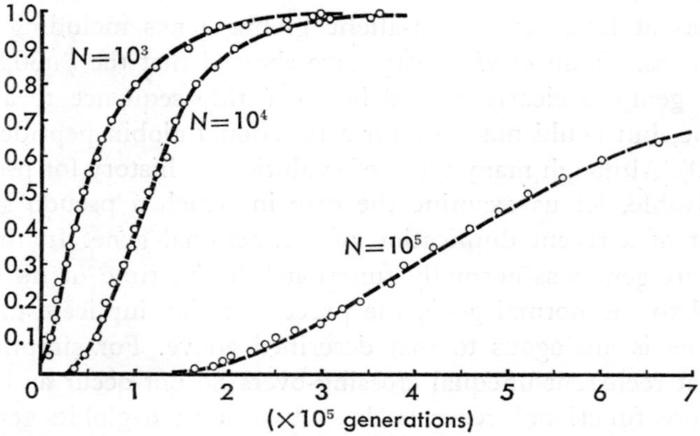

Fig. 6. Simulation results of the fixation probability of null alleles at one of duplicate loci as a function of time which is measured in units of 10^5 generations. Here, we assumed $c=0$, $v=10^{-5}$ and the complete recessive lethality of null alleles.

cisely, the effect is pronounced when Nc is small though N is large. However, the effect of linkage in small populations, say $N=10^3$ or 10^4, is not so much drastic. The finding of Li (1980), that tight linkage does not provide a strong sheltering effect as previously thought, holds true in this case (see Table II).

The case of $c=0$ and $N=10^6$ in Fig. 5 shows a rather striking pattern of trajectories. In this case, most trajectories travel along the line $p=q$ to the point $p=q=0.5$, and then turn to either $p=1$ or $q=1$ along the line $p+q=1$. This can be explained by noting that heterozygotes with at least one normal gene are completely normal, and that relatively strong mutation pressure compared with random genetic drift increases the frequencies of *Ab* and *aB* gametes up to about 0.5, and then the fixation can proceed along the line determined by $p+q=1$. Consequently, the trajectories are restricted to the regions near lines determined by $p=q<0.5$ and $p+q=1$. In small populations, however, strong random genetic drift makes the trajectories cover the entire regions surrounded by lines $p=0$, $q=0$, and $p+q=1$. Since the selection is inefficient under this condition, the rapid fixation of null alleles occurs. On the other hand, the prolonged retention of normal duplicates is caused by the selection which acts strongly against null alleles when Nc and Ns are large.

Finally, I would like to discuss our theoretical results for linked duplicate loci in the context of the widespread occurrence of pseudo genes (see Vanin *et al.*, 1980 and references therein). For example, the mouse genome has at least seven non-allelic globin genes including two adult α and β genes. Vanin *et al.* (1980) have showed that the $\psi\alpha 30.5$ fragment (pseudo α gene) is clearly related in nucleotide sequence to an adult α-globin gene, but could not code for a functional globin peptide (Nishioka *et al.*, 1980). Although many types of evolutionary history for pseudo genes are conceivable, let us examine the case in which a pseudo gene is the descendant of a recent duplication of a functional gene. In this case, as the duplicate gene was normally functional at the time of its occurrence and linked to the normal gene, the process of the duplicate to become a pseudo gene is analogous to that described above. For simplicity, if we assume that recurrent unequal crossing-overs do not occur and we regard two or more functional genes such as two adult α-globin genes as one gene, the process is equivalent to the case of linked loci where irreversible null mutations are considered to introduce frameshifts in DNA sequence. When the mutation rate v is about 10^{-5} and the heterozygous effect on fitness is weak, it can be seen from Table II and Fig. 6 that duplicate gene expression is disappeared rapidly and non-functional genes are more easily to be found at closer regions to the normal gene in genome. In actual cases, unless the effective population size is much larger than 10^5, we can expect that a pseudo gene will be able to fix in almost all populations in several million years.

Throughout the present study, we ignored the effect of recurrent unequal crossing-overs which can be a mechanism responsible for coincidental (concerted) evolution of duplicate genes (Hood *et al.*, 1975; Ohta, 1980; Zimmer *et al.*, 1980). Particularly, it is likely that unequal crossing-over has played a vital role in evolution of multigene families (Smith, 1974, 1975; Ohta, 1976, 1977). Theoretically it was still uncertain to what extent we must consider unequal crossing-over to account for coincidental evolution of duplicate genes with small size such as globin genes. Recently, Ohta (1981) studied this problem and concluded that the unequal crossing-over rate between α-globin genes in human may be as high as 10^{-4}. Although this value depends on the assumptions that two duplicates are almost identical, and that variants are selectively neutral, her study suggests that unequal crossing-over can occur at a rather high rate even between tightly linked duplicate loci (two adult α-globin genes are about 3–4 Kilobases apart in distance). Takahata (1981) also analysed a model which can account for a distribution of the number of repeated genes per chromosome under the joint effects of sister chromatid exchange, interchromosomal unequal crossing-over and selection, and reached the same conclusion. If this is the case, as pointed out by Ohta the effect of unequal crossing-over should be taken into account to investigate the history of pseudo genes. However, no one has studied theoretically on the process of the linked duplicates leading to non-functional genes with the condition that the number of genes in a genome undergoes construction and expansion due to unequal crossing-over. This problem should be solved before we discuss the effect of unequal crossing-over on the disappearance of duplicate gene expression such as is found in pseudo genes linked closely to their normal genes.

SUMMARY

Based on a numerical method for integrating Kolmogorov backward equations and a simulation method for solving stochastic differential equations, we studied the process of fixation of null (non-functional) alleles at duplicate loci. Assuming that normal alleles irreversibly mutate to null alleles, and that the population is finite and panmictic, we obtained the mean and median times until fixation measured in unit of generation, the amount of heterozygosity at specified times and the trajectories of null allele frequencies under various types of selection. It is shown that duplicate genes may rapidly return to the diploid condition when only double homozygotes for null alleles are selected against and linkage is tight. On the other hand, a prolonged retention of duplicate gene expression is expected if the heterozygous effect on fitness is pronounced. We have related the results of these analyses to recent observations on amounts of polymorphism, loss of duplicate gene expression in tetraploid organisms, and pseudo genes in mammals. Evolutionary implications of duplicate genes are discussed quantitatively from the viewpoint of population genetics.

Acknowledgments

I would like to express my thanks to Drs. M. Kimura, T. Maruyama, and T. Ohta for their useful suggestions and constant encouragements. Dr. Y. Tateno kindly spent much of time to read the manuscript and improve the presentation. Especially, I am much indebted to Dr. G. S. Whitt for many useful comments from the biological perspective and Dr. W.-H. Li for his comment on our simulation method.

REFERENCES

Allendorf, F. W. (1979). Rapid loss of duplicate gene expression by natural selection. *Heredity* **43**, 247–258.

Allendorf, F. W. and Utter, F. M. (1973). Gene duplication within the family Salmonidae: Disomic inheritance of two loci reported to be tetrasomic in rainbow trout. *Genetics* **74**, 647–654.

Bailey, G. S. and Lim, S. T. (1977). Evolution of duplicate lactate dehydrogenase isozyme in Salmon. *J. Biol. Chem.* **252**, 5708–5715.

Bailey, G. S., Poulter, R. T. M., and Stockwell, P. A. (1978). Gene duplication in tetraploid fish: Model for gene silencing at unlinked duplicate loci. *Proc. Natl. Acad. Sci. U.S.* **75**, 5575–5579.

Bisbee, C. A., Baker, M. A., Wilson, A. C., Hadji-Azimi, I., and Fischberg, M. (1977). Albumin phylogeny for clawed frogs (*Xenopus*). *Science* **195**, 785–787.

Bridges, C. B. (1919). Duplication. *Anat. Rec.* **15**, 357.

Crow, J. F. and Temin, R. G. (1964). Evidence for the partial dominance of recessive lethal genes in natural populations of *Drosophila*. *Am. Nat.* **98**, 21–33.

Crow, J. F. and Kimura, M. (1970). An introduction to population genetics theory. Harper & Row, New York.

Christiansen, F. B. and Frydenberg, O. (1977). Selection-mutation balance for two nonallelic recessives producing an inferior double homozygote. *Am. J. Hum. Genet.* **29**, 195–207.

Engel, W., Schmidtke, J., Vogel, W., and Wolf, U. (1973). Genetic polymorphism of lactate dehydrogenase isozymes in the carp (*Cyprinus carpio*) apparently due to a " null allele." *Biochem. Genet.* **8**, 281–289.

Ferris, S. D. and Whitt, G. C. (1977). Loss of duplicate gene expression after polyploidization. *Nature* **265**, 258–260.

Ferris, S. D. and Whitt, G. C. (1979). Evolution of the differential regulation of duplicate genes after polyploidization. *J. Mol. Evol.* **12**, 267–317.

Ferris, S. D., Portnoy, S. L., and Whitt, G. S. (1979). The roles of speciation and divergence time in the loss of duplicate gene expression. *Theor. Popul. Biol.* **15**, 114–139.

Fischer, R. A. (1935). The sheltering of lethals. *Am. Nat.* **69**, 446–455.

Fritsch, E. F., Lawn, R. M., and Maniatis, T. (1980). Molecular cloning and characterization of the human β-like globin gene cluster. *Cell* **19**, 959–972.

Frota-Pessoa, O. and Aratangy, L. R. (1968). The degeneration of the Y chromosome. *Rev. Bras. Pesqui. Med. Biol.* **1**, 241–244.

García-Olmedo, F., Carbonero, P., Aragoncillo, C., and Salcedo, G. (1978). Loss of redundant gene expression after polyploidization in plants. *Experientia* **34**, 332–333.

Grant, V. (1971). *Plant Speciation.* Columbia Univ. Press, New York.

Haldane, J. B. S. (1933). The part played by recurrent mutation in evolution. *Am. Nat.* **67**, 5–19.

Hardison, R. C., Butler III, E. T., Lacy, E., and Maniatis, T. (1979). The structure and transcription of four linked rabbit β-like globin genes. *Cell* **18**, 1285–1297.

Harris, H. (1980). Multi-locus enzymes in man. *CIBA Foundation Symposium*, 66 (new series), pp. 187–204.

Hood, L., Cambell, J. H., and Elgin, S. C. R. (1975). The organization, expression, and evolution of antibody genes and other multigene families. *Annu. Rev. Genet.* **9**, 305–353.

Ito, K. (1944). Stochastic integral. *Proc. Imp. Acad.* **20**, 519–524.

Itoh, Y. (1979). Random collision process on oriented graph. *Inst. Stat. Math. Res. Mem.* **154**, 1–20.

Jeffreys, A. J. (1979). DNA sequence variants in the $^{G}\gamma$-, $^{A}\gamma$-, δ- and β-globin genes of man. *Cell* **18**, 1–10.

Kan, Y. W. and Dozy, A. M. (1978). Polymorphism of DNA sequence adjacent to human β-globin structure gene: Relationship to sickle mutation. *Proc. Natl. Acad. Sci. U.S.* **75**, 5631–5635.

Kimura, M. (1964). Diffusion models in population genetics. *J. Appl. Probab.* **1**, 177–232.

Kimura, M. (1968). Evolutionary rate at the molecular level. *Nature* **217**, 624–626.

Kimura, M. (1980). Average time until fixation of a mutant allele in a finite population under continued mutation pressure: Studies by analytical, numerical and pseudo-sampling methods. *Proc. Natl. Acad. Sci. U.S.* **77**, 522–526.

Kimura, M. and King, J. L. (1979). Fixation of a deleterious allele at one of two "duplicate" loci mutation pressure and random drift. *Proc. Natl. Acad. Sci. U.S.* **76**, 2858–2861.

Lacy, E., Hardison, R. C., Quon, D., and Maniatis, T. (1979). The linkage arrangement of four rabbit β-like globin genes. *Cell* **18**, 1273–1283.

Lauer, L., Shen, C.-K. J., and Maniatis, T. (1980). The chromosomal arrangement of human α-like globin genes: Sequence homology and α-globin gene deletions. *Cell* **20**, 119–130.

Li, W.-H. (1980). Rate of gene silencing at duplicate loci: A theoretical study and interpretation of data from tetraploid fishes. *Genetics* **95**, 237–258.

Little, P. F. R., Annison, G., Darling, S., Williamson, R., Camba, L., and Modell, B. (1980). Model for antenatal diagnosis of β-thalassaemia and other monogenic disorders by molecular analysis of linked DNA polymorphisms. *Nature* **285**, 144–147.

Markert, C. L., Shaklee, J. B., and Whitt, G. S. (1975). Evolution of a gene. *Science* **189**, 102–114.

Maruyama, T. (1977). Stochastic problems in population genetics. In *Lecture Notes in Biomathematics* (ed. S. Levin), Vol. 17. Springer-Verlag, Berlin.

Maruyama, T. and Takahata, N. (1981). Numerical studies of the frequency trajectories in the process of fixation of null genes at duplicated loci. *Heredity* **46**, 49–57.

Muller, H. J. (1914). A gene for the fourth chromosomes of *Drosophila*. *J. Exp. Zool.* **17**, 325–336.

Nei, M. (1969). Gene duplication and nucleotide substitution in evolution. *Nature* **221**, 40–42.

Nei, M. (1970). Accumulation of nonfunctional genes on sheltered chromosomes. *Am. Nat.* **104**, 311–322.

Nei, M. and Roychoudhury, A. K. (1973). Probability of fixation of nonfunctional genes at duplicate loci. *Am. Nat.* **107**, 362–372.

Nishioka, Y., Leder, A., and Leder, P. (1980). Unusual α-globin-like gene that has cleanly lost both globin intervening sequences. *Proc. Natl. Acad. Sci. U.S.* **77**, 2806–2809.

Ohno, S. (1970). *Evolution by Gene Duplication*. Springer-Verlag, Berlin.

Ohno, S., Wolf, U., and Atkin, N. B. (1968). Evolution from fish to mammals by gene duplication. *Heredity* **59**, 169–187.

Ohta, T. (1976). A simple model for treating the evolution of multigene families. *Nature* **263**, 74–76.

Ohta, T. (1977). On the gene conversion model as a mechanism for maintenance of homogeneity in systems with multiple genomes. *Genet. Res.* **30**, 89–91.

Ohta, T. (1980). Evolution and variation of multiplegene families. In *Lecture Notes in Biomathematics* (ed. S. Levin), Vol. 37. Springer-Verlag, Berlin.

Ohta, T. (1981). Genetic variation in small multigene families. *Genet. Res.* **37**, 133–149.

Pederson, D. G. (1973). Note: An approximate method of sampling a multinomial population. *Biometrics* **29**, 814–821.

Smith, G. P. (1974). Unequal crossover and the evolution of multigene families. *Cold. Spring Harbor Symp. Quant. Biol.* **38**, 507–513.

Smith, G. P. (1976). Evolution of repeated DNA sequences by unequal crossover. *Science* **191**, 528–535.

Spofford, J. B. (1969). Heterosis and the evolution of duplications. *Am. Nat.* **103**, 407–432.

Takahata, N. (1981). A mathematical study on the distribution of the number of repeated genes per chromosome. *Genet. Res.* **38**, 97–102.

Takahata, N. and Maruyama, T. (1979). Polymorphism and loss of duplicate gene expression: A theoretical study with application to tetraploid fish. *Proc. Natl. Acad. Sci. U.S.* **76**, 4521–4525.

Tonegawa, S., Hozumi, N., Matthyssenes, G., and Schuller, R. (1977). Somatic changes in the content and context of immunoglobulin genes. *Cold. Spring Harbor Symp. Quant. Biol.* **41**, 877–889.

Vanin, E. F., Goldberg, G. I., Tucker, P. W., and Smithies, O. (1980). A mouse α-globin-related pseudogene lasking intervening sequences. *Nature* **286**, 222–226.

Weatherall, D. J. and Clegg, J. B. (1979). Recent developments in the molecular genetics of human hemoglobin. *Cell* **16**, 467–479.

Wright, S. (1945). The differential equation of the distribution of gene frequencies. *Proc. Natl. Acad. Sci. U.S.* **31**, 382–389.

Zimmer, E. A., Martin, S. L., Beverley, S. M., Kan, Y. W., and Wilson, A. C. (1980). Rapid duplication and loss of genes coding for the α chains of hemoglobin. *Proc. Natl. Acad. Sci. U.S.* **77**, 2158–2162.

8

8

Molecular Evolution, Protein Polymorphism and the Neutral Theory (1982) (ed. M. Kimura), pp. 193–214.
Japan Scientific Societies Press, Tokyo / Springer-Verlag, Berlin

Allelic Diversity of Human Populations in the Asian-Pacific Area

KEIICHI OMOTO

Department of Anthropology, Faculty of Science,
The University of Tokyo, Tokyo 113, Japan

The peoples of the Asian-Pacific area attract the interest of human population geneticists and biological anthropologists in several respects. They are essentially descended from Southeast Asian populations through a number of successive waves, the oldest of which probably took place as early as 50,000 years BP (for example, see Bellwood, 1980). Apart from Indonesia, that is, the islands in the western Pacific, the vast Pacific area is usually divided into four geographical areas, Australia, Melanesia, Micronesia, and Polynesia, with the total population size estimate to have been roughly 3.2 million at the time of European contact (Howells, 1973). Archeological and linguistic evidence suggests that Australia and New Guinea were probably settled first during the last glacial stage, some 20,000–50,000 years ago, by peoples with Upper Paleolithic or Mesolithic cultures. Until the end of the Pleistocene times Australia and New Guinea formed a continuous land mass called Sahulland and they were separated by water only during the last 8,000 years (Chappel, 1976). On the other hand, Sumatra, Java, Borneo, and other islands in what is now Indonesia including parts of the Philippines were also connected by land bridges to

Southeast Asia during the last glacial age, forming the so-called Sundaland. The Negrito groups now found scattered in the Philippines, Malaysia, and the Andaman Islands are probably descendants of the indigenous inhabitants of Sundaland. Following the Neolithic revolution on the Asian mainland, successive waves of migrations of Austronesian speakers starting from about 5,000 years BP, began to spread from Southeast Asia to Melanesia, and later to Micronesia, Most of the Polynesian islands were settled in the period from about 3,000 to 1,500 years BP. Also, the fact that peoples were scattered in numerous small populations separated by seas or mountains may have promoted the action of random genetic drift although isolation has never been complete because of occasional migrations.

Recently, population genetic studies have been actively carried out among various human groups in this area. The aim of the present paper is to test whether the measures now available for assessing genetic diversity between populations obtained by analyses of genetic distances are compatible with those population histories outlined above. Also, some data concerning genetic variants with restricted distribution obtained in this area are presented which seem to have some relevance to the neutral-mutation random genetic drift theory (Kimura, 1968; Kimura and Ohta, 1971).

THE GENERAL BACKGROUND OF THE APPROACH

In classic genetics where research was mainly based on mating experiments it was generally considered that man was among the most inadequate materials for genetic research because of the long generation time, small number of offspring, and, above all, the impossibility of mating experiments. Contrary to this notion, however, it is well known that human material did contribute a great deal to the development of modern genetics. In particular, humans appear to be an adequate source of data for population genetic studies, since individual and historical records are available and biological samples in relatively large amounts can be obtained from geographically diverse populations.

Screening for allelic variation using starch-gel electrophoresis followed by specific enzyme staining developed by H. Harris and his colleagues culminated in the estimation of the extent of genic variation within human populations. By 1972, data on a total of 71 structural gene loci for enzymes in European populations had been accumulated. Among these 71 loci, 20

(28%) were found to show genetic polymorphism, *i.e.*, with the frequency for the most common allele being less than 0.99. The average heterozygosity per locus for alleles determining electrophoretic variants was estimated to be 0.067 (Harris and Hopkinson, 1972). By 1976, the number of loci screened had increased to 104, 24 (23%) of which exhibited electrophoretic polymorphism. The average heterozygosity per locus was estimated to be 0.063 (Harris and Hopkinson, 1976). Obviously, these figures may be gross underestimates of the true average heterozygosity per locus, mainly because of the limitation on the detectability of amino acid substitutions by electrophoresis. For other animal species, average heterozygosity estimates of similar magnitudes have been obtained although it was suggested that the invertebrate species tend to show slightly higher values than the vertebrates (Selander and Kaufman, 1973). In any case, there is no doubt that humans, like *Drosophila* and other living systems, contain a large amount of allelic diversity within populations. However, the biological mechanisms leading to such a large number of polymorphisms are, as is well known, the subject of one of the greatest debates in the history of evolutionary genetics: the neutralist-selectionist controversy (see Dr. Kimura's contribution to this volume).

The detection of enzyme variants by electrophoresis has also been

TABLE I

Genetic Polymorphism of Enzymes Detected by Electrophoresis in Japanese

Enzyme	Locus	NCA[a]	FCA[a]	H[a]
Red cell				
Glutamic-pyruvic transaminase	GPT_S	2	0.54–0.62	0.47
Esterase D	ESD	2	0.63–0.66	0.45
Acid phosphatase	ACP_1	2	0.76–0.80	0.35
Phosphoglucomutase-1	PGM_1	3	0.70–0.81	0.39
6-Phosphogluconate dehydrogenase	PGD	2	0.90–0.94	0.14
Glyoxalase-1	GLO	2	0.91	0.16
UMP kinase	UMPK	2	0.95	0.10
Adenosine deaminase	ADA	2	0.97–0.98	0.04
Glutamic-oxaloacetic transaminase	GOT_S	2	0.98–0.99	0.02
Placenta				
Alkaline phosphatase	PL	3	0.72–0.75	0.39
Phosphoglucomutase-3	PGM_3	2	0.71–0.81	0.41
Peptidase A	PEPA	2	0.82	0.29

[a] NCA, number of common alleles; FCA, frequency of the commonest allele; H, heterozygosity. The symbol S in the locus designation indicates the soluble form of the enzyme.

actively carried out in Japan (Table I). However, since the enzyme systems studied were mostly those shown to be polymorphic in Europeans, no accurate estimate for the average heterozygosity per locus is available.

Recently, Harris and his colleagues reported their analysis of enzyme polymorphisms in relation to the size and the number of subunits (Harris *et al.*, 1977). Their arguments were that if those allelic variants detected by electrophoresis are mostly derived from effectively neutral mutations and their incidence in a population depends entirely on random genetic drift, then one would expect the following observations. First, the average heterozygosity and incidence of polymorphism are higher at loci coding for monomeric enzymes than at loci coding for multimeric enzymes, since in the latter a significant portion of the surface amino acids is engaged in the intersubunit contacts, and mutations occurring there are likely to influence the structural and functional stability of the protein. Secondly, the subunit sizes of enzymes coded by polymorphic loci are on the average larger than those coded by monomorphic loci.

The observation of 87 enzymes with known subunit structures, however, indicated rather contradictory results. The first expectation was supported by the observation that both the incidence of polymorphism and the average heterozygosity among 27 loci coding for monomeric enzymes were significantly higher than those at other loci coding for multimeric enzymes. It was further noted that multimeric enzymes in which interlocus molecular hybrids occur show a particularly low incidence of polymorphism. On the contrary, the result concerning the second expectation was negative, for there was no difference in the size of polypeptide subunits when the products of 33 polymorphic loci and 54 monomorphic loci were compared, with the mean molecular subunit size being 45,678 daltons (Harris *et al.*, 1977).

As the authors admit, the data accumulated thus far are still far from complete. For example, large-scale surveys for allelic variation are still limited to European populations and it is not certain whether the results hold true for the entire human species. Also, the electrophoretic conditions used may not be optimal for many enzyme systems, leaving a significant proportion of variants undetected. Nevertheless, this study shows a promising way for future studies to test hypotheses with regard to the neutralist-selectionist controversy on protein polymorphism. Obviously, the advantage of the neutral theory of protein polymorphism lies in the simplicity of making the predictions testable by laboratory studies. On the other hand, the selection theory requires the presence of selective pressures related to the function of proteins, which in turn should differ from protein to protein, and hence the predictions it provides are difficult to test using the data usually accumulated in laboratories engaged in the detection of protein polymorphisms.

At present, electrophoresis still remains the most convenient screening method for allelic variation. It provides the simplest method for the detection of allelic variation mostly derived from single amino acid substitutions. Only a very small sample amount is needed which does not require purification, and hence the method is adequate for screening a large number of samples. It can be applied to virtually all kinds of proteins if they are adequately solubilized. It is claimed that only a fraction (25–30%) of possible amino acid substitutions, that is, those giving rise to change in the net charge of a protein, can be detected by electrophoresis. However, a recent study by J.A.M. Ramshaw and colleagues indicates that this electrophoresis detectability may be an underestimate. They used 20 samples of human hemoglobin variants with known amino acid substitutions and showed that the resolving power of polyacrylamide gel electrophoresis is such that even chemically equivalent variants with the same amino acid substitutions can be distinguished on the gel in terms of mobility differences. Thus, for example, seven hemoglobin variants with Asp $\rightarrow$ Asn substitutions at different positions were distinguishable from each other when three different electrophoretic conditions were used (Ramshaw *et al.*, 1979). Therefore, we may say that at least the postulated discriminating power of 25–30% of the possible amino acid substitutions may well be the minimal estimate. Moreover, recent developments in electrophoretic methods (*e.g.*, isoelectrofocusing and thermostability tests) have raised the resolving power of electrophoresis substantially. For example, 10 common phenotypes are now distinguishable for both red cell phosphoglucomutase (PGM) and serum Gc globulin by means of isoelectrofocusing in thin-layer polyacrylamide gels, while for both only 3, or sometimes 6, phenotypes could be separated by conventional electrophoresis in starch or agar gels (Bark *et al.*, 1976; Constans and Viau, 1977). Also, it was reported that 10 additional alleles of xanthine dehydrogenase in *Drosophila pseudoobscura* were found by gel electrophoresis when thermostability tests were combined with conventional electrophoresis (Singh *et al.*, 1976). These facts may indicate that electrophoresis with a variety of improvements remains the most powerful and suitable method in population genetics for the detection of allelic variation for a while, until new methods capable of screening polymorphisms at the nucleotide level will become widely applied in this field.

Recent developments in rapid RNA and DNA sequencing methods are certainly welcomed by population and evolutionary genetics, but they still seem to be too laborious for screening purposes. On the other hand, the detection of polymorphisms at the nucleotide level using restriction endonuclease mapping seems to be promising for future studies, provided proper DNA samples are obtained. Recently, W. M. Brown applied this method to a comparison of 21 humans belonging to diverse racial groups using mitochondrial DNA (mtDNA) samples. Samples were digested with 18 restriction endonucleases and subsequently the sizes of the resulting DNA fragments were compared by gel electrophoresis. He concluded that "the data indicated that individuals differ from a postulated ancestral mtDNA sequence at 0.18% of their base pairs. On the basis of an estimated rate for base substitution of 1% per 10^6 years, *Homo sapiens* could have speciated or passed through a severe population constriction as recently as 180,000 years ago" (Brown, 1980).

GENETIC DIFFERENTIATION OF HUMAN RACES

The classification of human races was one of the major objectives of classic anthropology until the earlier half of this century. A great many attempts have been made to divide living human populations into a number of subgroups called races or varieties, which were thought to correspond to subspecies in other animal species. The classification was based largely on the so-called racial characters, most of which were somatological characteristics showing sharp geographical differences such as skin color, hair form, stature, nose form, and so on.

It became evident, however, that a clearcut division of mankind into subspecies was a matter of extreme difficulty. Since there was no way of defining *subspecies* in biologically unequivocal terms, most racial classifications were unavoidably subjective, emphasizing differences rather than similarities. Thus, for example, the Ainu of northern Japan was often considered to belong to the white race simply because the Ainu look very different from Japanese and there are some similarities to Europeans in superficial characters such as hairiness and a relatively rugged face. The disadvantage of these "racial" characters in their application to phylogenetic studies is obvious, since it is by no means certain whether the similarity really reflects a close relationship by descent, or whether it merely resulted from convergence or parallel evolution due to adaptation. The Pygmy-Negrito problem, one of the classic anthroplogical controversies, is a good example of this. The Pygmies of Central Africa and the Negritos of Southeast Asia and the Western Pacific share characteristics such as short stature, dark skin, round head, kinky hair, and so on. While one would speculate that they both derive from a common ancestral population, it must be considered alternatively that they may not necessarily be closely related to each other, and that those characteristics they possess in common are probably due to adaptation to similar environmental factors.

The situation in anthropology has changed in the last 25 years. While we are no longer interested in the classification of races as such, we have now the data and statistical means to estimate the degree of human variation at the genic level, so that we are able to compare human populations more or less objectively in terms of genetic affinities. Thus, for example, population genetic studies and multivariate analysis have shown that at the genic level the Ainu have many more similarities to the Mongoloids than to Caucasoids or other races (Omoto, 1970, 1972, 1975; Nei, 1978).

An interesting problem is the origins of the present races. In anthropology, this still seems to be a matter of controversy despite the accumulation of fossil records and advances in dating techniques. It appears, however, that the majority of anthropologists consider that the modern form of mankinds, *H. sapiens sapiens*, to which all of the present races belong, appeared in Upper Paleolithic times, dating back as early as 50,000–70,000 years BP. It is not agreed, however, whether *H. s. sapiens* resulted from continuous transition from *H. s. neanderthalensis* which took place over a wide range covering the tropical to temperate zones of Eurasia and Africa, or alternatively whether *H. s. sapiens* originated from a relatively small population of (or even ancestral to) *H. s. neanderthalensis*, probably during the end of Middle Paleolithic times some 100,000–200,000 years BP, and later expanded as the predominant group, while Neanderthal man became extinct by about 40,000 years BP.

Recent developments in human population genetics have shed some light on this problem. Nei and Roychoudhury (1974) examined the overall gene differences between the three major human races, *i.e.*, Caucasoids, Negroids, and Mongoloids. They gathered gene frequency data from the literature based on electrophoresis of 74 protein loci for Caucasoids and 62 loci for Negroids. Data on only 35 loci were available for Mongoloids, mostly from Japanese. Also, blood group (red cell antigen) gene frequency data were available for 57, 34, and 22 loci for Caucasoids, Negroids, and Mongoloids, respectively. The data indicated that the average heterozygosity for protein loci was approximately 10% in all the three races, while it was 13% for blood group loci in Caucasoids and 16–24% in Negroids and Mongoloids, suggesting that the blood group loci examined in these two latter groups may deviate from a random sample of the genes. There were high correlations in heterozygosity at individual loci between the three races. The genetic distances measured in terms of the number of net codon differences per locus between populations were small, being 1–3% for protein loci and 2–14% for blood group loci. The largest distance was observed between Negroids and Mongoloids. Moreover, the authors estimated the effective (minimum) divergence time between the three races (t) using the equation $t=5\times10^6\times D$, where D was Nei's standard genetic distance. This equation was based on the assumption that the rate of nucleotide substitution leading to electrophoretically detectable protein variants is approximately 10^{-7} per year (Nei, 1975). The results indicated that Caucasoids and Negroids, as well as Negroids and Mongoloids, may have been separated for about 120,000 years, while Caucasoids and Mongoloids may have diverged at a much later period, that is, about 55,000 years (Nei and Roychoudhury, 1974).

Recently, R. L. Kirk compared dendrograms and the estimated time of population divergence obtained by various genetic distance measures, particularly those developed by Cavalli-Sforza and Edwards (1967) and Nei (1972), using gene frequency data for proteins obtained in a number of native ethnic groups of Australia and New Guinea, in order to test empirically the usefulness of dendrograms in the assessment of genetic differentiation (Kirk, 1976). It was shown that while different models devised for the computation of genetic distances currently available gave essentially similar results with respect to genetic affinities between populations separated for several hundred generations, they vary in their discrimination of populations separated for short periods. Furthermore, it was suggested that Nei's procedure for estimating the minimum divergence time mentioned above may overestimate the actual time by roughly a factor of four. Thus, it is possible that the separation time of Negroids-Caucasoids and Negroids-Mongoloids estimated by Nei and Roychoudhury (1974) may be reduced to approximately 30,000 years, while the Caucasoids-Mongoloids separation time may be only 13,000 years. He considers while the former value may be realistic, the latter appears too short (Kirk, *l.c.*).

THE PROBLEM OF THE ORIGINS OF AUSTRALOIDS

In classical anthropology Australian aborigines and the inhabitants of New Guinea and other Melanesian islands were called Oceanic Negroes belonging to the Negroids race, mainly because of their dark pigmentation (for example, see Hooton, 1956). However, it is now usually accepted that they are descended from the Upper Paleolithic population of Asia, although some morphological features on their skulls are suggested to be derived from the *Homo erectus* stage (Thorne and Macumber, 1972). According to recent archeological evidence, the first arrival of humans in Australia probably dates back as early as 50,000 years BP, while the populations of Australia and New Guinea may have been separated by water only during the last 8,000 years (Chappel, 1976; White and O'Connell, 1979).

A large body of data on the allelic variation of Australian aborigines are available thanks to the electrophoretic survey by Kirk and his colleagues (Kirk, 1965; Kirk *et al.*, 1971, 1972). In addition, the results of tests for 22 red cell enzyme systems on samples from more than 4,000 aboriginal Australians in various localities covering the wide area of the Northern Territory, Queensland, and Central and Western Australia were recently tabulated (Blake, 1979).

In order to examine whether genetic distance analysis yields affinities and estimated separation times compatible with those supported by anthropological evidence, attempts were made in this study to compare gene frequency data on Australian aborigines, which were considered to represent the Australoids, with those of the three major human racial groups, the Caucasoids, Negroids, and Mongoloids. Since Blake's data on red cell enzymes were tabulated separately for more than 30 localities, the average gene frequencies for the whole sample were calculated for each system. Twelve out of 22 systems (loci) were found to be polymorphic (ACP, ADA, CA_1, CA_2, ESD, GPT_S, GLO, PEPA, PEPB, 6PGD, PGM_1, PGM_2).* Incidentally, the proportion of polymorphic loci in this case is rather high, since five of the above systems are nonpolymorphic in most other populations. The rest (10 systems) were found to be monomorphic

* ACP, acid phosphatase; ADA, adenosine deaminase; CA_1, carbonic anhydrase 1; CA_2, carbonic anhydrase 2; ESD, esterase D; GPT_S, soluble glutamic pyruvic transaminase; GLO, glyoxalase 1; PEPA, peptidase A; PEPB, peptidase B; 6PGD, 6-phosphogluconate dehydrogenase; PGM_1, phosphoglucomutase 1; PGM_2, phosphoglucomutase 2.

(AK_1, DIA_1, GOT_S, GPI, ICD_S, LDH_A, LDH_B, MDH_S, PGK_1, SOD_A).* So far as these systems are concerned, the average heterozygosity per locus was estimated at 0.0539, which is considerably lower than that for Caucasoids (0.115), Negroids (0.0925), and Mongoloids (0.0953), estimated by using data on the same 22 systems. The data for Caucasoids were mostly taken from those on English population published by Harris and his group (*e.g.*, Harris and Hopkinson, 1972; Harris *et al.*, 1974). The Negroid data were from South African Bantu populations taken from a variety of published data (*e.g.*, Kirk *et al.*, 1971; Mourant *et al.*, 1976; Nurse and Jenkins, 1977), but data on U.S. Black populations were neglected because of the well-known accumulation of Caucasoid genes in them. Mongoloids were represented by recently accumulated data on Japanese populations (Ishimoto, 1974; Ueda *et al.*, 1977; Satoh *et al.*, 1977; Tanis *et al.*, 1978). In order to use more information data on three polymorphic serum protein systems (haptoglobin (Hp), transferrin (Tf), and Gc) and seven red cell antigen systems (ABO, MNS, Rh, Di, Fy, K, and Jk) were taken from various sources (see Omoto, 1972) and added to the data, so that data for 32 loci were finally available for comparison between four racial groups. The average heterozygosity per locus was then 0.2040, 0.1620, 0.1711, and 0.1181, for Caucasoids, Negroids, Mongoloids, and Australoids. Still the figure for Australoids was low, being only about 57% of that of Caucasoids. Of course, these estimates of average heterozygosities are largely biased, since a relatively small number of nonpolymorphic loci were available. The proportion of polymorphic loci among the randomly selected loci may be close to 30% so far as electrophoretic survey is concerned (Harris and Hopkinson, 1972). Therefore, a hypothetical correction was made by adding 28 monomorphic loci under the assumption that monomorphism should be the same for all the four racial groups under consideration. Then the data on the resulting 60 loci yielded average heterozygosity values of 0.1088, 0.0864, 0.0913, and 0.0630 for Caucasoids, Negroids, Mongoloids, and Australoids, respectively. The heterozygosity for Caucasoids is now very close to the value (0.099) obtained by Nei and Roychoudhury (1974) using 74 protein loci.

On the basis of these gene frequency data, Nei's standard genetic dis-

* AK_1, adenylate kinase 1; DIA_1, diaphorase 1; GOT_S, soluble glutamic oxalacetic transaminase; GPI, glucose-phosphate isomerase; ICD_S, soluble isocitrate dehydrogenase; LDH_A, lactate dehydrogenase A; LDH_B, lactate dehydrogenase B; MDH_S, soluble malate dehydrogenase; PGK_1, phosphoglycerate kinase 1; SOD_A, superoxide dismutase A.

TABLE II
Genetic Distances between Four Racial Groups Based on Data of 32 Loci

		Caucasoids	Negroids	Mongoloids
(1)	Negroids	0.0310 (0.0145)		
	Mongoloids	0.0248 (0.0122)	0.0616 (0.0140)	
	Australoids	0.0410 (0.0200)	0.0631 (0.0067)	0.0143 (0.0078)
(2)	Negroids	0.9011 (0.5875)		
	Mongloids	0.7369 (0.4431)	1.1455 (0.5943)	
	Australoids	1.1032 (0.6002)	1.2779 (0.5582)	0.7403 (0.4170)

Distances based on 22 red cell enzyme loci only are given in parentheses. (1) Standard genetic distances of Nei (1972); (2) genetic distances of Cavalli-Sforza and Edwards (1967).

tances were calculated for four groups as shown in Table II. The distances were calculated separately for 22 enzyme loci and for 32 loci including blood groups and serum proteins. Also given in this table are the genetic distances calculated by the method of Cavalli-Sforza and Edwards (1967). It was found that the two methods, as well as different sets of loci, yielded different interrelationships between the four groups. For example, when 32 loci are used, Nei's method emphasizes the closeness between Mongoloids and Australoids, while the method of Cavalli-Sforza and Edwards assigns almost an identical distance between Caucasoids and Mongoloids on the one hand, and between Mongoloids and Australoids on the other. Also, when 22 loci for red cell enzymes only are used, Nei's method brings Australoids very close to Negroids, while in Cavalli-Sforza and Edwards' method the resulting interrelationship is almost the same as in the case when 32 loci are used. The results are shown diagrammatically in Fig. 1. The reason for this deviation between the two methods is not clear, although it may partly be accounted for by the fact that the two methods have different sensitivity to relatively uncommon variants included in the gene frequency data.

Figure 2 shows a dendrogram constructed by the method of Nei (1975) using the genetic distances for 32 as well as 60 loci. Mongoloids and Australoids belong to one branch, while Caucasoids and Negroids belong to another branch, although the latter two groups have a distance from each other much larger than the Mongoloid/Australoid interrelation.

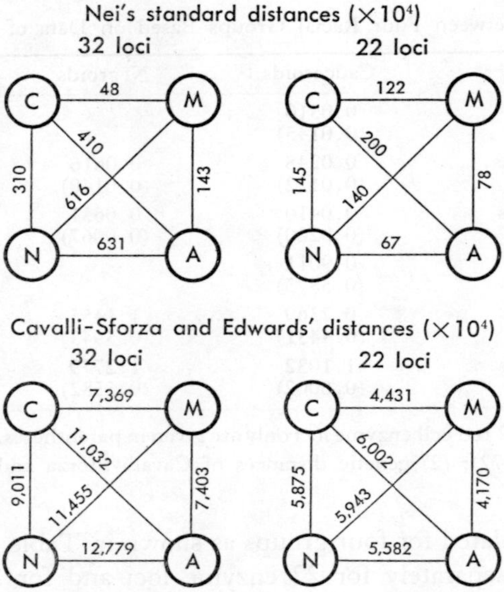

Fig. 1. Diagrammatic representation of various genetic distance estimates between four racial groups: Caucasoids (C), Negroids (N), Mongoloids (M), and Australoids (A). The data on 22 loci were taken exclusively from red cell enzymes, while blood group and serum protein loci were included in the 32 loci.

This dendrogram is clearly compatible with hypothesis of the Asiatic origin of Australoids mentioned earlier.

In Fig. 2, a relative measure of the minimum divergence time estimated by the method of Nei (1975) is given in the scale, together with the scale for standard genetic distances. It is suggested that the divergence time for Mongoloids and Australoids is roughly 37,500 years, while that for Caucasoids and Negroids is roughly 75,000 years. On the other hand, the estimated divergence time for the Mongoloid/Australoid cluster as against the Caucasoid/Negroid cluster is roughly 110,000 years. Although these figures must be subject to large standard errors, the overall relationships obtained in the present analysis seems to be compatible with anthropological findings mentioned above concerning the origin of the Australoid racial groups.

It should be pointed out that the dendrogram shown in Fig. 2 differs from that of Nei and Roychoudhury (1974) with respect to the relative position of Negroids. Among the three major races studied by Nei and

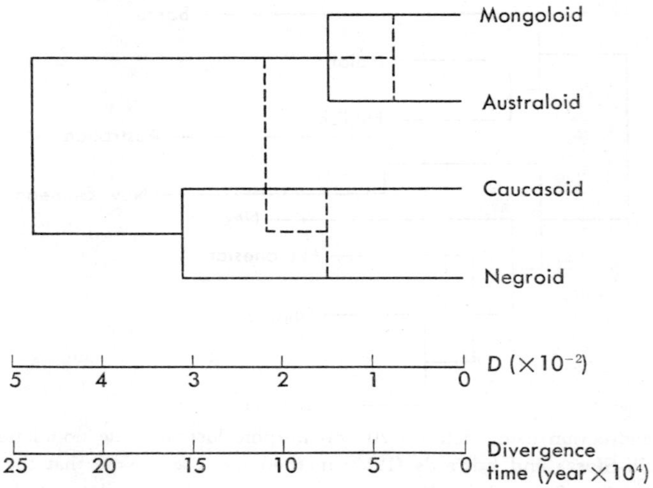

Fig. 2. Dendrogram constructed by Nei's method using data on 32 loci (solid line). Dashed line indicates hypothetical situation for 60 loci, among which about 20 are polymorphic.

Roychoudhury (1974) Negroids are equally far from Caucasoids and Mongoloids, with the estimated divergence time being about 120,000 years. In the present study, however, Negroids and Caucasoids form a cluster which is separated from the Mongoloid/Australoid cluster. The separation time for these two clusters is 110,000 years, corresponding well with the value of Negroid divergence estimated by Nei and Roychoudhury (1974). The reason for this discrepancy is unclear, but probably it is partly due to the fact that data on hemoglobin and glucose-6-phosphate dehydrogenase (G6PD) variants as well as on the *Fy* allele of the Duffy blood groups were not included in the present study, since the distribution of these is known to be positively correlated with that of the endemism of malaria (Livingston, 1967; Miller *et al.*, 1976).

THE NEGRITO PROBLEM

Since 1976, the present author and his colleagues have carried out population genetic studies of the Philippine Negritos. Thus far, blood samples totalling more than 600 have been obtained from the Negrito (Aeta) of west-central Luzon and of northeastern Mindanao (Mamanwa tribe). Investigations of genetic markers of red cell antigens, red cell en-

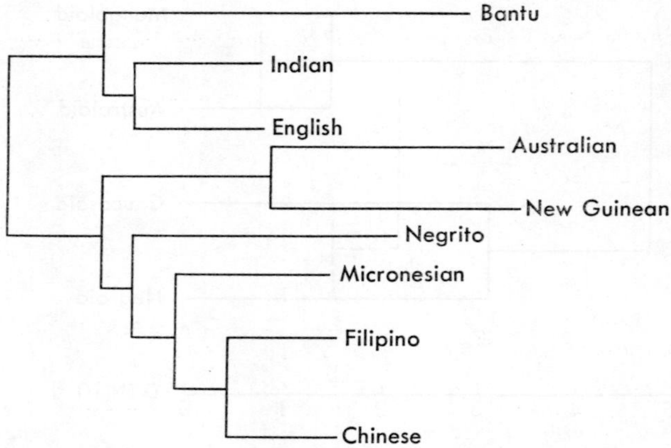

Fig. 3. Dendrogram using data on 20 polymorphic loci for nine populations (Omoto, 1981). Cavalli-Sforza and Edwards (1967) method was used. Note that the Negrito has a close genetic affinity to Asian-Pacific populations.

zymes, serum proteins, and human lymphocyte antigens (HLA) have shown a rather high level of genetic diversity in the Negrito populations and several unusual genetic variants with high frequency have been discovered (Omoto *et al.*, 1978, 1981; Matsumoto *et al.*, 1979; Horai *et al.*, 1981).

Figure 3 shows a dendrogram for nine populations based on the genetic distances computed by the method of Cavalli-Sforza and Edwards (1967). Allele frequency data on a total of 20 genetic loci were used: red cell antigens ABO, MNS_s, Rh, Fy, Di, Jk, P, Le, K; serum proteins Hp, Tf, Gc; and red cell enzymes ACP, PGM_1, PGM_2, PGD, ADA, AK, GPT, ESD. The genetic distances between populations are tabulated in Table III. So far as these data are concerned, the Negrito population appears to be genetically more closely related to the Asian-Pacific group than to the African group (Omoto, 1981). Unfortunately, the data published for African Pygmies with whom direct comparisons should be made are insufficient for this comparison. However, it has been shown that they have rather close genetic affinities to the Forest Negroes belonging to the Bantu group (Cavalli-Sforza *et al.*, 1968). Therefore, it may be speculated that the African Pygmies and the Asian Negritos have different origins and that similarities at the phenotypic level such as short stature, dark skin color, and kinky hair are probably the reflection of adaptation to similar environmental conditions. However, it is not known whether

TABLE III
Comparisons among Nine Population Groups Including the Negrito Using Data of 20 Polymorphic Loci

	Negrito	Filipino	Chinese	Indian	African	English	Australian	Micronesian
Filipino	0.7125							
Chinese	0.6948	0.3888						
Indian	0.7944	0.5288	0.5784					
African	1.3396	1.0537	1.1503	0.9373				
English	0.9158	0.6941	0.7137	0.4562	0.8335			
Australian	1.0333	0.9635	0.9064	1.1212	1.2966	1.1721		
Micronesian	0.7507	0.5059	0.5010	0.7766	1.1779	0.8770	0.7820	
Melanesian	1.0925	0.9677	0.9918	1.0779	1.3389	1.1234	0.8434	0.8426

Method of Cavalli-Sforza and Edwards (1967) was used.

these characteristics represent convergence or parallel evolution between the Pygmies and the Negritos, or alternatively, whether they are ancient human characters which remained relatively unchanged due to strong selective factors maintaining them only among certain groups of mankind including the two populations in question, while in other racial groups evolution at the phenotypic level with respect to these characters has proceeded quite rapidly and diverged into various directions leading to the marked variations now known as racial characters.

PRIVATE POLYMORPHISM

Among human polymorphisms, one may distinguish "universal" and "local" or "private" polymorphisms with respect to the distribution of specific alleles. Private polymorphisms are defined as the occurrence of a unique allele in a very restricted geographical population often corresponding to a tribal group with frequencies greater than 1%. Among peoples in the Pacific in particular, numerous examples of this phenomenon have been reported. Thus, as to electrophoretic variants of red cell enzymes alone, at least 11 different alleles were reported to have polymorphic frequency in restricted distribution limited to a tribal group (Table IV). Among these, three examples are of CA_1, and two are of PGM_2, while others are all of different enzymes. Thus there are seven different enzyme loci at which at least one allele was found to be locally polymorphic: 6PGD, PGM_1, PGM_2, PGK_1, ESD, CA_1, CA_2, and PEPB. Also, Neel (1978) listed eight red cell enzyme private alleles with polymorphic frequency among

TABLE IV
Red Cell Enzyme Electrophoretic Variants Showing Restricted Distribution (Private Polymorphism) in the Asian-Pacific Area

Enzyme (Locus)	Allele	Population	Locality	Highest frequency (%)	Local or private[a]	Reference
6PGD	PGD^E	Australian Aborigines	Northern Territory, Australia	6.1	l	Kirk *et al.*, 1971
PGM_1	PGM_1^3	Micronesian and Melanesian	W. Carolines and New Guinea	6.8	l	Blake and Omoto, 1975
PGM_2	PGM_2^3	Australian Aborigines	Australia	6.7	l	Blake and Omoto, 1975
	PGM_2^9	Melanesian	New Guinea	13.8	l	Blake and Omoto, 1971
PGK_1	PGK^2	Micronesian and Melanesian	W. Carolines, New Guinea, and Samoa	8.0	l	Omoto and Blake, 1972
ESD	ESD^{3N}	Negrito	Luzon Is., Philippines	10.1	p	Omoto *et al.*, 1978
CA_1	CA_1^3	Negrito	Mindanao Is., Philippines	26.2	p	Omoto, 1979
	CA_1^9	Australian Aborigines	Australia	12.3	l	Blake, 1979
	CA_1^{10}	Australian Aborigines	Northern Territory, Australia	2.3	p	Blake, 1979
CA_2	CA_2^4	Australian Aborigines	Australia	8.3	l	Blake, 1979
PEPB	$PEPB^7$	Australian Aborigines	Central Australia	6.1	p	Blake, 1979

[a] l, local; p, private, *i.e.*, found in only one tribal group.

South Amerindian tribal groups. From this data, four further enzyme loci (ESA (esterase A), PEPA, LDH_B, ICD_s) can be added to the list. Many of them (PGM_2, PGK_1, CA_1, ICD_s, ESA, LDH_B) are known to be monomorphic in the majority of human populations. It appears that this phenomenon is most readily explained by the neutral theory for protein polymorphism put forward by Kimura and Ohta (1971). On the other hand, it seems to be difficult to consider the selective factors leading to polymorphisms, which are different from loci to loci and specific in only a small fraction of human populations.

At least three different alleles were found to be highly polymorphic in some but not all of the Negrito populations thus far examined (Table V) (Omoto *et al.*, 1978, 1981; Omoto, 1979, 1980). ESD^{3N} and Gc^{1N} are both peculiar to the Negrito group of Luzon (Aeta), with the average allele frequency of both being approximately 0.10. They are absent in the Negrito group of Mindanao (Mamanwa) and the non-Negrito Philippine

TABLE V
Private Polymorphisms among Two Negrito Populations in the Philippines

System	Allele	Frequencies: Aeta (Luzon)	Frequencies: Mamanwa (Mindanao)	Occurrence elsewhere	Reference
ESD	ESD^{3N}	0.10 (0.02–0.23)	0	No	Omoto *et al.*, 1978
AK	AK_1^2 (?)	0.07	0.004	S.E. Asia (Low frequency)	Omoto *et al.*, 1978
Gc globulin	Gc^{1N}	0.11 (0.05–0.21)	0	Filipino (Low frequency)	Omoto *et al.*, 1978
CA_1	CA_1^{3N} ($=CA_1^{3Guam}$)	0	0.21	Western Pacific (Low frequency)	Omoto, 1980; Omoto *et al.*, 1981

populations with the exception of the Ifugao of Mountain Province, north Luzon, and the Tagalog in Manila, where the Gc^{1N} variant was detected at a low frequency, probably due to ancient gene flow from the Negritos to these populations (unpublished data). On the other hand, CA_1^{3N} is highly polymorphic only among the Mamanwa, with the allele frequency being 0.217±0.017. The variant allele also occurs in the non-Negrito Manobo population living adjacent to the Mamanwa in low but polymorphic frequency, while it is completely absent in the Aeta. Recently, peptide mapping and amino acid analysis were carried out on this variant protein, showing that this variant is identical to the "Guam" variant (253Gly → Arg) which has been known to have a wide distribution among Western Pacific populations as a rare variant (Omoto, 1980; Omoto *et al.*, 1981; Tashian *et al.*, 1980).

These results suggest that if these alleles are considered as essentially neutral in an evolutionary sense, ESD^{3N} and Gc^{1N} represent relatively recent mutations, possibly with an age similar to each other, while CA_1^{3N} is derived from a mutation of a much older age. According to Kimura and Ohta (1973), a neutral mutant whose current frequency is 10% has an expected age roughly equal to the effective population size, and a standard deviation of 1.4 times this (in generations), assuming that the mutant has increased by random genetic drift alone from a very low frequency. The observed frequencies for ESD^{3N} and Gc^{1N} coincide with this situation.

The effective population size of the Negrito population is unknown, but it seems likely to be small. The Negrito (Aeta) of west-central Luzon from whom the present material was taken now number approximately 6,000. If one takes 2,000 as a tentative estimate of the effective population size, the age for the two mutants would be roughly 40,000 years, assuming

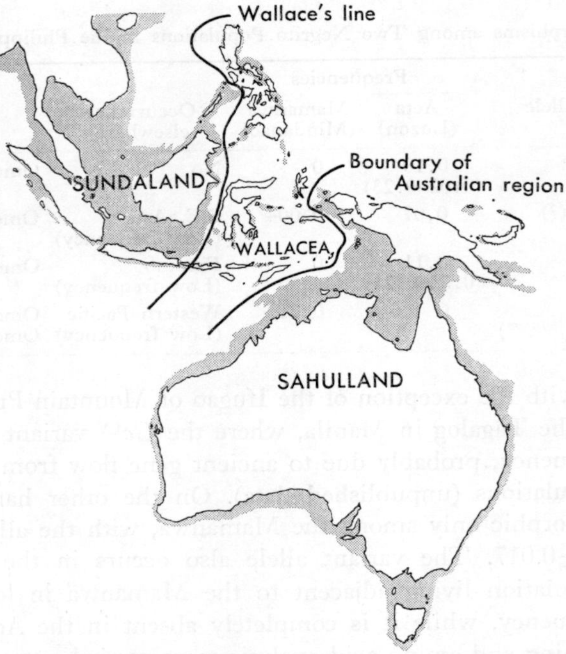

Fig. 4. Land areas in Southeast Asia and Western Pacific during Late Pleistocene, assuming a fall of 200 m in sea level (adapted from J. P. White and J. F. O'Connell, 1979).

20 years for an average generation time. While this estimated age is compatible with the time since the Negrito first inhabited the Philippines, it seems to be too long in view of the fact that these variants are confined to the Negrito of west-central Luzon and seem not to have spread to the Negrito group of Mindanao. If the actual ages of these mutants were younger, the observed high gene frequencies may be partly explained by the bottleneck effect, provided that these alleles are essentially neutral and no selection has acted on them.

The absence of these variant genes in the Mamanwa, the Negrito group of Mindanao, and the fact that $CA_1{}^{3N}$ ($=CA_1{}^{3Guam}$) which is common in the Mamanwa does not occur in the Aeta of Luzon raise an interesting question as to the origin of these two groups, which are usually combined as the Negritos. It is likely that they represent two different migrations to the Philippine Islands, the Aeta of Luzon arriving *via* Palawan Island from northern Borneo, while the Mamanwa migrated from the populations who inhabited somewhere in Wallacea during the end of the Pleistocene times (Fig. 4).

SUMMARY

In recent years, it has become possible to estimate the degree of gene diversity within and between human populations on the basis of detection of protein variants using electrophoresis. In this paper, gene frequency data on blood genetic markers obtained from various population groups of the Asian-Pacific area were analyzed using genetic distances and dendrograms in order to shed some light on the origins of two indigenous groups of this area, the Australian aborigines and the Philippine Negritos. It was shown that the Australian aborigines are genetically more closely related to Mongoloids than to Negroids or Caucasoids. The estimated time of divergence of the Australian aborigines from the Asian population (35,000–40,000 years) is not inconsistent with the dates of first arrival of humans in Australia as recently confirmed by archeological findings. Also, the Negrito group was shown to be descended from the Asian group rather than from the African group, suggesting that their superficial resemblance to African pygmies may be due to environmental adaptation or parallelism. Furthermore, some data on enzyme variants with polymorphic frequency in the restricted geographical area are presented and discussed in relation to the neutral theory of protein polymorphism and molecular evolution.

Acknowledgments

The author is deeply indebted to Dr. Kenichi Aoki, the National Institute of Genetics, Mishima, and to Mr. Naruya Saito, the University of Tokyo, for valuable discussions and assistance during the preparation of this manuscript. Thanks are also due to Miss Naoko Yokota and Miss Asami Ishihara for excellent clerical and laboratory assistance.

REFERENCES

Bark, J. E., Harris, M. J., and Firth, M. (1976). Typing of the common phosphoglucomutase variants using isoelectric focusing—a new interpretation of the phosphoglucomutase system. *J. Forens. Sci. Soc.* **16**, 115–120.

Bellwood, P. S. (1980). The peopling of the Pacific. *Sci. Am.* **243**, 138–147.

Blake, N. M. (1979). Genetic variation of red cell enzyme systems in Australian aboriginal populations. *Occas. Pap. Hum. Biol.* **2**, 39–82.

Brown, W. M. (1980). Polymorphism in mitochondrial DNA of humans as revealed by restriction endonuclease analysis. *Proc. Natl. Acad. Sci. U.S.* **77**, 3605–3609.

Cavalli-Sforza, L. L. and Edwards, A. W. F. (1967). Phylogenetic analysis: models and estimation procedures. *Am. J. Hum. Genet.* **19**, 233–257.

Cavalli-Sforza, L. L., Zonta, L. A., Nuzzo, F., Bernini, L., Jong, W. W. W., De Khan, P., Meera, Ray, A. K., Went, L. N., Siniscalco, M., Nijenhuis, L. E., van Loghem, E., and Modiano, G. (1968). Studies on African Pygmies in the Central African Republic. *Am. J. Hum. Genet.* **21**, 252–274.

Chappel, J. M. A. (1976). Aspects of late quartenary paleogeography of the Australian-East Indonesian region. In *The Origin of the Australians* (eds. R. L. Kirk and A. G. Thorne), pp. 11–22. Australian Institute of Aboriginal Studies, Canberra.

Constans, J. and Viau, M. (1977). Group-specific component: evidence for two subtypes of the Gc^1 gene. *Science* **198**, 1070–1071.

Harris, H. and Hopkinson, D. A. (1972). Average heterozygosity per locus in man: an estimate based on the incidence of enzyme polymorphisms. *Ann. Hum. Genet. Lond.* **36**, 9–20.

Harris, H. and Hopkinson, D. A. (1976). *Handbook of Enzyme Electrophoresis in Human Genetics*. North-Holland, Amsterdam.

Harris, H., Hopkinson, D. A., and Robson, E. B. (1974). The incidence of rare alleles determining electrophoretic variants: data on 43 enzyme loci in man. *Ann. Hum. Genet. Lond.*, **37**, 237–253.

Harris, H., Hopkinson, D. A., and Edwards, Y. H. (1977). Polymorphism and the subunit structure of enzymes: a contribution to the neutralist-selectionist controversy. *Proc. Natl. Acad. Sci. U.S.* **74**, 698–701.

Hooton, E. A. (1956). *Up from the Ape*, 2nd ed. MacMillan, New York.

Horai, S., Omoto, K., Juji, T., Sonozaki, H., Mitsui, H., Misawa, S., Sumpaico, J. S., and Mercado, A. D. (1981). The HLA antigens of two Negrito populations in the Philippines. *Tissue Antigens* **17**, 343–348.

Howells, W. W. (1973). *The Pacific Islanders*. Reed, Wellington, Sydney, and Auckland.

Ishimoto, G. (1975). Red cell enzymes. In *Anthropological and Genetic Studies on the Japanese* (eds. S. Watanabe, S. Kondo, and E. Matsunaga), JIBP Synthesis, Vol. 2, pp. 109–139. Univ. Tokyo Press, Tokyo.

Kimura, M. (1968). Evolutionary rate at the molecular level. *Nature* **217**, 624–626.

Kimura, M. and Ohta, T. (1971). Protein polymorphism as a phase of molecular evolution. *Nature* **229**, 467–469.

Kirk, R. L. (1965). The distribution of genetic markers in Australian Aborigines. In *Occasional Papers in Human Biology*. Australian Institute of Aboriginal Studies, Canberra.

Kirk, R. L. (1976). Genetic distance—a pragmatic approach. *Excerpta Med. Int. Congr. Ser.* **411**, 244–254.

Kirk, R. L., Blake, N. M., Cooke, D. R., and Jose, D. G. (1972). The distribution of some serum protein and enzyme groups among aboriginal populations in Queensland. *Hum. Biol. Oceania* **1**, 207–214.

Kirk, R. L. and Thorne, A. G. (eds.) (1976). *The Origin of the Australians*. Australian Institute of Aboriginal Studies, Canberra.

Kirk, R. L., Blake, N. M., and Vos, G. H. (1971). The distribution of enzyme group systems in a sample of South African Bantu. *South Afr. Med. J.* **45**, 69–72.

Kirk, R. L., Blake, N. M., Moodie, P. M., and Tibbs, G. J. (1971). Population genetic studies in Australian Aborigines of the Northern Territory. The distribution of some serum protein and enzyme groups among populations at various localities in the Northern Territory of Australia. *Hum. Biol. Oceania* **1**, 54–76.

Livingston, F. B. (1967). *Abnormal Hemoglobins in Human Populations.* Aldine, Chicago.

Matsumoto, H., Miyazaki, T., Omoto, K., Misawa, S., Harada, S., Hirai, M., Sumpaico, J. S., Medado, P. M., and Ogonuki, H. (1979). Population genetic studies of the Philippine Negritos. II Gm and Km allotypes of three population groups. *Am. J. Hum. Genet.* **31**, 70–76.

Miller, L. H., Mason, S. J., Clyde, D. F., and McGinniss, M. H. (1976). The resistance factor to plasmodium vivax in blacks. *N. Engl. J. Med.* **295**, 302–304.

Mourant, A. E., Kopeć, A. C., and Domaniewska-Sobczak, K. (1976). *The Distribution of the Human Blood Groups and Other Polymorphisms.* Oxford Univ. Press, London.

Neel, V. (1978). Rare variants, private polymorphisms, and locus heterozygosity in Amerindian populations. *Am. J. Hum. Genet.* **30**, 465–490.

Nei, M. (1972). Genetic distance between populations. *Am. Nat.* **106**, 283–292.

Nei, M. (1975). *Molecular Population Genetics and Evolution.* North-Holland, Amsterdam and Oxford.

Nei, M. (1978). The theory of genetic distance and evolution of human races. *Japan. J. Hum. Genet.* **23**, 341–369.

Nei, M. and Roychoudhury, A. K. (1974). Genic variation within and between the three major races of man, Caucasoids, Negroids, and Mongoloids. *Am. J. Hum. Genet.* **26**, 421–443.

Nurse, G. T. and Jenkins, T. (1977). *Health and the Hunter-Gatherer.* S. Karger, Basel.

Omoto, K. (1970). The distribution of polymorphic traits in the Hidaka Ainu. I. Defective colour vision, PTC taste sensitivity and cerumen dimorphism. *J. Fac. Sci. Univ. Tokyo,* **V**, 337–355.

Omoto, K. (1972). Polymorphisms and genetic affinities of the Ainu of Hokkaido. *Hum. Biol. Oceania* **1**, 278–288.

Omoto, K. (1975). Genetic affinities of the Ainu as assessed from data on polymorphic traits. In *Anthropological and Genetic Studies on the Japanese* (eds. S. Watanabe, S. Kondo, and E. Matsunaga), JIBP Synthesis, Vol. 2, pp. 296–307. Univ. Tokyo Press, Tokyo.

Omoto, K. (1979). Carbonic anhydrase-1 polymorphism in a Philippine aboriginal population. *Am. J. Hum. Genet.* **31**, 747–750.

Omoto, K. (1980). Genetic variants of red cell enzymes as potential anthropological markers in the Western Pacific. *Hemoglobin* **4**, 755–769.

Omoto, K. (1981). Genetic origins of the Philippine Negritos. *Curr. Anthropol.* **22** (in press).

Omoto, K. and Blake, N. M. (1972). Distribution of genetic variants of erythrocyte phosphoglycerate kinase (PGK) and phosphohexose isomerase (PHI) among some population groups in south-east Asia and Oceania. *Ann. Hum. Genet. Lond.* **36**, 61–67.

Omoto, K., Misawa, S., Harada, S., Sumpaico, S., Medado, M., and Ogonuki, H. (1978). Population genetic studies of the Philippine Negritos. I. A pilot survey of red cell enzyme and serum protein groups. *Am. J. Hum. Genet.* **30**, 190–201.

Omoto, K., Ueda, S., Goriki, K., Takahashi, N., Misawa, S., and Pagaran, I. G. (1981). Population genetic studies of the Philippine Negritos. III. Identification of the carbonic anhydrase-I variant with Ca_1 Guam. *Am. J. Hum. Genet.* **33**, 105–111.

Ramshaw, J.A.M., Coyne, J. A., and Lewontin, R. C. (1979). The sensitivity of gel electrophoresis as a detector of geneic variation. *Genetics* **93**, 1019–1037.

Satoh, C., Ferrell, R. E., Tanis, R., Ueda, N., Kishimoto, S., Neel, J., Hamilton, H. B., and Baba, K. (1977). The frequency in Japanese of genetic variants of 22 proteins III. Phosphoglucomutase-1, phosphoglucomutase-2, 6-phosphogluconate dehydrogenase, adenylate kinase, and adenosine deaminase. *Ann. Hum. Genet. Lond.* **41**, 169–183.

Selander, R. K. and Kaufman, D. W. (1973). Genetic variability and strategies of adaptation in animals. *Proc. Natl. Acad. Sci. U.S.* **70**, 1875–1877.

Singh, R. S., Lewontin, R. C., and Felton, A. A. (1976). Genic heterogeneity within electrophoretic "allele" of xanthine dehydrogenase in *Drosophila pseudoobscura*. *Genetics* **84**, 609–629.

Tanis, R. J., Ueda, N., Satoh, C., Ferrell, R. E., Kishimoto, S., Neel, J. V., Hamilton, H. B., and Ohno, N. (1978). The frequency in Japanese of genetic variants of 22 proteins. IV. Acid phosphatase, NADP-isocitrate dehydrogenase, peptidase A, peptidase B, and phosphohexose isomerase. *Ann. Hum. Genet. Lond.* **41**, 419–428.

Tashian, R. E., Kendall, A. G., and Carter, N. D. (1980). Inherited variants of human red cell carbonic anhydrases. *Hemoglobin* **4**, 635–651.

Thorne, A. G. and Macumber, P. G. (1972). Discoveries of late Pleistocene man at Kow Swamp, Australia. *Nature* **238**, 316–319.

Ueda, N., Satoh, C., Tanis, R. J., Ferrell, R. E., Kishimoto, S., Neel, J. V., Hamilton, H. B., and Baba, K. (1977). The frequency in Japanese of genetic variants of 22 proteins. II. Carbonic anhydrase I and II, lactate dehydrogenase, malate dehydrogenase, nucleoside phosphoryrase, triose phosphate isomerase, haemoglobin A and haemoglobin A_2. *Ann. Hum. Genet. Lond.* **41**, 3–52.

Watanabe, S., Kondo, S., and Matsunaga, E. (eds.) (1975). *Anthropological and Genetic Studies on the Japanese*, JIBP Synthesis, Vol. 2. Univ. Tokyo Press, Tokyo.

White, P. and O'Connell, J. F. (1979). Australian prehistory: new aspects of antiquity. *Science* **203**, 21–28.

9

Molecular Evolution, Protein Polymorphism and the Neutral Theory (1982) (ed. M. Kimura), pp. 217–229.
Japan Scientific Societies Press, Tokyo / Springer-Verlag, Berlin

Statistical Examination of Phylogenetic Tree Construction Methods by Computer Simulation

YOSHIO TATENO

Institute of Physical and Chemical Research, Wako 351, Japan

Evolutionary biology has recently made great advances thanks to the rapid accumulation of molecular data, *i.e.*, gene frequencies, amino acid sequences, cleavage patterns of restriction enzymes, and nucleotide sequences, on the one hand, and attempts to develop or modify the construction method of a phylogenetic tree so that it could be used for molecular data, on the other. In the past two decades or so, a number of construction methods have been proposed for studying the phylogenetic relationships of the organisms in question. Most of these methods were originally devised to deal with physiological and morphological data. It is well known that physiological and morphological characters change drastically in some periods and hardly do in other periods during the course of evolution; *cf.*, the evolutionary history from Eohippus to Equus. These characters are thus not appropriate measures for estimating divergence times of species, which is very important in the study of phylogeny.

The great advantage in using molecular data for the study of phylogenetic relationships was first pointed out empirically by Zuckerkandl and Pauling (1965), and theoretically by Kimura (1968), who proposed the neutral mutation theory, which states that if a gene causes neither selective advantage nor disadvantage, then its evolutionary rate is constant over time. It has been shown that the molecular data can be used safely for the construction of a phylogenetic tree without considering the influence of natural selection (*e.g.*, Nei, 1975; Wilson *et al.*, 1977). A phylogenetic tree based on such molecular data was first constructed by Fitch and Margoliash (1967) who used amino acid sequences of cytochrome *c* for 21 different organisms. After this pioneering work, many "molecular phylogenetic trees" have been produced for different molecular data. However, the accuracy of these constructed trees is unknown, because the actual trees can never be known in detail. One way to attack this difficulty is to study systematically the accuracy of the construction method used.

This paper presents a statistical study on the accuracy of the unweighted pair-group clustering (UPG) method (Sokal and Michener, 1958), the Farris method (Farris, 1972), and the modified Farris method (Tateno and Nei, submitted) which are used for the construction of a molecular phylogenetic tree. The study was carried out by conducting computer simulation which mimicked the evolutionary change of genes along a model tree.

MODEL AND METHOD

The basic idea of the computer simulation is as follows: the gene frequency change in a population (size N) is introduced by mutation and random genetic drift but not by natural selection. Mutation is created according to the Poisson distribution with a certain parameter in each generation, and sampling of genes is carried out by generating pseudorandom numbers introducing the effect of random genetic drift. The mutation thus created is assumed to be always different from the extant genes. Such a gene frequency change starts at the top (common ancestral population) of a model tree and proceeds all the way down to the terminal points (present populations, or operational taxonomic units (OTUs) as defined by Sokal and Sneath, 1963).

To set up the model tree, the number (n) of OTUs and topology must be determined. Since a large-scale simulation was planned, many OTUs were not affordable and n was finally determined to be eight. With respect to the topology, there are basically two types as shown in Fig. 1. In these trees the numbers 1 to 8 refer to the eight OTUs. Both are topologically extreme, and the topology of a tree is generally somewhere in between.

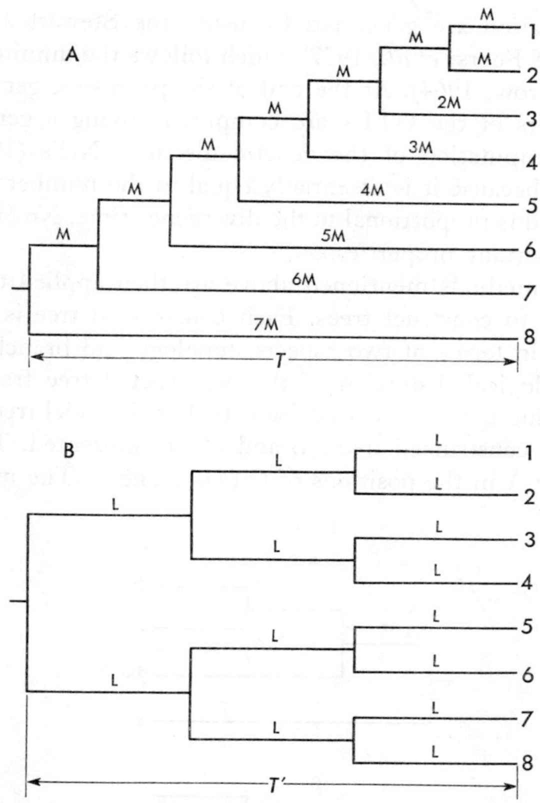

Fig. 1. Model trees considered in computer simulation. A: model tree A of 8 OTUs actually used in the simulation. Numbers 1 to 8 refer to 8 OTUs, M to the unit branch length, and T to the evolutionary time from the common ancestral population to each OTU. B: model tree B of 8 OTUs. Numbers 1 to 8 refer to 8 OTUs, L to the unit branch length, and T' to the evolutionary time from the common ancestral population to each OTU.

When the entire evolutionary time (T in tree A and T' in tree B) is fixed ($T=T'$), then tree A is subject to larger errors than tree B. This is because tree A has shorter branches than tree B, and the shorter the branch length the larger the coefficient of variation of an evolutionary rate. Since the study was concerned with the errors that would occur in tree construction, it was decided to use tree A as the model tree.

When a population splits into two at a branch point of the model tree, two identical populations are produced and reproductive isolation immediately operates on them. The initial gene frequencies of the common ancestral population are generated by using the Stewart algorithm (see the Appendix of Fuerst *et al.*, 1977) which follows the infinite allele model (Kimura and Crow, 1964). At the end of the processes, genetic distances between all pairs of the OTUs are computed, giving a genetic distance matrix. For computation of the genetic distance, Nei's (1972) measure was employed, because it is essentially equal to the number of nucleotide substitutions and is proportional to the divergence time. No other measures have these important properties.

The three methods mentioned above are then applied to the genetic distance matrix to construct trees. Each constructed tree is compared to the model tree in terms of two aspects, topology and branch lengths. To assess the topological distortion of the constructed tree from the model tree, the following measure was devised. In Fig. 2 model tree A with four OTUs and two constructed trees, B and C, are presented. Tree B is different from tree A in the positions of OTUs 3 and 4. The model tree can

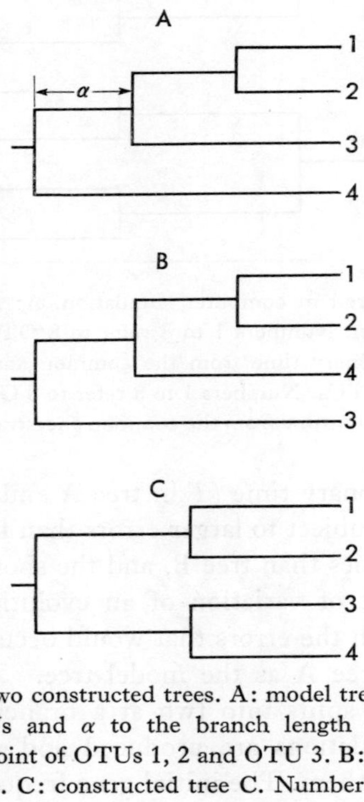

Fig. 2. Model tree and two constructed trees. A: model tree A of four OTUs. Numbers 1 to 4 refer to four OTUs and α to the branch length from the common ancestral population to the branch point of OTUs 1, 2 and OTU 3. B: constructed tree B. Numbers 1 to 4 refer to four OTUs. C: constructed tree C. Numbers 1 to 4 refer to four OTUs.

be obtained from tree B by exchanging OTU 3 with OTU 4. The degree of topological distortion is measured by the number of such exchanges that are required to obtain the model tree. This number is called the distortion index (DI). In the above example, the DI value is just one. In tree C, however, DI is defined in a different way. Since the order of the OTUs in this tree is the same as that in tree A, the topological distortion seems less serious than that in tree B. In fact, tree C can be regarded as a special case of tree A where the branch length α of tree A is zero, whereas tree B may be obtained when α is negative. The DI value of tree C is thus defined to be 0.5. To illustrate a more general case, another example is given in Fig. 3. In this figure the topology of the constructed tree (tree B) can be transformed into the model tree (tree A) by making exchanges between OTUs 3 and 4, 6 and 8, and 7 and 8 in that order. Even after these ex-

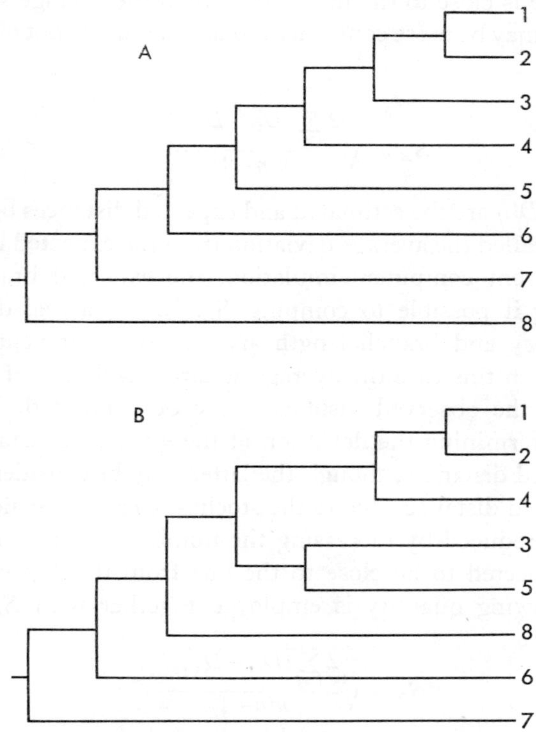

Fig. 3. Model tree and a constructed tree. A: model tree of 8 OTUs. Numbers 1 to 8 refer to 8 OTUs. B: constructed tree. Numbers 1 to 8 refer to 8 OTUs.

changes OTUs 4 and 5 are combined together like OTUs 3 and 4 of tree C in Fig. 2. The DI value of this constructed tree is thus 3.5.

As to measuring the deviation of the estimated branch length from the true one, the best way would be to compute the squared deviation of each branch length of a constructed tree from that of the model tree, and take the average over all branches. This is, however, not practical, because the topology of a constructed tree is often different from that of the model tree. The second choice is to use the matrices of the expected and estimated distances. In the present case the former can be obtained by adding the expected distances for all branches concerned, whereas the latter can be available from the constructed tree. Since the latter matrix is obtainable irrespective of the topology, it is possible to compute the average squared deviation of the elements of the estimated distance matrix from those of the expected distance matrix for any topology. It is expected that if a constructed tree is close to the model tree, then the average squared deviation is small. It may be more convenient to use the squar root of this quantity. Namely,

$$S_E = \sqrt{\frac{2\sum_{i>j}(D_{ij}-D'_{ij})^2}{n(n-1)}} \tag{1}$$

where D_{ij} and D'_{ij} are the estimated and expected distances between OTUs i and j. This is called the average deviation from the expected branch length.

In the present computer simulation, the expected branch length is known, making it possible to compute S_E. In the actual data, however, the true topology and branch length are unknown, preventing the computation of S_E. In this case the average squared deviation of the estimated distances from the observed distances may be computed. This quantity is useful for determining the deviation of the estimated distances from the actually observed distances, though the latter may be considerably different from the expected distances due to the stochastic error of gene substitution. If this error is reduced by increasing the number of genes used, then this deviation is expected to be close to the one from the expected distances. Thus, the following quantity is employed together with S_E.

$$S_O = \sqrt{\frac{2\sum_{i>j}(D_{ij}-D''_{ij})^2}{n(n-1)}} \tag{2}$$

where D''_{ij} is the observed distance between OTUs i and j. This is called the average deviation from the observed branch length. S_O is highly correlated with the Fitch and Margoliash (1967) percent standard deviation.

In the present simulation, the unit branch length M in Fig. 1, corresponds to vt_0 where v is the mutation rate and t_0 is the smallest time unit. The value of M was taken to be 0.002 which roughly corresponds to the gene differentiation of a population within a species (see Nei, 1975). Moreover, $4Nv$ was assumed to be 0.2, which is equivalent to an average heterozygosity of 0.18 in this case. Note that in many *Drosophila* species the average heterozygosity is of this order. With these conditions two sets of parameters of v, t_0, and N were used, *i.e.*, $v=0.002$, $t_0=1$, and $N=25$, and $v=0.001$, $t_0=2$, and $N=50$. These two sets were employed to study the effect of the population size on tree construction.

In each of the above two sets, the number of loci was as many as 100, and this was repeated 10 times (replications). In each replication, D_N was computed by using the gene frequency data for the first 10 loci, the first 20 loci, . . . , and all 100 loci.

DISTORTION OF TOPOLOGY OF A CONSTRUCTED TREE

1. *Effect of Population Size*

Li and Nei (1975) and Nei (1978) have shown that the mean and variance of D_N are largely determined by the product of N and v rather than each separately. These authors' results were obtained by using the diffusion approximation which may not be suitable to a small population. To elucidate this point, the effect of the population size on tree construction was examined using the results for $N=25$ and $N=50$. The relationship of the average DI and the number of loci is presented in Fig. 4. Figures 4A and B show the relationships for $N=25$ and $N=50$, respectively. It is clear that the effect of the population size is not crucial in the construction of a tree, because the difference in the relationship between the two figures are small to all number loci used except for the case where 10 loci were used. In this exceptional case the Farris method constructed less accurate topology compared with the other two, but this could be due to a large stochastic error rather than to an inherent deficiency of this method. At any rate, these results indicate that the above authors' conclusion is valid even for an N as small as 25.

2. *Effect of Number of Loci*

Studying the relative magnitude of the intralocus and interlocus variances of D_N, Nei and Roychoudhury (1974) have concluded that a

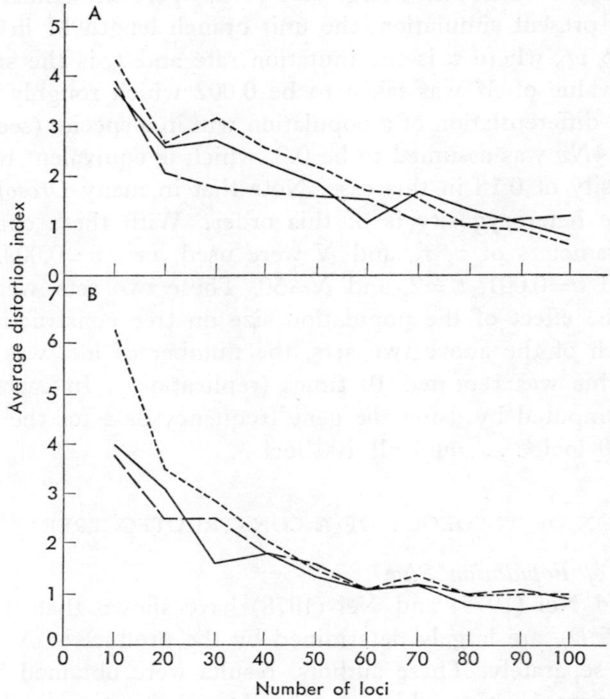

Fig. 4. Relationship between the number of loci and average DI. Results were obtained using Nei's genetic distance. The broken line refers to the UPG method, dotted line to Farris method, and solid line to the modified Farris method. Number of replications is 10. A: number of individuals is 25 and M is 0.002. B: number of individuals is 50 and M is 0.002.

large number of loci should be used to obtain a reliable estimate of the genetic distance. Their study refered to the variance of the genetic distance rather than to the accuracy of a constructed tree. However, Fig. 4 shows that their conclusion concerning the number of loci is essentially correct even with respect to the accuracy of a constructed tree. It is clear from the figure that the average DI value gradually declines as the number of loci increases. To make the average DI smaller than 2, about 50 loci are required, whereas to make it less than 1 about 80 loci are necessary. This indicates that the stochastic error of gene substitution affects considerably the construction of a tree, and thus the importance of using a large number of loci for this purpose.

It is observed from Fig. 4 that there is not much difference in the accuracy of constructing the true topology among the three methods. This is different from the observation made by Tateno and Nei (submitted) that the Farris and modified Farris methods are superior to the UPG method when the number of nucleotide substitutions is used as the measure of genetic distance. It is known that a single locus D_N shows an inverse J-shaped distribution when the evolutionary time is short, but gradually becomes U-shaped as the time increases (Nei and Tateno, 1975). On the other hand, the number of nucleotide substitutions per locus initially shows an inverse J-shaped distribution, but gradually tends toward a normal distribution as the time increases. It seems that the Farris and modified Farris methods are disadvantageous if the distribution of a single locus distance is U-shaped.

DEVIATION OF BRANCH LENGTHS OF A CONSTRUCTED TREE

1. *Deviation of Estimated Distances from Expected Ones*

Since Nei (1972) also showed the expectation of D_N, S_E can be computed for each constructed tree. The mean values of S_E for the cases of $N=25$ and $N=50$ are given in Table I. This table includes the means for the cases of the numbers of loci=10, 50, and 100. It is clear that the UPG method gives the smallest value, and the second best is the modified Farris method. The Farris method consistently shows poor performance in estimating the expected distance. Therefore, with the observation that the three methods give more or less the same DI values, it may be said that the UPG method is the best among the three when gene frequency

TABLE I
Average Deviation of Estimated Distances from Expected Distances

Number of loci	UPG method	Farris method	Modified Farris method
$N=25$			
10	0.0119	0.0152	0.0128
50	0.0060	0.0086	0.0068
100	0.0039	0.0058	0.0044
$N=50$			
10	0.0114	0.0173	0.0116
50	0.0060	0.0089	0.0067
100	0.0042	0.0062	0.0045

The number of replications is 10. N is the number of individuals in the population.

TABLE II
Average Deviation of Estimated Distances from Observed Distances

Number of loci	UPG method	Farris method	Modified Farris method
$N=25$			
10	0.0061	0.0066	0.0045
50	0.0048	0.0052	0.0027
100	0.0024	0.0040	0.0019
$N=50$			
10	0.0068	0.0084	0.0047
50	0.0037	0.0046	0.0021
100	0.0023	0.0039	0.0014

The number of replications is 10. N is the number of individuals of the population.

data are used for tree construction. Another advantage of the UPG method is that it does not give any branch with a negative length which could be produced by the other two.

It may be interesting to see if there is any correlation between S_E and DI, because if there is any DI can be used as a good substitute of S_E which is not computable for the actual data as mentioned above. The correlation coefficient between these two quantities was computed for each of the three methods using the results of 10 replications. None was, however, statistically significant.

2. Deviation of Estimated Distances from Observed Ones

The mean values of the S_O are given in Table II in the same format as the case of S_E. It is shown in the table that the modified Farris method gives the smallest S_O value among the three, whereas the Farris method gives the largest. This shows that the modification of the Farris method is successful in reducing the error in the estimation of the observed as well as expected distances. The correlation coefficient between S_O and DI was again not statistically significant for any of the three methods.

DISCUSSION

In the present study it is shown that the UPG method is the best among the three methods examined for estimating the expected genetic distance, whereas the modified Farris method is superior to the other two for estimating the observed genetic distance. The advantage of the UPG method over the others is rather anticipated, because this method is based on the assumption that the genetic distances of two populations from their common ancestor are the same. The corresponding expected genetic distances are, of course, identical, as shown in Fig. 1. The superiority of the modified Farris method may be reasonable, because this method (and the Farris method) is free from the above assumption, and the observed genetic distance is highly subject to the stochastic error creating the deviation from the expected genetic distance (see Tateno and Nei, submitted for details).

There are some other methods which have not been included in the present study. Particularly, the Dayhoff (1969) and the Moore *et al.* (1973) methods are quite different from those studied here in the sense that the former two methods construct trees based on amino acid sequences themselves, whereas the latter methods studied do so on the basis of a distance matrix obtained from such data as gene frequencies, amino acid sequences, cleavage patterns of restriction enzymes, and nucleotide sequences. The latter methods are in a wider usage than the former ones. At any rate, the Moore *et al.* method is a sort of trial-and-error procedure and it takes a considerable time to obtain a tree even if a high speed computer is employed. In this respect this method is not efficient at all. Moreover, the procedures of this method are unduly complicated, and it is almost impossible to make a close examination of it. This is a pity for serious wokers as well as for the method itself, because unless one knows a method well one would not be willing to use it.

Prager and Wilson (1978) used the Fitch and Margoliash (1967) percent standard deviation for the examination of constructed trees by the UPG, Fitch and Margoliash (1967), and Farris methods. They concluded that in general the Fitch and Margoliash method was better than the other two. As it is shown in the present study that no correlation was existed between S_O and DI, the percent standard deviation is not a good measure for the examination of a construction method. Thus, their conclusion is not justified. The finding that there is no correlation between S_E and DI, nor between S_O and DI suggests that both topology and branch lengths should be taken into account for the examination of a construction method.

Recently, Li (1981) has modified the UPG method. The basic idea behind the modification was to remove the undesirable property of this method that it was not devised in consideration of the stochastic error of gene substitution. He applied the modified UPG method to the genetic distances produced by a computer simulation (Tateno, 1978), and claimed that the modified method constructed better trees than the original method. The modification seems to be successful. However, his approach is not sufficient, because the examination was based only on the topological distortion ignoring the deviation of estimated branch lengths. The best way so far feasible to examine a construction method may be to carry out computer simulation taking into account the two aspects, topology and branch lengths.

The computer simulation performed in the present study may not be sufficient to draw a definite conclusion as to which one is the best among the three methods. A larger scale simulation should be carried out to obtain more solid results.

SUMMARY

The unweighted pair-group clustering method, the Farris method, and the modified Farris method were examined by computer simulation for their performances in constructing phylogenetic trees. In the simulation the gene frequency change along a model tree was considered. The examination was carried out in terms of two aspects: construction of topology and estimation of branch lengths. Under a few conditions it is shown that the unweighted pair-group clustering method is the best among the three methods in the estimation of expected distances, and the modified Farris method is superior to the others in the estimation of observed distances. The three methods show more or less the same performance in constructing the topology.

Acknowledgment

The author thanks Dr. M. Nei for stimulating discussions and encouragement throughout the course of this work, and Drs. N. Takahata, and K. Aoki for their valuable comments on the first draft of the manuscript.

REFERENCES

Dayhoff, M. O. (1969). *Atlas of Protein Sequence and Structure*, Vol. 4. National Biomedical Research Foundation, Silver Spring.

Farris, J. S. (1972). Estimating phylogenetic trees from distance matrices. *Am. Nat.* **106**, 645–668.

Fitch, W. M. and Margoliash, E. (1967). Construction of phylogenetic trees. *Science* **155**, 279–284.

Fuerst, P. A., Chakraborty, R., and Nei, M. (1977). Statistical studies on protein polymorphism in natural populations I. distribution of single locus heterozygosity. *Genetics* **86**, 455–483.

Kimura, M. (1968). Evolutionary rate at the molecular level. *Nature* **217**, 624–626.

Kimura, M. and Crow, J. F. (1964). The number of alleles that can be maintained in a finite population. *Genetics* **49**, 725–738.

Li, W.-H. (1981). Simple method for constructing phylogenetic trees from distance matrices. *Proc. Natl. Acad. Sci. U.S.* **78**, 1085–1089.

Li, W.-H. and Nei, M. (1975). Drift variances of heterozygosity and genetic distance in transient states. *Genet. Res.* **25**, 229–248.

Moore, G. W., Barnabas, J., and Goodman, M. (1973). A method for constructing maximum parsimony ancestral amino acid sequences on a given network. *J. Theor. Biol.* **38**, 459–485.

Nei, M. (1972). Genetic distance between populations. *Am. Nat.* **106**, 283–292.

Nei, M. (1975). *Molecular Population Genetics and Evolution.* North-Holland, Amsterdam.

Nei, M. (1978). Estimation of average heterozygosity and genetic distance from a small number of individuals. *Genetics* **89**, 583–590.

Nei, M. and Roychoudhury, A. K. (1974). Sampling variances of heterozygosity and genetic distance. *Genetics* **76**, 379–390.

Nei, M. and Tateno, Y. (1975). Interlocus variation of genetic distance and the neutral mutation theory. *Proc. Natl. Acad. Sci. U.S.* **72**, 2758–2760.

Prager, E. M. and Wilson, A. C. (1978). Construction of phylogenetic trees for proteins and nucleic acids: empirical evaluation of alternative matrix methods. *J. Mol. Evol.* **11**, 129–142.

Sokal, R. R. and Michener, C. D. (1958). A statistical method for evaluating systematic relationship. *Univ. Kansas Sci. Bull.* **38**, 1409–1438.

Sokal, R. R. and Sneath, P.H.A. (1963). *Principles of Numerical Taxonomy.* Freeman, San Francisco.

Tateno, Y. (1978). Statistical studies on the evolutionary changes of macromolecules. Ph. D. Dissertation at the University of Texas at Houston.

Tateno, Y. and Nei, M. Accuracies of several methods of constructing phylogenetic trees from molecular data (submitted).

Wilson, A. C., Carlson, S. S., and White, T. J. (1977). Biochemical evolution. *Annu. Rev. Biochem.* **46**, 573–639.

Zuckerkandl, E. and Pauling, L. (1965). Evolutionary divergence and convergence in proteins. In *Evolving Genes and Proteins* (eds. V. Bryson and H. J. Vogel), pp. 97–166. Academic Press, New York.

10

Molecular Evolution, Protein Polymorphism and the Neutral Theory (1982) (ed. M. Kimura), pp. 233–266.
Japan Scientific Societies Press, Tokyo / Springer-Verlag, Berlin

Evolutionary Changes and Functional Constraints in DNA Sequences

TAKASHI MIYATA

Department of Biology, Faculty of Science, Kyushu University, Fukuoka 812, Japan

Since a method was established for determining the amino acid sequences of proteins, a great deal of information has been accumulated on the primary structures of protein molecules (Dayhoff, 1978). From comparisons of amino acid sequences from various species sources for many protein families, much has become known about the features of amino acid substitutions during evolution (Zukerkandl and Pauling, 1965; Dickerson, 1971; Kimura and Ohta, 1974; Wilson *et al.*, 1977). Among these features, the following are particularly noteworthy. The rate of amino acid substitution of a given protein performing a well-established function is constant per residue per year. The rate of amino acid substitution varies with the nature of the specifications for a functioning protein. That is, protein molecules that are subject to fewer functional constraints evolve faster. By combining population genetic theory with the study of molecular evolution, Kimura (1968) developed a comprehensive theory of molecular evolution, the so-called neutral theory, by which most traits of evolutionary changes at the molecular level can be successfully understood. The neutral theory stresses that most evolutionary changes at the molecular level are caused not by selection but by random drift of mutant genes that are selectively equivalent. Although there have been extensive studies on gene evolution at the protein level, a comparative study of nucleotide sequences should be made to understand the more detailed picture of the process of gene evolution.

Recent advances in nucleotide sequencing techniques have made possible a rapid increase in our knowledge of the primary structure of many genes, and it is now possible to make a comparison between pairs of homologous nucleotide sequences of a wide variety of genes. Comparative study at the DNA level would provide much more knowledge on the process of gene evolution than study at the protein level (Miyata and Yasunaga, 1980; Miyata *et al.*, 1980a). Direct comparison between homologous DNA sequences enables us to evaluate the extent of sequence divergence or sequence difference per nucleotide site (K) in various functional or structural blocks as polypeptide coding regions, intervening sequences (introns), 5′- and 3′-transcribed non-coding regions and other untranscribed flanking regions. Furthermore, from comparison of the DNA sequences in the coding regions, it is also possible to evaluate the extent of sequence divergence for two types of substitutions, a nucleotide replacement leading to an amino acid change (amino acid substitution) and that leading to a synonymous codon change (synonymous substitution). For obtaining the usually defined evolutionary rate (*i.e.*, difference per site per year), the K value should be divided by the time since divergence of the two sequences compared. However, when a comparison is made between evolutionary rates of different functional blocks for the same pair of sequences, the corresponding K values can be treated directly as relative evolutionary rates, if closely related sequences are compared and no recombinations have taken place between them during evolution (Miyata *et al.*, 1980b). This is very useful, because paleontological estimates of the time of divergence are often subject to considerable uncertainty (Romero-Herrera *et al.*, 1973). Since the rate of substitution varies with the nature of specifications of a functioning protein and RNA and reflects the degree of functional constraint operating on the amino acid and nucleotide sequences, comparison of the rates for different functional blocks or segments within a single gene might provide some insight into the functional significance imposed on them and enable us to trace the evolutionary history of the gene in detail (Miyata *et al.*, 1980a, 1980b).

In this chapter, the extent of sequence divergence is shown for various parts of mRNAs of different genes and the relation between the rate of evolutionary change and the degree of functional constraint is examined in detail. From these data, some substitution patterns that are expected to be found commonly in many eukaryotic mRNAs are presented. Special substitution patterns that are correlated with the presence of unique structures and/or functions involved in segments of certain types of genes are also discussed. Especially, a substitution pattern due to homologous but unequal cross-overs within rapidly evolving segments, particularly introns, is discussed in rather greater detail, including the cases for immunoglobulin γ-type heavy chain genes and β-globin genes in the latter half of this chapter.

ESTIMATING THE EXTENT OF NUCLEOTIDE SEQUENCE DIVERGENCE

Sequence difference or simply "difference" (K) is defined as the number of nucleotide substitutions per site in a pair of aligned sequences. The calculation procedure is straightforword for the non-coding region. The case, however, is somewhat complicated for the protein coding region. Usually by counting the number of nucleotide base changes at every nucleotide position of codons from a pair of homologous nucleotide sequences and estimating the fraction of synonymous components of substitutions at the third position by the appropriate procedures, the number of synonymous codon substitutions and that of amino acid substitutions can be evaluated (Kimura and Ohta, 1972; Kafatos *et al.*, 1977; Salser, 1978; Lemedico *et al.*, 1979; Perler *et al.*, 1980; Jukes, 1980; Kimura, 1980a, b). Recently, Kmura (1980a, b) developed thistype of method into a more elaborate one which includes the effect of bias for transition and transversion mutations. A somewhat different approach was utilized by Miyata and Yasunaga (1980). By treating the *number of codon changes* caused by successive single-step nucleotide substitutions, rather than treating the *number of nucleotide changes* at the nucleotide positions of codons, they succeeded in evaluating the rate of synomynous and amino acid substitutions. Although this method has a rather complicated calculation procedure than the former type of method, the observed nucleotide changes within each codon can be discriminated accurately into two dstinct components, that is, synonymous substitution and amino acid substitution.

Miyata and Yasunaga's approach is briefly summarized as follows (for details, see Miyata and Yasunaga, 1980). A single nucleotide change (one-step change) within a certain codon α always causes an exchange of the codon for another α'. The resulting codon change $\alpha \rightarrow \alpha'$ is defined as "synonymous (codon) change," if the codons α and α' code for identical amino acids, and the change is an "amino acid change," if they code for different amino acids. Of the three possible one-step changes occurring at each nucleotide position of a codon, the fraction f leading to synonymous change is assigned for the number of "synonymous sites" and $1-f$ for the number of "amino acid sites." Thus, for a codon CUA, for example, the number of synonymous sites is 4/3 (=1/3+0/3+3/3) and that of amino acid sites is 5/3 (=2/3+3/3+0/3). The next step is to count the number of synonymous and amino acid substitutions from observation of the codon difference in the homologous sequences compared. For example, consider a codon change from UUU (Phe) to GUA (Val) or *vice versa* at a codon site of aligned sequences. The codon UUU reaches the GUA through a minimum of two successive one-step changes. For this case, the minimum substitution number (MSN) is two. On the basis of the assumption that no more substitutions than the MSN have occurred during evolution (the MSN assumption), all the possible "paths" from UUU to GUA or *vice versa* can be traced through successive one-step changes as

Path 1: UUU (Phe) ↔ GUU (Val) ↔ GUA (Val),
Path 2: UUU (Phe) ↔ UUA (Leu) ↔ GUA (Val).

It seems unlikely that all the potentially possible paths have contributed to the observed codon substitution UUU ↔ GUA equiprobably during evolution, but it is rather more probable that most evolutionary changes have occurred through the path that involves the array of codons coding for more similar amino acids (Miyata *et al.*, 1979a). Considering the above notion, Miyata and Yasunaga introduced a weight factor for each path, representing the relative degree of acceptance of the path. In the present example, since path 1 contains one synonymous change and one amino acid change, and on the other hand, in path 2, no synonymous changes appear and amino acid changes occur twice, the mean numbers of synonymous and amino acid substitutions are evaluated to be $1\times\omega_1+0\times\omega_2$ and $1\times\omega_1+2\times\omega_2$ for the observed change UUU ↔ UGA, respectively, where, ω_1 and ω_2 $(=1-\omega_1)$ are the weight factors of path 1 and path 2, respectively.

Carring out the above procedure over all the codon sites of the sequences compared, the desired synonymous (amino acid) site number ($N_S(N_A)$) and the synonymous (amino acid) substitution number ($M_S(M_A)$), and thus the synonymous (amino acid) difference ($K_S=M_S/N_S$ ($K_A=M_A/N_A$)) are obtained. The present method is based on the assumption that no more substitutions have occurred during evolution than the base changes observed within each codon (the MSN assumption). It is therefore necessary to correct K_S and K_A in order to get the correct order of evolutionary rates between sequence pairs with different divergence times. A computer simulation shows that Kimura and Ohta's formula (1972) is well suited for correction of the MSN assumption.

Procedure for calculating the difference of the non-coding region (K_N) is straightforward. Contrary to the coding region, however, considerable sequence divergence is found in non-coding segments, and mostly their lengths are somewhat different between the two genes compared. It is therefore necessary to introduce gaps (*i.e.*, deletions and insertions) into them to align for maximum homology. Since there is no solid backgrounds for treating gaps, it may be appropriate to estimate the K_N at the same time from different approaches which differ in gap treatments and also give the lower and upper boundaries of the actual sequence difference, respectively. For this perpose, Miyata *et al.* (1980a, 1980b) applied two methods, Method 1 and Method 2, for estimating the K_N of several non-coding sequences. When deletions or insertions are found in any one of the homologous sequences compared, the corresponding sites are excluded from the comparison (Method 1). In the other method (Method 2), a gap is counted as a "substitution." Clearly, Method 1 may give a value corresponding to the lower boundary of the true K_N and Method 2 gives the upper boundary of K_N.

EVOLUTIONARY CHANGES IN CODING REGIONS

In the last few years, comparative studies of nucleotide sequences in the protein coding regions have been made by several authors (Salser and Issacson, 1976; Grunstein *et al.*, 1976; Kimura, 1977; Kafatos *et al.*, 1978; Jukes and King, 1979; Lemedio *et al.*, 1979; Nichols and Yanofsky, 1979; Kedes, 1979; Miyata and Yasunaga, 1980; Miyata *et al.*, 1980a, 1980b; Perler *et al.*, 1980; Kimura, 1980; Jukes, 1980; Efstratiadis *et al.*, 1980) and have shown that substitution at the third position of codons (or synonymous substitution) has occurred at a relatively high rate as compared with the other two positions (or amino acid substitution). Indeed, applying the method mentioned in the preceding paragraph to about 50 pairs of homologous sequences for different genes, Myata *et al.* (1980a) have shown that in most genes the rate of synonymous substitution is much higher than that of amino acid substitution, except for the case of mouse immunoglobulin λ-type variable region genes. It is widely accepted that there is a selective force that prefers to preserve the correct tertiary structure of a protein molecule (Dickerson, 1971; Kimura and Ohta, 1974). On the other hand, as is well known, synonymous codons are used quite non-randomly in most genes (see Grantham *et al.*, 1980). Although the functional significance of specific codon utilization is still not fully understood (see Hasegawa *et al.*, 1979; Miyata *et al.*, 1979b), it might be related to selective constraint against synonymous changes. That K_S is much higher than K_A in a wide variety of genes suggests that mutations leading to synonymous codon changes are subject to much weaker constraint than amino acid-altering mutations. A striking example comes from the DNA sequences of histone genes. Histone is known to be the most slowly evolving protein. Even in these genes, synonymous substitutions occurred at a very high rate, almost comparable with the rate in the fibrinopeptide molecule (Grunstein *et al.*, 1976; Kimura, 1977). It is remarkable that this gene has undergone synonymous codon changes at the highest known rate, which is contrasted with the fact that the amino acid-altering changes have occurred at the lowest known rate.

UNIFORM RATE OF SYNONYMOUS SUBSTITUTION

To compared the evolutionary rates of different genes, it is usually necessary to know the time since divergence of the genes compared. However, when the comparison is made among genes on the same pair of DNA sequences or among genes with the same divergence times, the values of sequence differences can be compared directly between them without knowing the divergence times which are often subject to considerable uncertainty. On the basis of this idea, Miyata and his colleagues (Miyata and Yasunaga, 1980; Miyata *et al.*, 1980a, 1980b) compared K_Ss among closely linked genes on the same DNA sequences and among distantly located genes, and pointed out that the rate of synonymous substitution is not only very high but also approximately uniform among different genes (see Table I). The rate of amino acid substitution is known to vary with the nature of the specification of a functioning protein (Dickerson, 1971; Dayhoff, 1978). Indeed, as Table I shows, the K_As appear to have a wide range in value within each gene group. On the contrary, the values of K_S appear to be somewhat limited in range. It is remarkable that the value of K_A is about 2-fold greater in the growth hormone gene than in the preproinsulin gene, whereas the K_Ss are approximately the same in value between them. A similar trend is also found in histone genes and ϕX174 and G4 genes. The high and approximately uniform rate of synonymous substitution, as compared with amino acid substitution, would imply that the majority of mutations leading to synonymous codon changes are much less subject to selective constraint than amino acid-altering changes.

EVOLUTIONARY RATE OF SYNONYMOUS SUBSTITUTION AND THE MOLECULAR CLOCK

As mentioned above the rate of synonymous substitution is a very high and approximately uniform, regardless of different types of genes. Although at present sequence data are not available to enable us to draw a firm conclusion, it seems very likely that the accumulation of changes in synonymous sites is approximately proportional to the time since divergence of the sequences compared, at least within a short period of time. This characteristic feature of synonymous substitution might be useful for determining the evolutionary relationships between closely related genes which were derived from a common ancestor by gene duplication in relatively recent times. The rate of amino acid substitution is known to be approximately constant for a given protein performing a well-established function, and its value varies with the degree of functional constraints imposed on a protein molecule (Dickerson, 1971; Kimura and Ohta, 1974; Wilson *et al.*, 1977). The constant rate of amino acid substitution of a given protein is useful as a "clock" for determining the phylogenetic branching order of different species. However, it may be unsuitable for estimating the divergence time of duplicated genes coding for proteins with different functions, because both protein sequences have been evolving independently at different rates in each species. On the other hand, the use of synonymous substitution as a evolutionary clock might be adequate for such purpose, because its rate is approximately constant, regardless of the types of genes.

To obtain the absolute rate (evolutionary rate) of synonymous substitution (v_S) accurately, it is necessary to correct for the MSN assumption. This may partially be attained by using the formula $v_S=-(3/4)\ln[1-(4/3)K_S]/2T$, where T stands for the divergence time of the sequences compared. Assuming that $T=7.5\times10^7$ year, the time since divergence of the primates and rodents, we obtain $v_S=(5.3\pm0.4)\times10^{-9}$ (per nucleotide per year).

TABLE I

Comparison of Synonymous Differences (K_S) among Different Genes

	K_A	K_S	Mean (S.D.)
a) Comparison of K_Ss of closely linked genes or regions			
Histone genes			
(clone *h22* (1) *vs.* *pSp2/17* (2, 3))			
H4	0.010	0.488	
H2B	0.053	0.520	
H3	0.007	0.453	
H2A	0.020	0.501	
(*h22* *vs.* *h19* (4))			0.473 (0.035)
H4	0.0	0.408	
H2B	0.050	0.504	
H3	0.007	0.455	
H2A	0.020	0.491	
ϕX174 (5, 7) *vs.* G4 (6)			
F-gene	0.231	0.677	
G-gene	0.400	0.719	0.704 (0.029)
H-gene	0.188	0.744	
A-gene[a]	0.201	0.676	
SV40 (8, 9) *vs.* BKV (10, 11) (early genes)			
t-antigen	0.178	0.731	
T1-antigen	0.164	0.755	0.742 (0.001)
T2-antigen	0.148	0.740	
Mouse Ig $\gamma 1$ (12) *vs.* $\gamma 2b$ (13)[b]			
Hinge	0.311	0.552	
CH2	0.178	0.495	0.525 (0.023)
CH3	0.269	0.528	
b) Comparison of K_Ss of distantly located genes[c]			
β-Globin[d]			
(human (14) *vs.* mouse (15, 16))	0.147	0.459	
(rabbit (17, 18) *vs.* mouse)	0.157	0.507	
α-Globin			
(human (19) *vs.* mouse (20))	0.083	0.518	
(rabbit (21) *vs.* mouse)	0.105	0.486	0.487 (0.020)
Growth hormone			
(human (22) *vs.* rat (23))	0.181	0.471	
Preproinsulin			
(human (24) *vs.* rat II (25, 26))	0.098	0.483	

[a] Non-overlapping region of A-gene.

[b] The CHl domain gene is evidently reduced in the values of K_S as well as K_A compared with the other three, suggesting that it might be derived from recombination events occurring within the intervening sequences during evolution (Miyata *et al.*, 1980b; for details, see the following paragraph).

[c] Considering the phylogenetic relationships among humans, rabbits, mice, and rats,

it seems reasonable to assume that these six pairs of sequences have the same divergence time.

[d] In the β-globin gene, there are unique segments (codons 22–35 and 92–107) where the nucleotide sequences are strongly conserved, and it is suggested that this is due to other extra constraints, presumably related to the presence of rigid base-pairing structures (Miyata and Yasunaga, 1980; for details, see the following paragraph). Thus the K_S values shown were estimated by excluding the corresponding regions.

The Average was taken of the K_Ss of gene pairs with the same divergence time.

Data were taken from (1) Schaffner *et al.*, 1978; (2) Sures *et al.*, 1978; (3) Grunstein and Grunstein, 1978; (4) Busslinger *et al.*, 1980; (5) Sanger *et al.*, 1977; (6) Godson *et al.*, 1978; (7) Sanger *et al.*, 1978; (8) Reddy *et al.*, 1978; (9) Fiers *et al.*, 1978; (10) Dhar *et al.*, 1979; (11) Yang and Wu, 1979; (12) Honjo *et al.*, 1979; (13) Yamawaki-Kataoka *et al.*, 1980; (14) Marotta *et al.*, 1977; (15) Konkel *et al.*, 1978; (16) Konkel *et al.*, 1979; (17) Efstratiadis *et al.*, 1977; (18) Van Ooen *et al.*, 1979; (19) Wilson *et al.*, 1980; (20) Nishioka and Leder, 1979; (21) Heindell *et al.*, 1978; (22) Martial *et al.*, 1979; (23) Seeberg *et al.*, 1977; (24) Bell *et al.*, 1980; (25) Lemedico *et al.*, 1979; Cordell, *et al.*, 1979.

This rate is very high, almost comparable to the rate of amino acid substitution in fibrinopeptides (Kimura, 1977) which is known to be the most rapidly evolving protein.

Using the average rate of synonymous substitution, $v_S=5.3\times10^{-9}$, as a clock, and knowing the K_S value of a pair of sequences compared, the divergence time T of the sequences is evaluated as $T=K_S{}^c/2v_S=-(3/4)\ln[1-(4/3)K_S]/2\times5.3\times10^{-9}$ (year). Estimated divergence times of several pairs of duplicated genes and species are listed in Table II. Miyata *et al.* (1981) showed that mouse β-globin major and minor genes had undergone at least two recombinations during their evolution, one of which occurred within the large intron and the other somewhere at the 3′ termini of the 5′-non-coding region, resulting in the transfer of a long continuous segment including the complete first and second *exons* from one gene to another. Considering that the K_S value of the third exon is larger than that of the remaining exons, it may be adequate to estimate T, the time when immediate gene duplication had taken place, from the K_S value of exon 3. That is, the two non-allelic mouse β-globin genes might have diverged from a common ancestor about 12 million years ago. Recently Miyata and Hayashida (1981b) applied this idea to human leukocyte interferon multigene family and determined the phylogenetic branching order of the eight distinct genes.

To determine the phylogenetic branching order of different species or strains from amino acid sequence data, it is usually necessary to use a protein performing the same function as that with a known evolutionary

TABLE II
Estimated Divergence Times (T) of Duplicated Genes (A) and Species Compared (B)

A) Paralogous comparison	Species	K_S	K_S^c	T (MY)
GH (1) *vs.* CS (2)	Human	0.105	0.113	10.7
Insulin I (3) *vs.* insulin II (3)	Rat	0.186	0.214	20.2
Ig kappa V-gene *κ2* (4) *vs. κ3* (5)	Mouse	0.080	0.085	8.0
κ2 (4) *vs. κ41* (6)	Mouse	0.459	0.710	67.0
Ig lambda V-gene *λI* (7) *vs. λII* (7)	Mouse	0.023	0.023	2.2
Ovalbumin (8) *vs. X* gene (8)	Chicken	0.444	0.671	63.3
β-Globin major (9) *vs.* minor (9)	Mouse	0.067	0.070	6.6
1st and 2nd exons		0.041	0.042	4.0
3rd exon		0.123	0.134	12.6
B) Orthologous comparison	**Gene**	K_S	Kc_S	T (MY)
P. milialis (10) *vs. S. purupuratus* (11)	Histone[a]	0.048	0.050	4.7
E. coli (12) *vs. S. marcescens* (13)	Lipoprotein	0.290	0.367	34.6
E. coli (14) *vs. S. typhimurium* (14)	*TrpA*	0.726	—	—

[a] The K_S values of four histone genes *H4*, *H2B*, *H3*, and *H2A* were averaged. $K_S^c = -(3/4)\ln[1-(4/3)K_S]$. Averaged evolutionary rate of synonymous substitution was assumed to be 5.3×10^{-9} (per site per year).

GH, growth hormone; CS, chorionic somatomammotropin.

Sequence data were taken from (1) Martial *et al.*, 1979; (2) Shine *et al.*, 1977; (3) Lemedico *et al.*, 1979; (4) Nishioka and Leder, 1980; (5) Seidman *et al.*, 1978; (6) Seidman *et al.*, 1979; (7) Bernard *et al.*, 1978; (8) Heiling *et al.*, 1980; (9) Konkel *et al.*, 1979; (10) Busslinger *et al.*, 1980; (11) Sures *et al.*, 1978; Grunstein and Grunstein, 1978; (12) Nakamura and Inoue, 1979; (13) Nakamura and Inoue, 1980; (14) Nichols and Yanofsky, 1979.

rate. In lower organisms, however, amino acid sequence data are often unavailable for such purposes. The characteristic feature that the rate of synonymous substitution is approximately constant regardless of the types of genes would be quite useful for determining such phylogenetic relationships in lower organisms. Though there is at present no firm evidence, it seems likely that synonymous sites in prokaryotic genes evolved at approximately the same rate as in genes of higher animals, except for some special genes and possibly most virus genes. Assuming the mean evolutionary rate of synonymous substitution be 5.3×10^{-9} per site per year in lower organisms as well as in higher organisms, divergence times T of different species or strains can be determined (Table IIB). A comparison of *trpA* gene sequences from *Escherichia coli* and *Salmonella typhimurium* shows an extremely large K_S value of 0.726. Though it is difficult to estimate the correct divergence time from such a large K_S

value, this result implies that *E. coli* and *S. typhimurium* diverged from a common ancestor at a very remote time.

A problem arises when the DNA sequence of histone genes from clone *pSp2/17* of *Strongylosentus purpuratus* is compared with that from clone *h22* of *Psammechinus milialis*. It is apparent that the K_S value between clone *pSp2/17* and clone *h22* from different species is far greater than that between clones *h19* and *h22* from the same species (see also Table I). This indicates, as Busslinger *et al.* (1980) have noted, that the evolutionary relationship between *h19* and *pSp2/17* appears to be orthologous (Fitch and Margoliash, 1970). Thus, the divergence time of *P. milialis* and *S. purpuratus* is estimated to be 4.7 million years. In spite of such recent divergence, they are each quite different in morphological shape. A morphological consideration of the sea urchins suggests that they were separated about 27–65 million years ago (Shigei, personal communication). If this time estimation is correct, the evolutionary rate of synonymous substitution in sea urchin histone genes is estimated to be $0.07-0.18\times10^{-9}$ (per site per year), which is approximately $1/5-1/10$ times lower than the value estimated for mammalian genes. There is evidence that the presence of a strong bias in degenerate codon utilization might have influence upon the variability at synonymous sites, thereby reducing the rate of synonymous substitution considerably (see the following paragraph). Sea urchin histone genes, however, appear not to be significantly different from globin genes and insulin genes in their codon utilization pattern, but rather are similar to each other. Which interpretation is correct remains to be determined.

EVOLUTIONARY CHANGE IN NON-CODING REGIONS

Miyata *et al.* (1980a) calculated the sequence differences (K_N) of various non-coding parts of mRNA for α-globin, β-globin, insulin, and growth hormone genes and showed that, except for the amino acid-altering sites of coding regions, each functional block of non-coding regions as well as the synonymous sites of coding regions are approximately uniform in their K_N value, regardless of the types of genes, if they are subject to no more additional constraints. They also showed that non-coding regions are classified into two distinct blocks with respect to the extent of sequence divergence. The intron and 5′ portion of the 3′-non-coding region show marked sequence divergence, being almost comparable to the extent of divergence at the synonymous sites, whereas the nucleotide sequences

TABLE III
Comparison of the Sequence Differences (K_N) of Various Noncoding Parts with That (K_S) in the Synonymous Sites of Coding Regions

	K_N					K_S
	Small intron	Large intron	5′-Non-coding	3′-Non-coding 5′ portion	3′-Non-coding 3′ portion	Synonymous site
β-Globin[a]						
Rabbit *vs.* mouse	0.313 (0.408)	0.529 (0.547)	0.231 (0.245)	0.480 (0.567)	0.225 (0.225)	0.507
Human *vs.* mouse	— (—)	— (—)	0.160 (0.192)	0.457 (0.506)	0.196 (0.196)	0.459
α-Globin						
Rabbit *vs.* mouse	— (—)	— (—)	0.235 (0.278)	0.509 (0.581)	0.219 (0.219)	0.486
Human *vs.* mouse	— (—)	— (—)	0.206 (0.270)	0.419 (0.419)	0.188 (0.188)	0.518
Preproinsulin						
Human *vs.* rat	0.398 (0.408)	0.493 (0.526)	0.327 (0.393)	0.438 (0.500)	0.191 (0.191)	0.483
Growth hormone						
Human *vs.* rat	— (—)	— (—)	0.321 (0.345)[b]	0.400 (0.451)	0.152 (0.177)	0.471
Mean	0.356 (0.408)	0.511 (0.537)	0.247 (0.287)	0.451 (0.504)	0.195 (0.199)	0.487
±S.D.	0.043 (0.000)	0.018 (0.011)	0.060 (0.066)	0.037 (0.058)	0.024 (0.017)	0.020

[a] Highly conserved segments consisting of codons 22–35 and 92–107, each of which involves within it a splicing point of an intron, are excluded from the analysis.

[b] The nucleotide sequence of this region is not established completely for both species and thus comparison is possible only for limited nucleotides near the initiation codon. Values estimated by Method 2 are shown in parentheses. When consecutive gaps with more than ten nucleotide sites were found in any one of the sequences compared, the corresponding sites were excluded from the calculation. The 3′-non-coding region was divided into two portions, the 5′ portion and the 3′ portion, depending on the extent of divergence. It should be noted that the divergence time of sequences compared is the same for the six pairs of genes listed above and therefore all the values could be compared directly between them.

of the 5′-non-coding region and the 3′ portion of the 3′-non-coding region are strongly conserved, with the K_N values being approximately one-half or so of the K_S value (see Table III; it should be noted that, considering the phylogenetic relationships among human, rabbit, and mouse (rat), the divergence time of the sequences compared is just the same among the six pairs and therefore all the values listed in this table can be compared

TABLE IV
Ratio of Difference (K_N) at the Non-coding Block to Difference (K_S) at Synonymous Site in Several Genes

	Small intron		Large intron		3′-Non-coding (5′ portion)		5′-Non-coding		3′-Non-coding (3′ portion)		Synonymous site
	K_N	K_N/K_S	K_N	K_N/K_S	K_N	K_N/K_S	K_N	K_N/K_S	K_N	K_N/K_S	K_S
Globin and hormone genes[a]											
Method 1	0.356	0.73	0.511	1.05	0.451	0.93	0.247	0.51	0.195	0.40	0.487
Method 2	0.408	0.84	0.537	1.10	0.504	1.03	0.287	0.59	0.199	0.41	
Ig C region genes[b] ($\gamma 1$ *vs.* $\gamma 2b$)											
Method 1	0.357	0.68			0.361	0.69			0.223	0.42	0.525
Method 2	0.465	0.89			0.371	0.71			0.263	0.50	
Preproinsulin genes[c] (gene *I* *vs.* *II*)											
Method 1							0.088	0.47	0.096	0.52	0.186
Method 2							0.088	0.47	0.113	0.61	
Histone genes (*H4* and *H2A*)[d] (clone *h22* *vs.* *pSp2/17*)											
Method 1							0.258	0.54			0.481
Method 2							0.334	0.69			
Mean Method 1		0.71		1.05		0.81		0.51		0.45	
Mean Method 2		0.87		1.10		0.87		0.58		0.51	

[a] The average values of K_N and K_S for genes listed in Table III were used.
[b] For the K_N of small introns, the values for the 5′-flanking intron, the 3′ half of the first intron, and the complete second and third introns were averaged, and the average value shown in Table I was used for the K_S.
[c] When the sequences of the 3′-non-coding regions are compared between rat preproinsulin *I* and *II* genes, this region appears to involve no diverging regions; thus the entire 3′-non-coding region was classified in the "3′ portion" in this table.
[d] K_Ns values for the 5′-non-coding region of *H4* and *H2A* were estimated by Method 1 (Method 2) as 0.258 (0.343) and 0.257 (0.325), respectively, and they were averaged. The K_S value corresponds to the mean value among the four genes (*i.e.*, *H4*, *H2b*, *H3*, and *H2A*), each of which is listed in Table I.

directly between them). This strong sequence preservation might be related to functional requirements for transcription and modification of mRNA.

To confirm the above results, further analysis was carried out for other genes including mouse immunoglobulin constant domain genes ($\gamma 1$ *versus* $\gamma 2b$), rat preproinsulin genes (gene *I versus II*) and histone genes (clone *h22 versus pSp2/17*). Since for each pair of homologous sequences the ratio K_N/K_S would be insensitive to the divergence time of the two sequences compared, if they are closely related each other, the value of K_N/K_S for a non-coding block of a certain gene pair can be compared directly with the corresponding value for another pair, regardless of the divergence time. The results clearly support the above findings (Table IV).

The 5′-non-coding region of globin genes (α-globin and β-globin) appears to be somewhat different in K_N value from hormone genes (preproinsulin and growth hormone) (Table III). This may be related to the effect of statistical fluctuation due to the short nucleotide length of this region. Alternatively, and presumably more plausibly, the nucleotide sequences around the "cap" are almost always conserved, while a short segment bordering the coding region shows considerable divergence. The nucleotide sequence of the diverging segment appears to differ in length from gene to gene. Thus the diverse values of K_N in the 5′-non-coding region may be due to the difference in size of the diverging segment involved.

Both globin and insulin genes contain two introns which are different in size. The small intron is about 150 nucleotides or so in length, while the large intron consists of more than 500 nucleotides. The large intron appears to be larger in K_N value than the small intron (see Table III). This may be due to at least two reasons. The large intron involves several long simple sequences like pyrimidine-rich or purine-rich sequences, which are not found in the small intron, and there is a large number of deletions in it, some of which are more than 100 nucleotides in length. It seems likely that the large intron is subject to constraint against sequence variation to a lesser extent than the small intron. Furthermore, as shown in globin genes (Van den Berg *et al.*, 1978; Leder *et al.*, 1978), sequence homology in introns is found only in the region bordering the structural gene sequence, which may be important in splicing (Lerner *et al.*, 1980), and the central portion diverges. Considering the fraction of the length of this homology region compared to that of the overall region, it is expected that the small intron is estimated to be smaller in the value of difference per site than the large intron.

As noted above, introns and the 5′ portion of the 3′-non-coding region show marked sequence divergence and considerable length heterogeneity, suggesting that, even if any functions are associated with them, they do not depend on a unique primary structure of the sequences and there is little constraint on the sequences except for some very limited nucleotides. Considering the large K_N values in these regions, which are nearly equal to or slightly lower than the K_S value in synonymous sites, the degree of constraint operating on these sequences might be almost comparable to that in the synonymous sites of coding regions. On the other hand, the strong sequence preservation found in the 5′-non-coding region and the 3′ portion of the 3′-non-coding region suggests the presence of important functions related to transcription and modification of mRNA and others. That the pattern of substitution and the extent of sequence divergence are similar among different genes, except for the amino acid-altering sites of protein coding regions, might imply that most functions associated with conserved sequences are very similar among different genes. Such sequence preservation and functional constraint relationship may be well explained by the neutral theory proposed by Kimura (1968).

RAPIDLY EVOLVING α-GLOBIN PSEUDOGENES

An interesting and unexpected finding has come from DNA sequencing of gene families that code for globins. The nucleotide sequence of an α-globin-like gene ($\alpha 3$) from BALB/c mouse has been determined by Nishioka *et al.* (1980). Furthermore a gene ($\psi\alpha 30.5$) equivalent to the $\alpha 3$ was also isolated from outbred Swiss mice and sequenced by Vanin *et al.* (1980). These α-globin-like genes (α-globin pseudogenes) are apparently related to the adult α-globin mouse gene, but they clearly lack two introns which are commonly present in other globin genes and moreover, they have several deletions and insertions relative to adult α-globin genes, resulting in frame shifts which prevent them from producing a functioning α-globin polypeptide (Nishioka *et al.*, 1980; Vanin *et al.*, 1980). Nishioka *et al.* (1980) proposed a plausible model for a mechanism that would permit the clean loss of either or both introns independently. On the basis of the complete absence of both introns and the widespread occurrence of globin pseudogenes in other species, Vanin *et al.* (1980) proposed an interesting hypothesi on the possible function of the α-globin pseudogene. Similar α- and β-globin pseudogenes which cannot code for normal polypeptides have been isolated and sequenced in several other mammalian species including human (Proundfoot and Maniatis, 1980), rabbit (Lacy and Miniatis, 1980), mouse (Jahn *et al.*, 1980), and goat (Haynes *et al.*, 1980).

The globin pseudogene would provide much insight into the processes of gene evolution, particularly the relationship between the extent of sequence divergence and functional constraint, since it lacks at least a function coding for an active polypeptide. Miyata and Yasunaga (1981) compared the nucleotide sequence of the mouse α-globin pseudogene $\psi\alpha 30.5$

with normal polypeptide-producing α-globin genes (productive genes) from mouse and rabbit, and showed that this gene evolves at an unusual rate which is higher than the rate of synonymous substitution in productive genes.

To investigate the properties of evolutionary changes in the globin pseudogene, it may be effective to consider a pair of genes both of which are productive genes from mouse and other species and another pair of genes involving the mouse pseudogene counterpart, and to compare the sequence differences between them. In this case, the difference in the K values between the two gene pairs directly reflects the difference of the extent of sequence variation between the productive gene and the corresponding mouse pseudogene. According to the idea, four α-globin genes including productive genes from mouse ($M\alpha$), rabbit ($R\alpha$), and human ($H\alpha$) and the mouse pseudogene ($M\psi\alpha$) were compared pairwise and the sequence differences at every functional block was calculated (Table V). The sequence difference $K(M\alpha\text{-}M\psi\alpha)$ between the $M\alpha$ and $M\psi\alpha$ is clearly smaller than the $K(R\alpha\text{-}M\psi\alpha)$ and $K(H\alpha\text{-}M\psi\alpha)$ along the entire region of the genes, suggesting that after the divergence of mouse, rabbit, and human, the $M\psi\alpha$ was derived from an ancestral α-globin gene by gene duplication in the mouse line. This also implies that, considering the phylogenetic relationship among $H\alpha$, $R\alpha$, $M\alpha$, and $M\psi\alpha$, the four gene pairs, $R\alpha\text{-}M\alpha$, $R\alpha\text{-}M\psi\alpha$, $H\alpha\text{-}M\alpha$, and $H\alpha\text{-}M\psi\alpha$ have the same divergence time as each other. This is very useful because the corresponding K values could be compared directly between the four pairs without using paleontlogical estimates of divergence times which are often subject to considerable uncertainty. Ratios $K(R\alpha\text{-}M\psi\alpha)/K(R\alpha\text{-}M\alpha)$ and $K(H\alpha\text{-}M\psi\alpha)/K(H\alpha\text{-}M\alpha)$ are therefore particularly suggestive for the present purpose, and they are also included in Table V.

Examination of the ratios reveals interesting features in the nucleotide substitution of the pseudogene. 1) The values of the ratios are always larger than unity, regardless of functional blocks, implying that the $M\psi\alpha$ evolves at a higher rate than the productive gene $M\alpha$. 2) As noted before, the eukaryotic mRNA sequence consists of two distinct blocks with respect to the extent of sequence divergence, that is, strongly constrained blocks including the amino acid sites, the 5′-non-coding region, and the 3′ portion of the 3′-non-coding region, and less constrained blocks including the synonymous sites, the 5′ portion of the 3′-non-coding region, and the intron. Table V clearly shows that the ratios are always larger in the strongly constrained blocks than in the less constrained blocks. This feature is

TABLE V
Comparison of the Sequence Difference of Pair: Productive Gene-Pseudo Gene with That of Pair: Productive Gene-Productive Gene

	Sequence difference per site				
	Strongly constrained blocks			Less constrained blocks	
	Amino acid site	5′-Non-coding	3′-Non-coding (3′ portion)	3′-Non-coding (5′ portion)	Synonymous site
1) $K(M\alpha\text{-}M\psi\alpha)$	0.148	0.059 (0.059)	0.219 (0.219)	0.210 (0.210)	0.298
2a) $K(R\alpha\text{-}M\psi\alpha)$	0.225	0.294 (0.333)	0.344 (0.344)	0.604 (0.661)	0.530
2b) $K(R\alpha\text{-}M\alpha)$	0.105	0.235 (0.278)	0.219 (0.219)	0.501 (0.581)	0.486
3a) $K(H\alpha\text{-}M\psi\alpha)$	0.203	0.265 (0.324)	0.313 (0.313)	0.500 (0.500)	0.558
3b) $K(H\alpha\text{-}M\alpha)$	0.083	0.206 (0.270)	0.188 (0.188)	0.419 (0.419)	0.518
4a) $K(RH\alpha\text{-}M\psi\alpha)$[a]	0.214	0.280 (0.329)	0.329 (0.329)	0.552 (0.581)	0.544
4b) $K(RH\alpha\text{-}M\alpha)$[b]	0.094	0.221 (0.274)	0.204 (0.204)	0.460 (0.500)	0.502
Ratio					
$K(R\alpha\text{-}M\psi\alpha)/K(R\alpha\text{-}M\alpha)$	2.14	1.25 (1.20)	1.57 (1.57)	1.21 (1.14)	1.09
$K(H\alpha\text{-}M\psi\alpha)/K(H\alpha\text{-}M\alpha)$	2.45	1.29 (1.20)	1.66 (1.66)	1.19 (1.19)	1.08
$K(HR\alpha\text{-}M\psi\alpha)/K(HR\alpha\text{-}M\alpha)$	2.28	1.27 (1.20)	1.61 (1.61)	1.20 (1.16)	1.08

[a] $K(RH\alpha\text{-}M\psi\alpha)$ is an average difference of $K(R\alpha\text{-}M\psi\alpha)$ and $K(H\alpha\text{-}M\psi\alpha)$, that is, $K(RH\alpha\text{-}M\psi\alpha)=(K(R\alpha\text{-}M\psi\alpha)+K(H\alpha\text{-}M\psi\alpha))/2$.

[b] $K(RH\alpha\text{-}M\alpha)=(K(R\alpha\text{-}M\alpha)+K(H\alpha\text{-}M\alpha))/2$.

$M\alpha$, $R\alpha$, and $H\alpha$, mouse, rabbit, and human α-globin gene, respectively; $M\psi\alpha$, mouse α-globin pseudogene. $K(M\alpha\text{-}M\psi\alpha)$, for example, stands for the sequence difference per site for gene pair $M\alpha$ and $M\psi\alpha$. It should be noted that since $M\psi\alpha$ was apparently derived from an ancestral α-globin gene in the mouse line, the four gene pairs, $R\alpha\text{-}M\alpha$, $H\alpha\text{-}M\alpha$, $R\alpha\text{-}M\psi\alpha$, and $H\alpha\text{-}M\psi\alpha$ have the same divergence time as each other. Values estimated by Method 2 are in parentheses.

particularly significant in the amino acid sites where the sequence is strongly conserved in many genes. 3) The values of the ratios are larger than unity even in the synonymous sites which evolve at the highest known rate in most productive genes. These three points strongly suggest that the α-globin pseudogene evolves at an unusual rate which is higher than the rate of synonymous substitution in productive genes and is almost uniform throughout the entire region regardless of functional blocks. The significant increase in the values of the ratios in the highly conserved blocks suggests the disappearence of functional constraint in the pseudogene sequence. It seems very likely that the entire nucleotide sequence of the pseudogene serves almost no important function, enabling it to escape the selective presures that operate to preserve the sequence and thereby allow the accumulation of several deletorious mutations which would otherwise be excluded from a population. Similar results are obtained for other globin pseudogenes (Kimura, 1980; Miyata and Hayashida, 1981a; Li *et al.*, 1981; Takahata and Kimura, 1981). Although the pseudogene appears to have no functional constraint against sequence variation throughout the entire region, the values of $K(R\alpha\text{-}M\psi\alpha)$ and $K(H\alpha\text{-}M\psi\alpha)$ are sill smaller in the strongly constrained blocks than in the less constrained blocks, for which the following three factors should be noted. 1) One of the genes compared is a productive gene (*i.e.*, $R\alpha$ or $H\alpha$). 2) As noted by Miyata and Yasunaga (1981), there is a possibility that the present-day pseudogene was not necessarily a functionless gene at the beginning, but functioned normally in some ancient period. 3) The $K(R\alpha\text{-}M\psi\alpha)$ of $K(H\alpha\text{-}M\psi\alpha)$ involves substitutions accumulated in a period between the time since mouse and rabbit or human separated and the time before $M\alpha$ and $M\psi\alpha$ divergene in the mouse lineage, at which substitution would be normal in pattern as in productive genes.

The above notion that the evolutionary rate of the globin pseudogene exceeds the rate of synonymous substitution in productive genes is particularly interesting in relation to the neutral theory proposed by Kimura (1968). According to the theory, the possibility of a mutational change being neutral depends on functional constraints. That is, the weaker the constraint, the larger the probability of random change being selectively neutral (Kimura, 1977). This leads to a conclusion that the maximum evolutionary rate is attained when functional constraints are completely absent. This limiting rate might be close to the rate of the pseudogene where functional constraint seems to have disappeared almost completely throughout the entire sequence.

BIAS IN CODE WORD UTILIZATION AND EVOLUTIONARY RATE OF SYNONYMOUS SUBSTITUTION

The accumulation of nucleotide sequence data for mRNAs from various sources has revealed the presence of distinct biases in the use of degenerate codons (see, *e.g.*, Grantham *et al.*, 1980). Several arguments have already been noted on the primary factors determining the bias of codon utilization (see Miyata *et al.*, 1979b). Although the functional significance of specific codon utilization is still not understood fully, it might be related to selective constraint against synonymous codon changes, thereby reducing the rate of synonymous substitution to some extent. There is some evidence showing that the evolutionary rate of synonymous substitution appears to be influenced by constraint due to bias in codon utilization directly or indirectly.

From the comparison of the codon utilization pattern between α- and β-globin genes from rabbit, Salser (1978) pointed out that there are significant differences in the codon utilization between them. That is, C-ending codons are used more frequently in the α-globin gene than in the β-globin gene. Miyata and Hayashida (1981a), however, showed that this difference does not necessarily correspond to the difference in gene types, but rather depends on species. In chickens, the β-globin gene is rather similar in codon utilization pattern to the α-globin gene and it shows a strong preference for the use of C and/or G in the third position of codons, which is a common feature in α-globin genes of mammals and chickens. This trend is most remarkable in human and rabbit α-globin genes. From the pairwise comparison of α-globin genes and β-globin genes from human, rabbit, mouse, and chicken, Miyata and Hayashida (1981a) also showed that the evolutionary rate of synonymous substitution is negatively correlated with the degree of bias in codon utilization. That is, the evolutionary rate of synonymous substitution is highest in the β-globin gene of mammals among the eight globin genes compared. Chicken α- and β-globin genes and mouse α-globin gene evolve at a moderate rate. The most slowly evolving genes are human and rabbit α-globin genes, which show the strongest preference for the use of C- and/or G-ending codons. Combining the above two results, they suggested that, though evolutionary rate of synonymous substitution is approximately constant, it is slightly influenced by the difference in codon utilization. This also implies the presence of constraint against synonymous codon changes due to the bias in codon utilization.

As noted previously, pseudogenes undergo accumulation of mutations at a rate higher than that at the synonymous sites of normally functioning genes, suggesting that, even if the sequences have some biological functions, they don not depend on the unique primary structure of the sequences. Comparison of codon utilization of the pseudogene with that of other α- and β-globin genes reveals that the pseudogene shows an increasing tendency to use U- and/or A-ending codons. This result supports the above notion. (Miyata and Hayashida, 1981a).

The yeast glyceraldehyde-3-phosphate dehydrogenase (G3PDH) gene

TABLE VI
Codon Usage in G3PDH *pgap491* (DH1), G3PDH *pgap63* (DH2), Cytochrome *c* (CT),

		DH1	DH2	CT	α	β			DH1	DH2	CT	α	β
Phe	UUU	0	0	5	3	12	Ser	UCU	13	11	4	11	7
	UUC	10	11	3	19	11		UCC	12	14	1	9	7
Leu	UUA	0	0	3	0	0		UCA	0	0	2	0	0
	UUG	21	20	8	3	2		UCG	0	0	1	0	0
Leu	CUU	0	0	1	2	1	Pro	CCU	0	1	4	5	10
	CUC	0	0	0	7	9		CCC	0	0	0	12	1
	CUA	0	1	1	1	1		CCA	12	10	5	0	3
	CUG	0	0	0	39	40		CCG	0	0	0	3	0
Ile	AUU	9	7	5	1	2	Thr	ACU	12	10	4	2	8
	AUC	11	12	3	4	2		ACC	12	13	4	26	8
	AUA	0	0	1	0	1		ACA	0	0	5	0	1
Met	AUG	6	7	5	4	4		ACG	0	0	4	0	0
Val	GUU	22	23	2	2	9	Ala	GCU	25	24	5	10	20
	GUC	15	10	1	5	6		GCC	7	10	6	32	24
	GUA	0	0	1	2	0		GCA	0	0	2	1	3
	GUG	0	0	2	24	35		GCG	0	0	1	11	1

provides clear evidence for the presence of a relationship between the degree of codon selectivity and the evolutionary rate of synonymous substitution. Genomic DNA sequencing of two non-allelic yeast G3PDH genes (*pgap491* and *pgap63*) reveals unusual codon utilization in these genes (Holland and Holland, 1979; 1980). Table VI shows the codon usage in the two G3PDH genes together with those in three other genes for comparison. It is apparent that yeast G3PDH genes are heavily biased in the use of codons compared with the other three. For instance, all amino acids are coded by at most two codons, and eleven amino acids almost exclusively use only one codon out of two, four, or six in both *pgap491* and *pgap63*. It is very likely that the rate of synonymous substitution is considerably reduced in such heavily biased genes. This is really the case. Comparison between the DNA (sequences of *pgap491* and *pgap63*) revealed an unusually low K_S value (Table VII). Although the value of K_S is still larger than that of K_A in the coding region, it is, however, remarkably small compared with other flanking non-coding regions. At present, the cap site and poly(A) addition site have not been determined for both genes. However, considering the extent of sequence divergence, a region corresponding to the tentatively defined "3′ portion of the 5′-flanking region" (see Table VII) may be a candidate for the trans-α-Globin (α), and β-Globin (β) mRNAs

		DH1	DH2	CT	α	β			DH1	DH2	CT	α	β
Tyr	UAU	0	0	5	4	3	Cys	UGU	2	2	4	0	4
	UAC	11	10	5	5	6		UGC	0	0	1	3	1
Stp	UAA	—	—	—	—	—	Stp	UGA	—	—	—	—	—
	UAG	—	—	—	—	—	Trp	UGG	3	3	2	3	6
His	CAU	0	0	5	4	8	Arg	CGU	0	0	0	4	0
	CAC	8	8	2	28	19		CGC	0	0	0	0	0
Gln	CAA	5	6	2	0	0		CGA	0	0	0	0	0
	CAG	0	0	3	3	10		CGG	0	0	0	2	1
Asn	AAU	0	0	3	1	8	Ser	AGU	0	0	1	2	7
	AAC	12	14	11	9	12		AGC	0	0	0	14	3
Lys	AAA	1	2	13	5	7	Arg	AGA	11	11	5	0	0
	AAG	25	24	19	29	27		AGG	0	0	1	3	8
Asp	GAU	9	7	5	5	12	Gly	GGU	25	26	16	6	11
	GAC	16	13	4	19	8		GGC	0	0	3	18	21
Glu	GAA	12	15	7	7	7		GGA	0	0	2	1	3
	GAG	2	0	5	8	16		GGG	0	0	3	2	2

scribed 5′-non-coding region where sequences usually show strong conservation for most genes (see Tables III and IV). If one assigns this segment to the 5′-non-coding region, then the ratio $K_{5'noncoding}/K_S$ nearly equals unity, which is in contrast with the standard case in which the corresponding value is about one-half or so.

More qualitatively, using the mean K_N values 0.247 (Method 1) and 0.287 (Method 2) for the 5′-non-coding region listed in Table III, and assuming $T=75$ million years, the time since the separation of primates and rodents, one obtains the standard evolutionary rate ($\bar{v}_{5'noncoding}$) at the 5′-non-coding region as $\bar{v}_{5'noncoding}=2.0-2.4\times10^{-9}$ (per site per year). Assuming that the 5′-non-coding region of yeast G3PDH genes evolves at the same rate as the $\bar{v}_{5'noncoding}$, the divergence time of *pgap491* and *pgap63* genes can be estimated from the $K_{5'noncoding}$ value in Table VI, and thus the evolutionary rate of synonymous substitution in yeast G3PDH genes is obtained to be $1.8-2.4\times10^{-9}$ (per site per year). This value is about 1/3–1/2 of the standard rate of synonymous substitution (*i.e.*, 5.3×10^{-9}). It is apparent that such a large reduction in the evolutionary rate of synonymous substitution could be attributed to the heavily biased utilization of codons in yeast G3PDH genes.

It is noteworthy that such a strong bias in codon usage might reduce

TABLE VII
Sequence Divergence between Two Non-allelic G3PDH Genes (*pgap491* and *pgap63*) of Yeast

	K
Coding region	
Synonymous site	0.142
Amino acid site	0.026
Non-coding region	
5′-Flanking sequence (overall)	0.341 (0.360)
5′ portion	0.463 (0.482)
3′ portion	0.140 (0.157)
3′-Flanking sequence (overall)	0.325 (0.364)

The 5′-flanking sequence was divided into two portions, the 5′ portion and the 3′ portion, with respect to the extent of sequence divergence. K, sequence difference per site. For non-coding regions, K was estimated by two different methods, Method 1 and Method 2 (enclosed in parentheses).

the variability not only at the synonymous sites but also at the amino acid altering sites, though to a lesser extent. For instance, single-step nucleotide changes at the argine codon AGA lead to codons coding for different amino acids, *i.e.*, AUA(Ile), ACA(Thr), AAA(Lys), UGA(Stop) and GGA(Gly), whose use is prohibited completely or is infrequent in both G3PDH genes. This is also the case with the prolin codon CCA. A model calculation shows that a hypothetical gene with the same codon utilization pattern as G3PDH yeast genes is appreciably lower in its degree of variability both at the synonymous and amino acid sites than other genes with a pattern similar to cytochrome *c* and α- and β-globin genes (Miyata *et al.*, unpublished results).

In conclusion, although the rate of synonymous codon substitution is very high and approximately constant among different types of genes, it is slightly influenced by the codon utilization pattern, depending on the degree of bias in degenerate code word usage.

UNIQUE HOMOLOGY REGION RELATED TO SPECIFIC STRUCTURE; AND FUNCTION INVOLVED

We have seen in the preceding paragraphs that synonymous codon substitution occurs at an approximately constant rate among different types of genes, irrespective of their relative location on the chromosome. This naturally leads us to expect that when homologous nucleotide sequences are compared, codon positions occurring with synonymous changes are usually found to be distributed uniformly throughout the entire coding region. For instance, when the entire coding sequence is divided into several segments each of which consists of 50 codons or so, the K_S value of each segment is expected to be almost the same as that of other segments, somewhat fluctuating around the mean. In some genes, however, a certain segment R within the coding region appears to differ appreciably in the extent of sequence divergence from the remaining ones ($\bar{R}$). If amino acid difference is found to be reduced significantly in region R compared with other regions $\bar{R}$, whereas synonymous differences are almost identical between them (*i.e.*, $K_S(R)=K_S(\bar{R})$ and $K_A(R)<K_A(\bar{R})$, where $K_S(R)$ and $K_S(\bar{R})$ stand for synonymous differences of R and $\bar{R}$, respectively, and similarly, for amino acid differences), it is highly probable that segment R codes for a polypeptide region that is critical for the functioning protein. A case where both the K_S and K_A are reduced to a larger extent in segment R than in $\bar{R}$ (*i.e.*, $K_S(R)<K_S(\bar{R})$ and $K_A(R)<K_A(\bar{R})$) is especially interesting. This substitutional trend can be understood at the mRNA level or DNA level but not at the protein level alone (Salser, 1978; Miyata and Yasunaga, 1980).

Usually the extent of sequence divergence between homologous nucleotide sequences may be determined by at least two distinct factors: functional constraint imposed on the sequences and the time since divergence of the two sequences compared. The mosaic structure in the substitution pattern along the sequence may be derived from the above factors. The presence of a unique sequence, base-pairing structure, and overlapping genes in the R segment actually reduces sequence divergence, depending on the degree of selective pressure operating on the structures. On the other hand, a recombination event which has taken place between non-allelic homologous genes during evolution gives rise to a difference in divergence time of segments before and after the cross-over point, resulting in a unique mosaic pattern in sequence divergence when the non-allelic genes are compared. In this paragraph, the former case was discussed and the latter case will be discussed in detail in the following paragraph.

1. *Secondary Structure in Coding Region of mRNA and Substitution Pattern*

It has been shown that MS2 phage RNA has an extensive secondary structure throughout the genome (Min-Jou *et al.*, 1972; Fiers *et al.*, 1975; Fiers *et al.*, 1976). Miyata and Yasunaga (1980) compared the nucleotide sequence of the MS2 phage with that of closely related phage R17. On the basis of the secondary structure proposed by Fiers *et al.*, they calculated K_S and K_A in the base-pairing sites and those in the non-pairing sites separately and showed that both substitutions leading to amino acid change and synonymous codon change are observed less frequently in the pairing sites than in the non-pairing sites. Furthermore, in the base-pairing sites, amino acid substitution is reduced to a larger extent than synonymous substitution.

A simple interpretation of this pattern of substitution may be (Miyata and Yasunaga, 1980) that a segment involving a secondary structure would be subject to a constraint that prefers to conserve the already existing base pairings. Transversion mutations always disrupt base-pairing and therefore would be excluded. On the contrary, transition mutations preserve base-pairing more than 50%, if wobble pairings are accepted. According to the genetic code, of the transversion mutations at the third codon position, on the average about one-half result in synonymous codon changes. On the other hand, most transition mutations (95%) at the same position cause synonymous codon changes. As most mutations at the first and second positions lead to amino acid changes, a single transition mutation per codon results in synonymous changes with a probability of $\sim 1/3$, and a single transversion mutation per codon results in the corresponding changes with a probability of $\sim 1/6$. Thus, the transversion mutations contribute to the synonymous changes with a weight of $\sim 1/2$ $(=(1/6)/(1/3))$ relative to the transition mutations, whereas the corresponding weight for amino acid changes exceeds unity $(=(5/6)/(2/3)=5/4)$. Assuming that transversion mutations are totally excluded in the base-pairing sites, both K_S and K_A are reduced in these sites and K_A is reduced to a larger extent than K_S.

From the above considerations, it is apparent that if the degree of functional constraint other than that due to the presence of a secondary structure is the same between regions R and $\bar{R}$ and if region R involves within it an extensive secondary structure, then, the relation $1 > K_S(R)/K_S(\bar{R}) > K_A(R)/K_A(\bar{R})$ is expected to hold. That is, in region R, amino acid-altering sites are reduced more significantly in sequence divergence than synonymous codon sites, relative to the corresponding divergence in the other $\bar{R}$ regions where no secondary structures are present. From a comparison between human and rabbit β-globin sequences, Salser (1978) pointed out the presence of conserved segments where a significant lack of base substitutions is observed between them. Miyata and Yasunaga (1980) tentatively assigned these segments to regions involving codons

14–39 and 80–107, both of which interestingly involve splicing points of introns, and showed that in these segments the value of K_A is reduced more significantly than that of K_S, suggesting the presence of extensive secondary structures.

2. *Pattern of Substitution in Overlapping Genes*

Genomes of ϕX174 and G4 phages contain several overlapping genes (Barrell *et al.*, 1976; Sanger *et al.*, 1977; Godson *et al.*, 1978), in which the same stretch of nucleotide sequence can code for two proteins which are translated in different reading frames. The gene-within-gene phenomenon was also observed in several other viruses including single-stranded RNA phages MS2 (Atkins *et al.*, 1979) and f2 (Beremand and Blumenthal, 1979), λ phage (Davies, 1980), papova viruses SV40 (Contreras *et al.*, 1977), BKV (Yang and Wu, 1979) and polyoma (Deininger *et al.*, 1979; Soeda *et al.*, 1980), and influenza virus (Porter *et al.*, 1980; Lamb and Lai, 1980). Presumably such a genetic coding economy system may be found commonly in viruses where the total amount of DNA is severely limited by the physical size of the capsid in which the DNA is packaged.

The nucleotide sequence of overlapping segments is subject to both selective constraints operating on the two proteins which are translated in different reading frames. The sequence divergence would therefore be reduced more significantly in overlapping genes than in non-overlapping genes in which the sequence is read in only one reading frame. Miyata and Yasunaga (1978) developed a theory for treating the evolutionary rate in overlapping genes in which a parameter $\rho_S(\rho_A)$ was introduced, representing the rate of synonymous (amino acid) substitution in overlapping genes, relative to the corresponding rate in non-overlapping genes, and showed that the relation $1>\rho_A>\rho_S$ holds in most genes. If the same stretch of sequence of region R carries the information for producing two proteins with entirely different amino acid sequences and the sequence of the remaining $\bar{R}$ regions can code for only one protein, then, this result leads to the relation: $1>K_A(R)/K_A(\bar{R})>K_S(R)/K_S(\bar{R})$. That is, in region R containing overlapping genes the extent of reduction is larger in K_S than in K_A, which is quite the reverse of the case where region R involves an extensive secondary structure. This pattern of substitution in overlapping genes is confirmed from a comparison between ϕX174 and G4 sequences (Miyata and Yasunaga, 1980) and between papova viruses SV40 and BKV (Yasunaga and Miyata, 1981).

In conclusion, when the rates of synonymous and amino acid substitutions in a certain region are compared with the corresponding rates in other regions, it is expected that, if the region contains overlapping genes, the rate of synonymous substitution is reduced to a larger extent than that of amino acid substitution, and conversely, if the sequence of the region forms an extensive secondary structure, the rate of synonymous substitution is reduced to a lesser extent than that of amino acid substitution.

RECOMBINATION WITHIN AN INTRON BETWEEN NON-ALLELIC DUPLICATED GENES DURING EVOLUTION

In the preceding paragraphs, we have examined the sequence divergence-functional constraint relationship in some detail. A stretch of a DNA sequence carrying an important function preserves its sequence throughout evolution. Thus when a pair of homologous nucleotide sequences are compared, the sequence of the corresponding region is observed to be highly homologous compared with other functionally less important regions. The reverse is not always true. That is, even if some region shows strong sequence preservation, it is not necessarily implied that the region carries some important function. Usually, the extent of sequence divergence is determined by at least two factors: functional constraint and the time since divergence of the sequences compared. Consider a case where a comparison is made between non-allelic duplicated genes which were derived from a common ancestor by gene duplication and furthermore have undergone recombination between them during evolution. It is apparent that in this case the extent of sequence divergence between them varies discontinuously at a point where a homologous but unequal cross-over has taken place, since the divergence time between the sequences of genes compared is different before and after the cross-over point.

A way to determine whether the duplicated genes, say a_1 and a_2, of a species A have undergone recombination during evolution may be to introduce a third gene, say b, from a different species B which is equivalent to a_1 or a_2 in function and to compare the sequences between the three pairwise. Except for some very special cases, recombination between genes from different species can be excluded from consideration. This leads to the suggestion that if sequence homology is found in some region in the comparison between b and a_1 or a_2, it could be attributed to the effect due to functional constraint. Consider a case where when the duplicated genes a_1 and a_2 are compared a unique sequence homology is found in some region compared with other regions. If a similar sequence homology is also observed in the equivalent region in comparison between a_1 or a_2 with the other gene b from a different species, then, the unique homology found between a_1 and a_2 might be derived from functional constraint operating on the sequence of the region. On the contrary, if the region in question shows no similar homology between a_1 and b or a_2 and b, then the unique homology region in a_1 and a_2 might be derived from recombination between them during evolution.

As noted previously, each eukaryotic gene is divided into two distinct blocks with respective to the extent of sequence divergence. If no recombinations occurred between a_1 and a_2, the substitution pattern between them must be quite similar to that between a_1 and b or a_2 and b. If the duplicated genes a_1 and a_2 were derived from gene duplication in the lineage of species A after the divergence of species A and B, the value of $K(a_1-a_2)$, or the sequence difference between a_1 and a_2, must be small relative to the value of $K(a_1-b)$ or $K(a_2-b)$, though their absolute values vary with the different blocks considered. It may be especially useful for detection of recombination points occurring between a_1 and a_2 to introduce a parameter $R(=K(a_1-a_2)/K(a_1-b)$ or $K(a_1-a_2)/K(a_2-b))$. It should be noted that in this case both the ratios are almost the same in value. It is important that, though each component of R (*i.e.*, $K(a_1-a_2)$, $K(a_1-b)$, and $K(a_2-b)$) varies its value depending on the different blocks, the R value itself is expected to be insensitive to the degree of functional constraint operating on the sequence of each block. Thus if no recombinations occurred between a_1 and a_2, the R value would be almost uniform throughout the entire region, somewhat fluctuating around the mean. A clear-cut example comes from a comparison among two non-allelic insulin gene sequences from the rat and a human insulin gene sequence. The estimated R values are almost constant along the entire region of these genes, suggesting that no recombinations have taken place between the non-allelic rat insulin genes during evolution (Miyata *et al.*, 1981). Quite the contrary, if R is found to change its value significantly at some point, it seems likely that a_1 and a_2 have undergone recombination near the point between them during evolution. From a close examination of the R values calculated from sequences of non-allelic mouse β-globin major and minor genes and rabbit β-globin genes, Miyata *et al.* (1981) suggested that more than two unequal cross-overs (or a double unequal cross-over) have taken place during the evolution of mouse β-globin major and minor genes, which give rise to a significant sequence homology in the stretch of sequence from the termini of the 5′-non-coding region to the first half of the large intron between them.

Immunoglobulin γ subclass genes consist of four related genes in mouse (Honjo and Kataoka, 1978). At present, the three genes, $\gamma 1$, $\gamma 2b$, and $\gamma 2a$ have already been cloned and sequenced (Honjo *et al.*, 1979; Yamawaki-Kataoka, 1980; Yamawaki-Kataoka, 1981). The nucleotide sequences of the three genes have demonstrated that three introns interrupt the coding sequence at a precisely equivalent position among the three, separating three domains (CH1, CH2, and CH3) and a hinge region, each of which codes for a polypeptide that is a functional constituent of the entire immunoglobulin molecule (Sakano *et al.*, 1979). The nucleotide sequences of γ subclass genes share considerable homology, suggesting that they have evolved from a common ancestor by gene duplication.

The $\gamma 1$ and $\gamma 2b$ genes were the first in which the phenomenon of a double unequal cross-over within different introns was recognized, which gives rise to the transfer of a DNA segment involving a complete CH1 exon and its neighborhood from one gene to another (Miyata *et al.*, 1980b). Miyata *et al.* (1980b) compared the nucleotide sequences of the constant portion of mouse $\gamma 1$ and $\gamma 2b$ genes and found a remarkably conserved segment including the entire CH1 and the first half of the first intron. In the CH1 domain (exon), both K_S and K_A are appreciably smaller in value than those of the other exons and the K_S of CH1 and the $K_{5'half\ of\ 1st\ intron}$ are almost the same in value, whereas the synonymous sites of the other domains and introns together with the 3′ half of the first intron show considerable divergence with values of K being almost the same between them, which is consistent with the observed pattern in other eukaryotic genes. To determine whether such a long homology region was derived by recombination or functional constraint, they carried out a further comparison using an amino acid sequence of the γ subclass chain from different species, human, and concluded that during the evolution of $\gamma 1$ and $\gamma 2b$ genes a double unequal cross-over event took place in the different introns, resulting in the complete transfer of CH1 and its neighborhood. From a comparison of nucleotide sequences of mouse $\gamma 1$, $\gamma 2b$, and $\gamma 2a$ genes, Yamawaki-Kataoka *et al.* (1981) suggested that such an "intron-mediated exon transfer" event took place more than twice during evolution. As noted above, mouse β-globin major and minor genes present a clear example of such a process. Similar events were also observed in human fetal $^{G}\gamma$- and $^{A}\gamma$-globin genes (Slightom *et al.*, 1981) and in mouse immunoglobulin κ-type variable domain genes (Hayashida *et al.*, 1981). Tsuzukida *et al.* (1979) reported the amino acid sequence of human immunoglobulin $\alpha 2$ allotype *A2m(1)* which appears to be a hybrid that is identical to $\alpha 1$ in the CH3 domain and identical to *A2m(2)* in the CH1 and CH2 domains. This suggests a recombination occurring within the third intron. Similar hybrid immunoglobulin molecules produced by recombination between human $\gamma 3$ and $\gamma 1$ heavy chains (Werner and Steinberg, 1974) and between $\gamma 4$ and $\gamma 2$ chains (Natvig and Kunkel, 1974) have already been reported. These data suggest that recombination within introns is not an extremely rare event but one which could be reasonable anticipated during the evolution of duplicated genes.

Recombination within segments showing strong sequence divergence such as the intron and the 5′ portion of the 3′-non-coding region may play an important role in the evolution of duplicated genes. The presence of an intron may be of special importance for such processes, because it is considerably large in length, with some introns exceeding 500 nucleotides (Konkel *et al.*, 1978; Konkel *et al.*, 1979; van Ooyen *et al.*, 1979; Lomedico *et al.*, 1979; Bell *et al.*, 1980), thereby increasing the recombination rate. A single recombination event results in a hybrid gene which is different in many sites from the original one. If recombination takes place within a functionally important block, such as a polypeptide coding region, the resulting saltatory change might largely be disadvantageous for the structure and function of the protein molecule and would necessarily be eliminated from the population. Furthermore, such an event would rarely be accepted, because it would ordinarily be required to preserve the same reading frame as before. The above two points however would not be a problem when recombination takes place within an intron. As noted previously, introns show strong sequence divergence and considerable length heterogeneity, suggesting that even if any function is associated with them, they do not depend on a unique primary structure of the sequence except for very limited nucleotides. Furthermore, recombination within introns results in the complete transfer of the exon(s) from one of the duplicated genes to the corresponding position of the other. As already shown in immunoglobulin heavy chain genes (Sakano *et al.*, 1979) and β-globin genes (Eaton, 1980; Craik *et al.*, 1980), each exon codes for a polypeptide that is a functional constituent of the entire protein molecule. Thus the exchange of functional units of one protein for the corresponding homology units of another might be possible without seriously influencing the functions of the translation product of the gene. Through the process of recombination within introns, it might therefore be possible to reach a new gene with a multiply-changed structure by one event or so, accompanied by little influence upon both gene activity and protein function. These considerations lead to the inference that, at least in duplicated genes, recombination within introns might largely be selectively neutral (Kimura, 1968; Kimura and Ohta, 1974) and thus hybrid genes would be produced at a considerable frequency during evolution (Miyata *et al.*, 1980b; Miyata *et al.*, 1981).

Recent advances in two related technologies, gene cloning and rapid DNA sequencing, are now revealing the presence of many duplicated genes hitherto unknown. Our growing knowledge on the structure and function of eukaryotic genes together with the finding of drastic changes in gene structure mediated by recombination within introns provide an opportunity to open up new areas for investigating the processes of more dynamic and flexible evolution of genes.

SUMMARY

Growing knowledge of the primary structure of genes has provided an opportunity to investigate the processes of gene evolution at the DNA level. Here some recent work on the evolution of eukaryotic genes has been reviewed. The extent of sequence divergence was examined for structurally and/or functionally different parts of genes, and the relation between the evolutionary rate and the degree of functional constraint was discussed in detail. It was pointed out that the rate of change between synonymous codons is not only very high but also uniform among different types of genes and that the uniform rate for different genes is particularly useful for determining the evolutionary branching order of duplicated genes. From a comparison of a mouse α-globin pseudogene sequence with its functional counterparts, it was shown that the pseudogene evolves at an extraordinary rate which is higher than the rate of change between synonymous codons in functional genes. This strongly suggests the presence of functional constraint against changes between synonymous codons. Based on examinations of codon utilization patterns and the extent of divergence at synonymous codon altering sites of coding regions, it was pointed out that even synonymous codon changes are not completely free from selective pressure but are partially constrained, although presumably weakly in most cases, depending on the degree of bias in code word usage. Special substitution patterns that are correlated with the presence of a unique structure and/or function involved in segments of certain types of genes were also discussed. Finally, an unusual substitution pattern due to homologous but unequal cross-overs within introns was considered in some detail.

REFERENCES

Atkins, J. F., Steitz, J. A., Anderson, C. W., and Model, P. (1979). *Cell* **18**, 247–256.

Barrell, B. G., Air, G. M., and Hutchison III, C. A. (1976). *Nature* **264**, 34–41.

Bell, G. I., Pictet, R. L., Rutter, W. J., Cordell, B., Tischer, E., and Goodman, H. M. (1980). *Nature* **284**, 26–32.

Beremand, M. N. and Blumenthal, T. (1979). *Cell* **18**, 257–266.

Bernard, O., Hozumi, N., and Tonegawa, S. (1978). *Cell* **15**, 1133–1144.

Busslinger, M., Portmann, R., Irminger, J. C., and Birnstiel, M. L. (1980). *Nucleic Acids Res.* **8**, 957–977.

Contreras, R., Rogiers, R., Van de Voorde, A., and Fiers, W. (1977). *Cell* **12**, 529–538.

Craik, C. S., Buchman, S. R., and Beychok, S. (1980). *Proc. Natl. Acad. Sci. U.S.* **77**, 1384–1388.

Cordell, B., Bell, G., Tischer, E., DeNote, F. M., Ullrich, A., Pictet, R., Rutter, W. J., and Goodman, H. M. (1979). *Cell* **18**, 533–543.

Davies, R. W. (1980). *Nucleic Acids Res.*, **8**, 1765–1782.

Dayhoff, M. O. (1978). In *Atlas of Protein Sequence and Structure* (ed. M. O. Dayhoff), Vol. 5, Suppl. 3, National Biomedical Research Foundation, Silver Spring.

Deininger, P., Esty, A., La Porte, P., and Fviedman, T. (1979). *Cell* **18**, 771–779.

Dhar, R., Seif, I., and Khoury, G. (1979). *Proc. Natl. Acad. Sci. U.S.* **76**, 565–569.

Dickerson, R. E. (1971). *J. Mol. Evol.* **1**, 26–45.

Eaton, W. A. (1980). *Nature* **284**, 183–185.

Efstratiadis, A., Kafatos, F. C., and Maniatis, T. (1977). *Cell* **10**, 571–585.

Efstratiadis, A., Posakony, J. W., Maniatis, T., Lawn, R. M., O'Connell, C., Spritz, R. A., DeRiel, J. K., Forget, B. G., Weissman, S., Slighton, J. L., Blechl, A. E., Smithies, O., Barralle, F. E., Shoulders, C. C., and Proudfoot, N. J. (1980). *Cell* **21**, 653–668.

Fiers, W., Contreras, R., Duerinck, F., Haegeman, G., Merregaert, J., Min-Jou, W., Raeymaekers, A., Volckaert, G., Ysebaert, M., Van de Kerckhove, J., Nolf, F., and Van Mantagu, M. (1975). *Nature* **256**, 273–278.

Fiers, W., Contreras, R., Duerinck, F., Haegeman, G., Iserentant, D., Merregaert, J., Min-Jou, W., Molemans, F., Raeymaekers, A., Van den Berghu, A., Volckaert, G., and Ysebaert, M. (1976). *Nature* **260**, 500–507.

Fiers, W., Contreras, R., Haegeman, G., Rogiers, R., Van de Voorde, A., Van Heuverswyn, H., Van Herreweghe, J., Volckaert, G., and Ysebaert, M. (1978). *Nature* **273**, 113–120.

Fitch, W. M. and Margoliash, E. (1970). *Evol. Biol.* **4**, 76.

Godson, G. N., Barrell, B. G., Staden, R., and Fiddes, J. C. (1978). *Nature* **276**, 236–247.

Grantham, R., Gautier, C., and Gouy, M. (1980). *Nucleic Acids Res.* **8**, 1893–1912.

Grunstein, M., Schedle, P., and Kedes, L. (1976). *J. Mol. Biol.* **104**, 351–369.

Grunstein, M. and Grunstein, J. E. (1978). *Cold Spring Harbor Symp. Quant. Biol.* **42**, 1083–1092.

Hasegawa, M., Yasunaga, T., and Miyata, T. (1979). *Nucleic Acids Res.* **7**, 2073–2079.

Hayashida, H., Honjo, T., and Miyata, T. (1981). *J. Mol. Evol.* (submitted).

Haynes, J. R., Rosteck Jr., P. R., Schon, E. A., Gallagher, P. M., Burks, D. J., Smith, K., and Lingrel, J. B. (1980). *J. Biol. Chem.* **255**, 6355–6367.

Heiling, R., Perrin, F., Gannon, F., Mardel, J. L., and Chambon, P. (1980). *Cell* **20**, 625–637.

Heindell, H. C., Liu, A., Paddock, G. V., Studnicka, G. M., and Salser, W. A. (1978). *Cell* **15**, 43–54.

Holland, J. P. and Holland, M. J. (1979). *J. Biol. Chem.* **254**, 9839–9845.

Holland, J. P. and Holland, M. J. (1980). *J. Biol. Chem.* **255**, 2596–2605.

Honjo, T., Obata, M., Yamawaki-Kataoka, T., Kawakami, T., Takahashi, N., and Mano, Y. (1979). *Cell* **18**, 559–568.

Jahn, C. L., Hutchison III, C. A., Phillips, S. J., Weaver, S., Haigwood, N. L., Voliva, C. F., and Edgell, M. H. (1980). *Cell* **21**, 159–168.

Jukes, T. H. and Cantor, C. R. (1969). Evolution of protein molecules. In *Mammalian Protein Metabolism* (ed. Munro, H. N.), Vol. II, pp. 21–123. Academic Press, New York.

Jukes, T. H. and King, J. L. (1979). *Nature* **281**, 605–606.

Jukes, T. H. (1980). *Science* **210**, 973–978.

Kafatos, F. C., Efstratiadis, A., Forget, B. G., and Weissman, S. M. (1977). *Proc. Natl. Acad. Sci. U.S.* **74**, 5618–5622.

Kimura, M. (1968). *Nature* **217**, 624–626.

Kimura, M. and Ohta, T. (1972). *J. Mol. Evol.* **2**, 87–90.

Kimura, M. and Ohta, T. (1974). *Proc. Natl. Acad. Sci. U.S.* **71**, 2848–2852.

Kimura, M. (1977). *Nature* **267**, 275–276.

Kimura, M. (1980a). *J. Mol. Evol.* **16**, 111–120.

Kimura, M. (1980b). *Proc. Natl. Acad. Sci. U.S.* **78**, 454–458.

Konkel, D. A., Tilghman, S. M., and Leder, P. (1978). *Cell* **15**, 1125–1132.

Konkel, D. A., Maizel Jr., V., and Leder, P. (1979). *Cell* **18**, 865–873.

Lacy, E. and Maniatis, T. (1980). *Cell* **21**, 545–553.

Lamb, R. A. and Lai, C. (1980). *Cell* **21**, 475–485.

Leder, A., Miller, H. I., Hamer, D. H., Seidman, J. G., Norman, B., Sullivan, M., and Leder, P. (1978). *Proc. Natl. Acad. Sci. U.S.* **75**, 6187–6191.

Lemedico, P., Rosenthal, N., Efstratiadis, A., Gilbert, W., Kolodner, R., and Tizard, R. (1979). *Cell* **18**, 545–558.

Lerner, M. R., Boyle, J. A., Mount, S. M., Wolin, S. L., and Steiz, J. A. (1980). *Nature* **283**, 220–224.

Li, W., Gojobori, T., and Nei, M. (1981). *Nature* **292**, 237–239.

Marotta, C. A., Wilson, J. T., Forget, B. G., and Weissman, S. M. (1977). *J. Biol. Chem.* **252**, 5040–5053.

Martial, J. A., Hallewell, R. A., Baxter, J. D., and Goodman, H. M. (1979). *Science* **205**, 602–607.

Min-Jou, W., Haegeman, G., Ysebaert, M., and Fiers, W. (1972). *Nature* **237**, 82–88.

Miyata, T. and Yasunaga, T. (1978). *Nature* **272**, 532–535.

Miyata, T., Miyazawa, S., and Yasunaga, T. (1979a). *J. Mol. Evol.* **12**, 219–236.

Miyata, T., Hayashida, H., Yasunaga, T., and Hasegawa, M. (1979b). *Nucleic Acids Res.* **7**, 2431–2438.

Miyata, T. and Yasunaga, T. (1980). *J. Mol. Evol.* **16**, 23–36.

Miyata, T., Yasunaga, T., and Nishida, T. (1980a). *Proc. Natl. Acad. Sci. U.S.* **77**, 7328–7332.

Miyata, T., Yasunaga, T., Yamawaki-Kataoka, Y., Obata, M., and Honjo, T. (1980b). *Proc. Natl. Acad. Sci. U.S.* **77**, 2143–2147.

Miyata, T. and Yasunaga, T. (1981). *Proc. Natl. Acad. Sci. U.S.* **78**, 450–453.

Miyata, T. and Hayashida, H. (1981a). *Proc. Natl. Acad. Sci. U.S.* **78**, 5739–5743.

Miyata, T. and Hayashida, H. (1981b). *Nature* (in press).

Miyata, T., Hayashida, H., and Honjo, T. (1981) (manuscript in preparation).

Natvig, J. B. and Kunkel, H. G. (1974). *J. Immunol.* **112**, 1277–1284.

Nakamura, K. and Inouye, M. (1979). *Cell* **18**, 1109–1117.

Nakamura, K. and Inouye, M. (1980). *Proc. Natl. Acad. Sci. U.S.* **77**, 1369–1373.

Nichols, B. P and Yanofsky, C. (1979). *Proc. Natl. Acad. Sci. U.S.* **76**, 5244–5248.

Nishioka, Y. and Leder, P. (1979). *Cell* **18**, 875–882.

Nishioka, Y., Leder, A., and Leder, P. (1980). *Proc. Natl. Acad. Sci. U.S.* **77**, 2806–2809.

Nishioka, Y. and Leder, P. (1980). *J. Biol. Chem.* **255**, 3691–3694.

Perler, F., Efstratiadis, A., Lemedico, P., Gilbert, W., Kolonder, R., and Dodgson, J. (1980). *Cell* **20**, 555–566.

Proudfoot, N. J. and Maniatis, T. (1980). *Cell* **21**, 537–544.

Porter, A. G., Smith, J. C., and Emtage, J. S. (1980). *Proc. Natl. Acad. Sci. U.S.* **77**, 5074–5078.

Reddy, V. P., Thimmappaya, B., Dhar, R., Subramanian, K. N., Zain, B. S., Pan, J., Ghosh, P. K., Celma, M. L., and Weissman, S. M. (1978). *Science* **200**, 494–502.

Romero-Herrera, A. E., Hehman, H., Joysey, K. A., and Friday, A. E. (1973). *Nature* **246**, 389–395.

Salser, W. and Issacson, J. S. (1976). *Prog. Nucleic Acids Res.* **19**, 205–220.

Salser, W. (1978). *Cold Spring Harbor Symp. Quant. Biol.* **42**, 985–1002.

Sanger, F., Air, G. M., Barrell, B. G., Brown, N. L., Coulson, A. R., Fiddes, J. C., Hutchison III, C. A., Slocombe, P.M.S., and Smith, M. (1977). *Nature* **265**, 689–695.

Sanger, F., Coulson, A. R., Friedmann, T., Air, G. M., Barrell, B. G., Brown, N. L., Fiddes, J. C., Hutchison III, C. A., Slocombe, P. M., and Smith, M. (1978). *J. Mol. Biol.* **125**, 225–246.

Schaffner, W., Knuz, G., Daetwyler, H., Telford, J., Smith, H. O., and Birnstiel, M. L. (1978). *Cell* **14**, 655–671.

Seeburg, P. H., Shine, J., Martial, J. A., Baxter, J. D., and Goodman, H. M. (1977). *Nature* **270**, 486–494.

Seidman, J. G., Leder, A., Edgell, M. H., Polsky, F., Tilghman, S. M., Tiemeier, D. C., and Leder, P. (1978). *Proc. Natl. Acad. Sci. U.S.* **75**, 3881–3885.

Seidman, J. G., Max, E. E., and Leder, P. (1979). *Nature* **280**, 370–375.

Shine, J., Seeburg, P. H., Martial, J. A., Baxter, J. D., and Goodman, H. M. (1977). *Nature* **270**, 494–499.

Slightom, J. L., Blechl, A. E., and Smithies, O. (1980). *Cell* **21**, 627–638.

Soeda, E., Arrand, J. R., Smolar, N., Walsh, J. E., and Griffin, B. E. (1980). *Nature* **283**, 445–453.

Sures, I., Lowry, J., and Kedes, L. H. (1978). *Cell* **15**, 1033–1044.

Takahata, N. and Kimura, M. (1981). *Genetics* (in press).

Tsuzukida, Y., Wang, C., and Putnam, F. W. (1980). *Proc. Natl. Acad. Sci. U.S.* **76**, 1104–1108.

Van Den Berg, J., Van Ooen, A., Mantei, N., Schambock, A., Grosveld, G., Flavell, R. A., and Weissmann, C. (1978). *Nature* **276**, 37–44.

Van Ooen, A., Van Den Berg, J., Mantei, N., and Weissmann, C. (1979). *Science* **206**, 337–344.

Vanin, E. F., Goldberg, G. I., Tucker, P. W., and Smithies, O. (1980). *Nature* **286**, 222–226.

Werner, B. G. and Steinberg, A. G. (1980). *Immunogenetics* **3**, 254–271.

Wilson, A. C., Carlson, S. S., and White, T. J. (1977). *Annu. Rev. Biochem.* **46**, 573–639.

Wilson, J. T., Wilson, L. B., Reddy, V. B., Covallesco, C., Ghosh, P. K., de Riel, J. K., Forget, B. G., and Weissman, S. M. (1980). *J. Biol. Chem.* **255**, 2807–2815.

Yamawaki-Kataoka, Y., Kataoka, T., Takahashi, N., Obata, M., and Honjo, T. (1980). *Nature* **283**, 786–789.

Yamawaki-Kataoka, Y., Miyata, T., and Honjo, T. (1981). *Nucleic Acids Res.* **9**, 1365–1381.

Yang, R.C.A. and Wu, R. (1979). *Science* **206**, 456–462.

Yasunaga, T. and Miyata, T. (1981). *J. Mol. Evol.* (in press).

Zuckerkandl, E. and Pauling, L. (1965). In *Evolving Genes and Proteins* (eds. V. Bryson and H. J. Vogel), pp. 97–116. Academic Press, New York.

11

Molecular Evolution, Protein Polymorphism and the Neutral Theory (1982) (ed. M. Kimura), pp. 269–283.
Japan Scientific Societies Press, Tokyo / Springer-Verlag, Berlin

Molecular Evolution of Papova Viruses and Their Host Species

TAKEO MARUYAMA AND EIICHI SOEDA

National Institute of Genetics, Mishima 411, Japan

The first major molecular data which became available for the study of evolution were the primary structures of various proteins. When the amino acid sequences of the proteins coded from the homologous genes are compared in taxonomically related species, the similarity is found to be approximately proportional to the taxonomical closeness of the species. Phylogenies based on amino acid sequence data are known to be in good agreement with the phylogeny based on classical taxonomy (Fitch and Margoliash, 1967). Amino acid sequence data also enable us to count the number of gene substitutions which have occurred among the species being compared. A large amount of data on primary protein structure has been compiled and used in various research areas (Dayhoff, 1972). Although the amino acid sequence data have provided a unique means to study evolution, the ultimate and most direct kind of data which we want is the nucleotide sequence of genes and genomes. But this was not available until the new powerful techniques of DNA sequencing were developed.

In the past few years, a large amount of DNA sequence data has been made available for molecular evolutionary study. For several viral species the entire genomes have been sequenced, and many genes of both prokaryotic and eukaryotic species have also been sequenced. Usually, DNA sequence data provide direct evidence for evolutionary studies and therefore have a strong impact. For instance, the discovery of a high evolutionary rate at the third positions of genetic codons and a high incidence of synonymous substitutions have provided rather decisive evidence for the validity of the neutral hypothesis of molecular evolution (Kimura, 1968, 1977). Using such DNA sequence data, we intend to elucidate several interesting phenomena in the evolution of viral genomes and their host species.

MOLECULAR EVOLUTION OF THE PAPOVA VIRUSES AND THEIR HOSTS

The molecular biology of the three papova viruses, BKV, SV40, and polyoma virus (Py) has been extensively studied (Fried and Griffin, 1976; Tooze, 1973; Takemoto and Mullardey, 1973; Shah *et al.*, 1977). Earlier work on the genomic organizations of these species revealed a strong similarity and suggested that these papova viruses have diverged from a common ancestor. However, it was not possible to discuss their relationships in detail until the nucleotide sequences of the viral genomes had been elucidated. Comparisons of the nucleotide sequences of these viral genomes revealed striking homologies among the corresponding genes (Reddy, 1978; Soeda *et al.*, 1980; Yan and Wu, 1979; Seif *et al.*, 1979). It is now beyond doubt that these viral species have diverged from a common origin.

Each of these viral genomes consists of two coding regions called the early and late regions, and there are a total of five common genes. The Py genome appears to have one additional gene designated as middle *T* in the early region. A graphic presentation of homologies in terms of amino acid sequence among genomes is given in Fig. 1, where each of the five genes common to the three species is compared. In the figure the homology is viewed from the polyoma genes, and the dark areas indicate strong homology and the dotted areas indicate the intervening sequences. The areas where lines are missing indicate inserted gaps.

BKV, SV40, and Py are easily isolated from healthy human, African green monkey, and mouse tissues, respectively (Tooze, 1973; Gardner *et al.*, 1971). They grow *in vitro* to high titers on cells from these animals. Therefore, we can regard these mammals as their natural hosts. Comparing

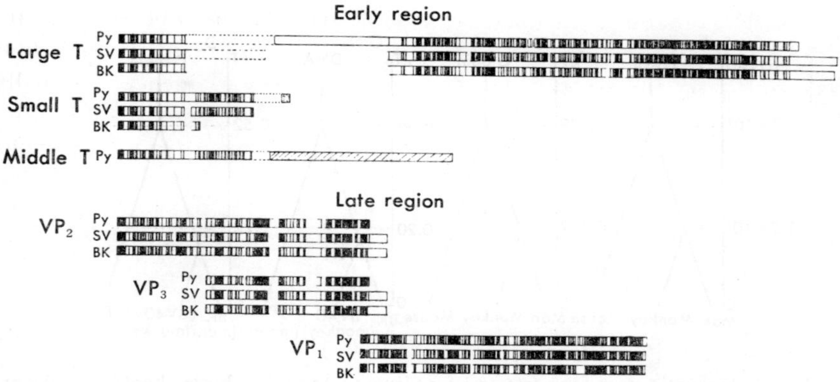

Fig. 1. A comparison of amino acid homology in individual genes of Py and SV40, and of Py and BKV. Black areas represent homologies and colorless areas nonhomologies. The genes have been linearized about their respective origins of replications and divided into early and late regions (Soeda *et al.*, 1980a).

the DNA sequences of homologous genes, we can calculate linearized relative evolutionary distances and construct a phylogenetic tree for the viral species. For the hosts, the evolutionary time and relationships between species have been well documented from paleontological studies based on the fossil record.

According to paleontology, the order Rodentia, which includes rats, mice, and squirrels, diverged about 80 million years ago from the order Primates to which various monkeys, the apes, and humans belong. Later in mammalian evolution, various forms of Primates came into existence. In particular, the evolution of the suborder Catarrhini, a well-defined group of Old World Primates including the living monkeys of Asia and Africa, the anthropoid apes, and humans occurred some 35 million years ago in the late Eocene and early Oligocene (Romer, 1974; Colbert, 1955). Therefore, applying these paleontological facts to the evolution of the host organism, it seems quite reasonable to suggest that the mouse separated from the African green monkey and humans some 80 million years ago and the green monkey and humans diverged from each other 35 million years ago. Figure 2A shows these phylogenetic relationships.

For DNA sequence comparisons, data were used from four different genes: those coding for small T-antigen, large T-antigen, and the virus capsid proteins $VP_{2,3}$, and VP_1. In the comparisons (made by computer) the corresponding nucleotide sites of DNAs, or amino acid residues of

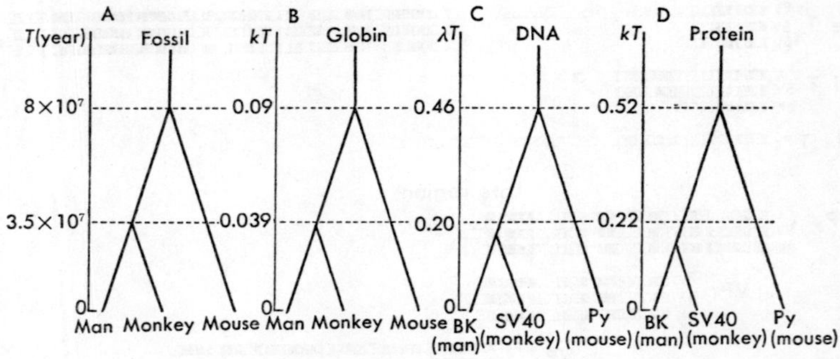

Fig. 2. Phylogenetic trees for hosts and for viruses. A: for the hosts, based on paleontological evidence; each branch of the tree represents evolutionary divergence time. B: for the homologous globin amino acids; each branch represents the evolutionary distance measured by kT. C: for the viruses based on nucleotide sequence; each branch represents the evolutionary distance measured by λT. D: for the viruses based on the presumed amino acid sequence comparisons; each branch is measured by kT. The dotted horizontal lines indicate the relative points where the branch should occur if the molecular evolutionary distances among the viruses agreed completely with the branchings of the hosts. Since the dotted lines show only relative locations, we can place one of the two lines at an arbitrary location. We have adjusted the branchings between Py and SV40 (BKV) to the same height of the primates-rodents branching in the figures (Soeda *et al.*, 1980a).

proteins, were compared and the number of sites or residues at which the two were identical was counted. We then calculated the ratio of the number of sites having identical nucleotides (or amino acid residues) at corresponding positions to the total number of sites (or residues) examined. We denote this ratio by P_n in the case of the DNA sequence and P_a in the case of the primary structure (predicted) of the protein. Thus, P_n is the fraction of the DNA sequence which has the same nucleotides, and P_a the fraction of amino acid sites having identical residues. We made this calculation for parts of genes for which the homology was clear and therefore a meaningful comparison was possible. We have thus excluded from the calculations those parts in which gaps have been inserted in either of the two species being compared. Such gaps appear to exist between Py and SV40 or BKV, but not between SV40 and BKV. With three species to be compared for each of four different genes there are a total of 12 pairwise comparisons. The results are presented in Table I.

The probability of having identical residues can be translated into a measure of the evolutionary distance between the species compared.

TABLE I
Comparison of Nucleotide and Amino Acid Sequences among the Three Virus Species BKV, SV40, and Py

Compared species and genes		Probability of identity		Linearized evolutionary distance; divergence time (T) times substitution rate (λ or k)		Number of nucleotides examined
		P_n DNA		$2\lambda T$ DNA	$2kT$ Protein	
Small T	SV40/Py	0.453	0.306	0.98 (1)	1.25 (1)	519
	BK/Py	0.470	0.322	0.92 (0.94)	1.19 (0.95)	513
	BK/SV40	0.709	0.709	0.37 (0.38)	0.35 (0.28)	516
Large T	SV40/Py	0.501	0.423	0.82 (1)	0.89 (1)	1,881
	BK/Py	0.425	0.407	1.09 (1.33)	0.93 (1.05)	1,380
	BK/SV40	0.721	0.753	0.35 (0.43)	0.29 (0.32)	1,535
VP_1	SV40/Py	0.574	0.564	0.63 (1)	0.56 (1)	1,080
	BK/Py	0.586	0.532	0.60 (0.95)	0.64 (1.19)	1,077
	BK/SV40	0.768	0.818	0.28 (0.44)	0.20 (0.36)	1,086
$VP_{2,3}$	SV40/Py	0.501	0.434	0.82 (1)	0.86 (1)	885
	BK/Py	0.469	0.341	0.92 (1.13)	1.12 (1.31)	510
	BK/SV40	0.764	0.758	0.28 (0.35)	0.28 (0.33)	534

The distance measure is linear with respect to time in units of years, and thus when two or more cases are compared, relative evolutionary times can be estimated without knowing the actual rate of evolution. This method, originally developed by Zuckerkandl and Pauling (1965) and Jukes and Cantor (1969) has been extensively used in molecular evolutionary studies (Kimura, 1969, 1977). The principle of the method is as follows: when we look at the evolution of genes at the molecular level, that is, the primary structure of proteins coded for by those genes, the rate at which amino acid residues become different is constant with time and the probability of identity between the corresponding homologous sites of two species declines exponentially with time. Since there is only a finite number of possible substitutions for both amino acids and nucleotides, this factor must be and has been included in the calculations. If we denote the rate of amino acid substitution by k and that for nucleotides by λ, and the divergence time by T, we have P_n=0.25+0.75 exp($-8kT/3$) for the nucleotide identity and P_a=(1/20)+(19/20) exp($-40\lambda T/19$) for the amino acid sequence identity. Since P_n and P_a are quantities determined from the observed data, substituting them into these formulas gives values for λT for the nucleotide comparison and kT for the amino acid comparison. The results are shown in Table I. If the divergence time is known, the rates λ and k can be determined explicitly. For the time being, however, it is important to realize that values of λT or kT, though they are the product of two quantities (rate and time), are of a linear nature and can be directly compared, while the values of P_n and P_a cannot.

Each comparison gives one value of the evolutionary distance and we have made a total of twelve such comparisons for three species. Although the evolutionary distances calculated from the data did not vary much from gene to gene, we have calculated a weighted mean distance between each pair of two species to construct a phylogeny. The result for the DNA sequences is presented in Fig. 2C and that for the amino acid sequences in Fig. 2D. A remarkable similarity is evident among the host phylogeny based on the fossil record (Fig. 2A) and the two virus phylogenies constructed from genome comparison at the molecular level. We take this agreement of the phylogenies as evidence for the coevolution of the hosts and viruses. That is, that these three virus species diverged from each other simultaneously with the divergence of their host species.

It is necessary however to examine whether this hypothesis is consistent with other known facts. From the data presented below, we have found no serious contradictions to invalidate our hypothesis. A phylogeny based on the amino acid substitutions among homologous proteins of different species has proved to be in good agreement with relationships inferred from classical taxonomy and paleontology. Proteins often used in constructing molecular evolutionary phylogenies are cytochrome *c* and hemoglobins (Fitch and Margoliash, 1967, 1970). The sequences of hemoglobin-α and -β are known for a number of species and, when the homologous chains are compared, identity indices (P_a) are found to be: human-monkey, 0.95; human-mouse, 0.85; and monkey-mouse, 0.83. If these P_a values are converted into the linearized evolutionary distance ($2kT$), we get: human-monkey, 0.074; human-mouse, 0.165; and monkey-mouse, 0.18. From these a phylogenetic tree of the host species can be constructed. One based on these data is presented in Fig. 2B. With different proteins, although the distance values change, the relative relationships remain approximately the same (Soeda *et al.*, 1980a).

Based on the assumption that these viral genomes have diverged with their host organisms, it is possible to estimate the rates at which the genes in these viral genomes are evolving. That is, assuming the divergence time between Py and SV40 or BKV to be about 8×10^7 years and that between SV40 and BKV to be about 3.5×10^7 years, we can get an estimate of λ or k from the linearized evolutionary distance discussed above. We have

TABLE II

Estimated Evolutionary Rates of Genes in Papova Viruses BKV, SV40, and Py: Rate Estimated from the Linearized Evolutionary Distance Divided by the Divergence Time

Gene	Species compared	Nucleotide position in codons				Amino acid ($\times 10^{-9}$)	Number of nucleotides or amino acid sites
		1st ($\times 10^{-9}$)	2nd ($\times 10^{-9}$)	3rd ($\times 10^{-9}$)	Average ($\times 10^{-9}$)		
Small *T*	Py/SV40	5.4	5.0	8.8	6.4	7.8	171
	Py/BK	5.4	4.9	7.3	5.9	7.4	171
	BK/SV40	4.3	3.1	9.3	5.6	5.0	172
Large *T*	Py/SV40	4.9	3.3	8.7	5.6	5.5	627
	Py/BK	4.8	3.7	8.4	5.6	5.8	460
	BK/SV40	3.1	2.2	10.0	5.1	4.1	512
VP_1	Py/SV40	3.0	2.2	9.4	4.7	3.6	360
	Py/BK	2.6	2.4	7.3	4.1	3.4	359
	SV40/BK	2.9	1.8	8.6	4.4	2.8	362
$VP_{2,3}$	Py/SV40	5.1	3.0	9.5	5.9	5.4	295
	Py/BK	5.4	3.7	10.5	6.5	7.0	170
	BK/SV40	3.3	1.4	8.7	4.5	4.0	178

estimated the rates for each nucleotide position of codons separately, and also the rate of the amino acid substitutions. The results are presented in Table II. As noted by various authors, the rates at the third positions of codons appear to be faster by a factor of two to three as compared with the other two positions (Kimura, 1977), and this supports the neutral theory of molecular evolution (Kimura, 1968). The average rates for the viral nucleotides were found to be about $4.5–6.5 \times 10^{9}$ per year per nucleotide site. This value is somewhat larger than the rates estimated for genes in other mammalian genomes, such as the hemoglobin sequences (Kimura, 1977).

The rates for the virus proteins were found to be $4.0–7.8 \times 10^{-9}$ per year per amino acid site, which is relatively fast. The rates are known for several proteins and range from 1×10^{-9} for hemoglobins to an exceptionally slow rate of 6×10^{-12} for histone-IV (Dayhoff, 1972).

EVOLUTIONARY PATTERN IN OVERLAPPING GENES IN BACTERIOPHAGES ϕX174 AND G4

The complete nucleotide of bacteriophages ϕX174 and G4 have been sequenced (Sanger *et al.*, 1977; Godson *et al.*, 1979). The two genomes have the same set of 10 genes, and each pair of corresponding genes shows

TABLE III
Percentage of Identical Nucleotides at Corresponding Sites between Genomes of ϕX174 and G4, for Non-overlapping or Underlying Genes

Gene	Position in codons 1st	2nd	3rd	Number of nucleotides compared at each position
A	0.783	0.871	0.583	420
C	0.694	0.812	0.494	85
D	0.810	0.928	0.516	153
F	0.708	0.795	0.452	425
H	0.756	0.824	0.432	324
J	0.808	0.731	0.462	26
G	0.497	0.594	0.352	175

TABLE IV
Incidence of Identical Nucleotides at Corresponding Sites in Overlapping Genes *B*, *E*, and *K* of ϕX174 and G4

Gene	Position in codons 1st	2nd	3rd	Number of nucleotides compared at each position
B	0.645	0.785	0.843	121
(*A*)	(3) 0.559	(1) 0.783	(2) 0.880	299
E	0.957	0.620	0.848	92
(*D*)	(2) 0.885	(3) 0.361	(1) 0.754	61
K(1–86)	0.897	0.759	0.821	28 or 29
(*A*)	(2) 0.871	(3) 0.583	(1) 0.783	392
K(83–174)	0.586	0.800	0.833	30 or 31
(*C*)	(3) 0.444	(1) 0.630	(2) 0.796	55

The values given for genes *A*, *D*, and *C* are the incidence calculated for the regions excluding the overlapping region; integers in parentheses indicate nucleotide positions in codons of *A*, *D*, or *C*.

strong homology, indicating a common origin of the two genomes. Godson *et al.* (1979) have provided some comparison of the two genomes, and have shown that there are considerable differences in the evolutionary rates among different genes.

Using the nucleotide sequence data of ϕX174 and G4 we examined the evolutionary rate of each codon position in these 10 genes. The results are presented in Tables III and IV. As noted by others, we also found that the third positions of codons appear to be the fastest evolving sites

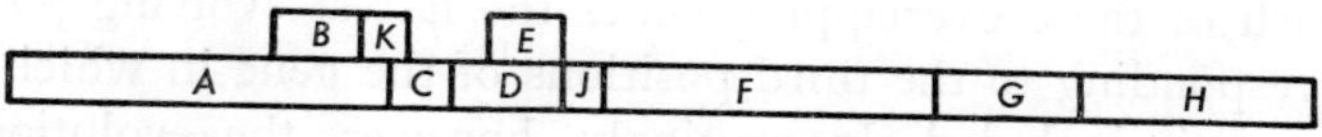

Fig. 3. Approximate physical map of the genomes of bacteriophages ϕx174 and G4. Letters represent the genes (after Godson *et al.*, 1978).

among the three positions. However, the three overlapping genes *B*, *K*, and *E* are totally different in this aspect. As illustrated in Fig. 3, gene *B* is contained in gene *A* and consists of 363 nucleotides, providing 121 site comparisons for each of the three nucleotide positions. Although the two genes use the same region of DNA they have different reading frames and therefore code entirely different proteins. The first nucleotide of every codon in *B* is the third position of *A*, and accordingly the second and third positions of *B* correspond to the first and second positions of *A*, respectively. A similar reading frame difference exists between genes *D* and *E* which is contained in *D* and consists of 276 nucleotides. The first, second, and third positions of *E* are, respectively, the second, third, and first positions of *D*. The situation in gene *K* is slightly more complicated. This gene consists of 174 nucleotides, and the first 86 sites are in *A*, while the last 72 sites are in *C*. The four nucleotides from the 83rd to 86th are therefore in both *A* and *C* where all three genes are coded using codons triply. In every case of these overlapping genes, *B*, *E*, and *K*, the rule that the third positions evolve fastest does not hold.

In the case of gene *B*, there are 121 nucleotides for each position which can be compared between ϕX174 and G4. In *B*, the incidence (*P*) of identical nucleotides at corresponding sites are 0.645, 0.785, and 0.843 for the first, second, and third positions, respectively. This is quite the contrary to the rule that the third position is the fastest evolving site. Interestingly the average of *P*'s, the percent difference, of the three positions are 0.758 for *B* and 0.746 for *A*. Thus *A* and *B* are changing at about the same rate. Now we compare the values of *P* for the first, second, and third positions of *B* to the *P*'s of the third, first, and second positions of *A*, excluding the part which codes for *B*. Then the corresponding *P*'s are quite similar. That is the first positions of *A* excluding the part coding for *B* have $P=234/299=0.783$ which is about the same as $P=0.785$ of the second positions of *B*. Likewise, the second and third positions of *A* have *P* values of 0.880 and 0.560 which are to be compared with the *P* values of the third and first positions of *B* which are 0.843 and 0.645. A similar comparison can be made between genes *D* and *E*.

In each of these overlapping genes the fastest evolving positions are those corresponding to the third positions of the gene in which the overlapping gene is included. Interestingly, however, the evolutionary rates of these nucleotides in the overlapping genes are somewhat slower than the rates observed in the part outside of the overlapping regions. Details of the comparisons are presented in Table IV. It is clear that the evolutionary patterns with respect to the nucleotide positions in codons in *B*, *E*, and *K* are very much determined by *A*, *D*, and *C*, though the constraints appear to be slightly more stringent in those overlapping regions. This observation seems to support the hypothesis that these overlapping genes *B*, *E*, and *K* came into existence long after the other genes were established. These overlapping genes must have been quite tolerable to nucleotide changes. This seems to be consistent with the fact that unless a gene is very flexible with respect to its primary structures it is probably difficult for it to arise by using a part of DNA which is already in use for other genes.

PATTERN OF NUCLEOTIDE SUBSTITUTION

While the third positions of codons are changing more rapidly than the first and second positions, the substitutions at the third positions of codons (informational strand of DNA) do not appear to be symmetrical between species. For instance, at the corresponding nucleotides between Py and SV40, there are 69 cases with T in Py and C in SV40, while there are 121 cases with C in Py and T in SV40. Therefore, substitutions increasing T in SV40 and C in Py appear to have been occurring. In fact, if we examine the numbers of T's and C's at the third positions of codons in Py and SV40, there are 441 T's and 374 C's in Py, and 594 T's and 212 C's in SV40. This difference seems to be very large. We may say that in the genome of SV40 the number of T's is increased by decreasing the number of C's, whereas the reverse situation holds in Py.

A similar pattern of asymmetric substitutions seems to exist between the genomes of ϕX174 and G4. In that case, the third position of ϕX174 is rich in T, but poor in A, as opposed to G4 which is rich in A but poor in T (in the informational strand of DNA). In Table V, comparisons of the numbers of different nucleotides found at each position of codons are presented, and Table VI provides a comparison of every possible substitution between species.

The asymmetric substitutions at the third positions of codons are

TABLE V
Relative Frequencies of T, C, A, and G Nucleotides at the First, Second, and Third Positions of Codons in Genomes of Py, SV40, ϕX174, and G4

Position	Organism	T	C	A	G
1st	Py	0.174	0.234	0.288	0.304
	SV40	0.209	0.161	0.309	0.321
2nd	Py	0.269	0.218	0.312	0.201
	SV40	0.277	0.215	0.344	0.164
3rd	Py	0.259	0.221	0.295	0.225
	SV40	0.389	0.125	0.286	0.199
Aver.	Py	0.234	0.224	0.299	0.243
	SV40	0.292	0.167	0.313	0.228
1st	ϕX174	0.202	0.215	0.268	0.315
	G4	0.200	0.222	0.277	0.301
2nd	ϕX174	0.270	0.251	0.316	0.163
	G4	0.266	0.267	0.326	0.141
3rd	ϕX174	0.432	0.191	0.164	0.213
	G4	0.344	0.274	0.218	0.164
Aver.	ϕX174	0.301	0.219	0.249	0.231
	G4	0.270	0.255	0.274	0.202

TABLE VI
Frequencies of All the Possible Nucleotide Combinations Found at Corresponding Sites of the Third Positions of Codons between Py and SV40, and between ϕX174 and G4

		Nucleotide at the third position in SV40			
		T	C	A	G
Nucleotide at the third position in Py	T	190	69	67	57
	C	121	55	82	45
	A	134	39	162	94
	G	94	27	104	115
		Nucleotide at the third position in G4			
		T	C	A	G
Nucleotide at the third position in ϕX174	T	435	216	105	55
	C	102	199	42	18
	A	40	37	178	52
	G	63	63	87	186

quite puzzling. The third positions are believed to be more tolerable to substitutions than other positions, because of the high rate of synonymous mutations. This seems to imply that, if these nucleotide compositions are in equilibrium, substitutions should be approximately symmetric and be independent of the mutation rates among different nucleotides.

Amino acid sequences of proteins have often been used in the molecular evolutionary study of homologous genes. It is generally believed that the evolutionary rate inferred from the primary structures of protein is not the same over the entire region of a gene. On the contrary the rate can differ considerably depending upon the part of the gene examined. Most genes have some part which appears to be extremely conservative and to have remained unchanged for a long time. This, in terms of nucleotide change, implies that the nucleotide substitutions may not be independent among different codons. This can be confirmed at the nucleotide level using the sequence data.

For each of the first, second, and third positions, we compute the probability of having identical nucleotides. Then the theoretical expectation of having identical or different nucleotides in various combinations can be calculated. If P_1, P_2, and P_3 represent, respectively, the probability of having identical nucleotides at the first, second, and third positions, then for example, the probability of having identical nucleotides at the first and second positions, but different nucleotides at the third is equal to $P_1P_2(1-P_3)$, and similarly that of having different nucleotides at all three positions is equal to $(1-P_1)(1-P_2)(1-P_3)$, and so forth. Of course, these theoretical expectations are probabilities based on the assumption that the substitutions occur independently of each other.

In both comparisons of Py-SV40 and of ϕX174-G4, we confirmed that the nucleotide substitutions are not independent. Rather striking is that the cases which have a substitution at either the first or second position alone are found only half as frequently as predicted by the expectations based on the independence hypothesis. On the other hand, the case which has substitutions at all three positions is found three to four times more often than expected. Similarly the case which has substitutions at the first and second positions, but with the third position being identical is also found about twice as often than expected. However, interestingly, the case which has a substitution at the third position alone is always in good agreement with the expectations. The numerical details are presented in Table VII.

However, this substitution pattern does not appear to apply to the overlapping genes. This is partly because in these genes the relative rates of nucleotide substitutions at different positions of codons are controlled by those genes which underly the overlapping ones. A typical example

TABLE VII
Three Examples of Frequencies of Nucleotide Substitutions in Different Combinations

		OOO	OOX	OXO	XOO	OXX	XOX	XXO	XXX
Py-SV40 Large *T*	Observ.	116	134	27	41	47	98	46	118
	Expect.	73.7	127.3	45.1	69.0	77.9	119.0	42.2	72.8
								$\chi^2=87.63$	
ϕX174-G4 Gene *A*	Observ.	195	111	13	26	10	35	11	21
	Expect.	166.8	119.1	25.1	46.1	18.0	33.0	7.0	5.0
								$\chi^2=77.6$	
ϕX174-G4 Gene *E*	Observ.	46	8	29	2	5	1	1	0
	Expect.	46.2	8.3	28.4	2.1	5.1	0.4	1.3	1.4
								$\chi^2=1.4$	

Symbols O and X indicate, respectively, sites having identical nucleotides (no substitution) and sites not having identical nucleotides, at corresponding sites between species compared. The symbols O or X indicate, from left to right, the first, second, and third positions in codons.

is the pattern of substitutions in gene *E* of ϕX174 and G4. Interestingly, in *E* the substitutions appear to be independent of each other, and the agreement between the observed values and the expectations is very good (see Table VII).

The third positions of codons have been shown to be changing considerably faster than the other two positions. Is this because the third positions are far more degenerate than the others, or because the third positions *per se* have a faster evolving property? Although the former seems to be the case, it is worth testing. This may be possible using the codons for the amino acids leucine and serine. Leucine has six degenerate codons, TTA, TTG, and CTN where N indicates any nucleotide of the four. The first positions of -TA and -TG are degeneracy for T and C, the third positions of TT- are degenerated for A and G, and the third positions of CT- are completely degenerated. Therefore, if all these six codons are equally in use, and if the mutation rates are the same for each of the twelve different combinations of the nucleotides, we would expect synonymous substitutions at the first and third positions to be in a 1 to 2.25 ratio. That is, we should find about four substitutions at the first positions for every nine at the third positions among the synonymous substitutions for the leucine sites. In fact, in the data on ϕX174 and G4, we find 27 synonymous substitutions at the first positions and 63 such cases at the third positions for the leucine sites. Therefore the observed ratio is 1: 2.33 which is in good agreement with the expectations based on the assumption that every position is equally substitutable if synonymous.

A similar kind of codon degeneracy exists for serine. In this case, all the three positions have degeneracy, *i.e.*, TCN, AGT, and AGC code for serine. If these codons are equally frequent, we would expect a 1 : 1 : 2.25 ratio for the first, second, and third positions. But, in this case, codons AGT and AGC are used considerably less frequently than the others. In the data of ϕX174 and G4, we find the actual numbers of synonymous substitutions to be 13 : 8 : 35 which gives a ratio of 1 : 0.6 : 2.7. Considering the uneven usage of codons for serine, this may also be in agreement with the expectations.

We interpret these observations as evidence that the third position evolves faster, because of high degeneracy.

SUMMARY

Comparing homologous genes of three papova viral genomes, we have shown that the relative phylogeny among the viral species is very close to that of their host species, and therefore that the viral species have evolved with their host organisms. Additionally, using the DNA sequence data of bacteriophage viruses ϕX174 and G4 the overlapping genes have been examined for their evolutionary pattern. It has been made evident that overlapping genes have a quite different substitutional pattern with respect to the position of nucleotides in codons compared to non-overlapping sequences. That is, in non-overlapping regions the third positions are usually substituted fastest, followed by the first positions, while the second positions are slowest in each gene, though different genes may have different rates of nucleotide substitution. With overlapping genes, this pattern does not apply, but rather is altered because of an interaction of the substitution rates in the two genes involved in an overlap.

Acknowledgments

This study was partially supported by Grants-in-Aids II-56213007 from the Ministry of Education, Science and Culture, Japan.

REFERENCES

Colbert, E. H. (1955). *Evolution of the Vertebrates*. Wiley, New York.

Dayhoff, M. O. (ed.) (1972). *Atlas of Protein Sequence and Structure*. National Biomedical Research Foundation, Washington, D.C.

Fiers, W., Contreras, R., Haegeman, G., Rogiers, R., Van de Voorde, A., Van Heuverswyn,

H., Van Herreweghe, J., Volckaert, G., and Ysebaert, M. (1978). Complete nucleotide sequence of SV40 DNA. *Nature* **273**, 113–120.

Fitch, W. M. and Margoliash, E. (1967). Construction of phylogenetic tress. *Science* **155**, 279–284.

Fitch, W. M. and Margoliash, E. (1970). The usefulness of amino acid and nucleotide sequences in evolutionary studies. *Evol. Biol.* **4**, 67–109.

Fried, M. and Griffin, B. E. (1977). Organization of the genomes of polyoma virus and SV40. *Adv. Cancer Res.* **24**, 67–113.

Gardner, S. D., Field, A. M., Coleman, D. V., and Hulme, B. (1971). New human papovavirus (B.K.) isolated from urine after renal transplantation. *Lancet* **i**, 1253–1257.

Godson, G. N., Barrell, B. G., Staden, R., and Fiddes, J. C. (1978). Nucleotide sequence of bacteriophage G4 DNA. *Nature* **276**, 236–247.

Jukes, T. H. and Cantor, C. R. (1969). Evolution of protein molecules. In *Mammalian Protein Metabolism* (ed. H. N. Munro), Vol. III. Academic Press, New York.

Kimura, M. (1968). Evolutionary rate at the molecular level. *Nature* **217**, 624–626.

Kimura, M. (1969). The rate of molecular evolution considered from the standpoint of population genetics. *Proc. Natl. Acad. Sci. U.S.* **63**, 1181–1188.

Kimura, M. (1977). Preponderance of synonymous changes as evidence for the neutral theory of molecular evolution. *Nature* **267**, 275–276.

Reddy, V. B., Thimmappaya, B., Dhar, R., Subramanian, K. N., Zain, B. S., Pan, J., Ghosh, P. K., Celma, M. L., and Weismann, S. M. (1978). The genome of simian virus 40. *Science* **200**, 494–502.

Romer, A. S. (1974). *Vertebrate Palaeontology*, 3rd ed. Univ. Chicago Press, Chicago.

Sanger, F., Air, G. M., Barrell, B. G., Brown, N. L., Coulson, A. R., Fiddes, J. C., Hutchison III, C. A., Slocombe, P. M., and Smith, M. (1977) Nucleotide sequence of bacteriophage ϕX174 DNA. *Nature* **263**, 687–695.

Seif, I., Khoury, G., and Dahr, R. (1979). The genome of human papovavirus BKV. *Cell* **18**, 963–977.

Shah, K. V., Ozer, H. L., Ghazey, H. N., and Kelly Jr., T. J. (1977). Common structural antigen of papovaviruses of the simian virus 40-polyoma subgroup. *J. Virol.* **21**, 179–186.

Soeda, E., Arrand, J. R., Smolar, N., Walsh, J. E., and Griffin, B. E. (1980). Coding potential and regulatory signal of the polyoma virus genome. *Nature* **283**, 445–453.

Soeda, E., Maruyama, T., Arrand, J. R., and Griffin, B. E. (1980a). Host dependent evolution of three papova viruses. *Nature* **285**, 165–167.

Takemoto, K. K. and Mullarkey, M. F. (1973). Human papovavirus, BK strain: biological studies including antigenic relationship to simian virus 40. *J. Virol.* **12**, 625–631.

Tooze, J. (1973). *Molecular Biology of Tumour Virus*, pp. 1–73. Cold Spring Harbor Laboratory, New York.

Yang, R.C.A. and Wu, R. (1979). BK virus DNA: Complete nucleotide sequence of a human tumor virus. *Science* **206**, 456–462.

Zuckekandl, B. and Pauling, L. (1965). Evolutionary divergence and convergence in proteins. In *Evolving Genes and Proteins* (eds. V. Bryson and H. Vogel), pp. 97-166. Academic Press, New York.

12

Molecular Evolution, Protein Polymorphism and the Neutral Theory (1982) (ed. M. Kimura), pp. 287–296.
Japan Scientific Societies Press, Tokyo / Springer-Verlag, Berlin

A Discussion of the Molecular Evolution of Proteins, Especially Hemoglobin

GENJI MATSUDA

Department of Biochemistry, Nagasaki University School of Medicine, Nagasaki 852, Japan

Hemoglobin is widely distributed in the animal kingdom. In 1961, the primary structures of the α- and β-chains of human hemoglobin were determined (Braunitzer *et al.*, 1961). After that the primary structure of the α-chain of horse hemoglobin was analyzed, and it was reported that there may be at least fifteen differences in the amino acid sequence of the α-chain in comparison with α-chain of human hemoglobin (Braunitzer and Matsuda, 1961). Referring to this report, it is calculated that 14.5 million years were needed for one amino acid substitution in hemoglobin evolution (Zuckerkandl and Pauling, 1962). For the sake of calculation, it was assumed that the evolutionary effective mutation rate fluctuated during the time of evolution of hemoglobin around a mean without showing a predominant trend to increase or to decrease.

After that, a theory was proposed that the chief cause of molecular evolution is the random fixation of selective neutral mutants (Kimura, 1968; King and Jukes, 1969). The most convincing basis for the neutral theory is that the rate of amino acid substitution per protein molecule is uniform. Matsuda *et al.* assumed the primary structure of a common ancestral hemoglobin of various kinds of mammlian hemoglobins by comparing the primary structures of these present-day animal hemoglobins, and obtained the rate of amino acid substitutions of these animal hemoglobins (Matsuda *et al.*, 1968). Goodman *et al.* constructed the phylogenetic tree of hemoglobin sequences by the maximum parsimony method, and calculated the rate of evolution of the hemoglobin molecule (Goodman *et al.*, 1975). Since 1962, we have been making comparative studies of the primary structures of various kinds of hemoglobins. Considering these results, I shall try here to discuss a few aspects of the molecular evolution of proteins, especially hemoglobin.

EVOLUTION OF LIVING ORGANISMS AND THEIR MOLECULAR EVOLUTION

According to the studies of geologists and geophysicists, the age of the earth is said to be 4,500–5,000 million years old. On the other hand, fossil-like microorganisms have been discovered in rock 3,000 million years old (Barghoorn and Schopf, 1966) and characteristic hydrocarbons which seem to have originated from those living organisms have also been discovered (Eglinton and Calvin, 1967). If these things are really remains of organisms which lived on earth, it proves that previous to this time the first life was born on earth. After that, these living organisms evolved gradually. It is widely known that the fundamental structures of proteins, lipids, glucides, and other compounds which compose the body of living organisms or the fundamental mechanisms of biochemical reactions in them are surprisingly similar throughout the bacterium, plant, and animal kingdoms.

It is assumed from these facts that the fundamental structures of constructed compounds of almost of all living present-day organisms or the fundamental mechanisms of biochemical reactions in them had already been formed before these living organisms separated in the process of evolution. This period seems to be very important in evolution at the molecular level and it could be called "the Completion period." When was the Completion period? I do not have enough data to answer this question. But, for example, according to Schopf it is conceivable that protozoa separated from bacterium and blue-green algae at least 1,000 million years ago (Schopf, 1967). As Fig. 1 shows, I tentatively assume that the Completion period was from 1,500 million years ago to 500 million years ago. For these approximately 1,000 million years, some genes seem to have formed their fundamental structure and functions at a comparatively earlier stage and others at a later stage.

Before and after the completion of genes, the evolution of living organisms at the molecular level might have differed remarkably. The period before the completion of genes could be called "the mainly qualitative adaptation era at the molecular level" as shown in Fig. 1. That is, in this era living organisms adapted themselves to circumstances at the molecular level, generally underwent natural selection, continued orthoevolution, and then the fundamental structures and functions of many genes in them seem to have been established in a qualitative manner. Accordingly it seems to also have been in this period that the structures and functions of many proteins encoded by these genes were decided. After that these genes may have survived to date, making good use of many living organisms as carriers. Of course, it cannot be denied that quantitative adaptation, to which I will refer later, would also have been partially carried out in this era.

On the other hand, as Fig. 1 shows, the era from the completion of genes to the present could be called "the mainly quantitative adaptation era at the molecular level." That is, in this era the fundamental structures and functions of many genes in living organisms which had been already decided in the Completion period seem to have been evolving mainly in a

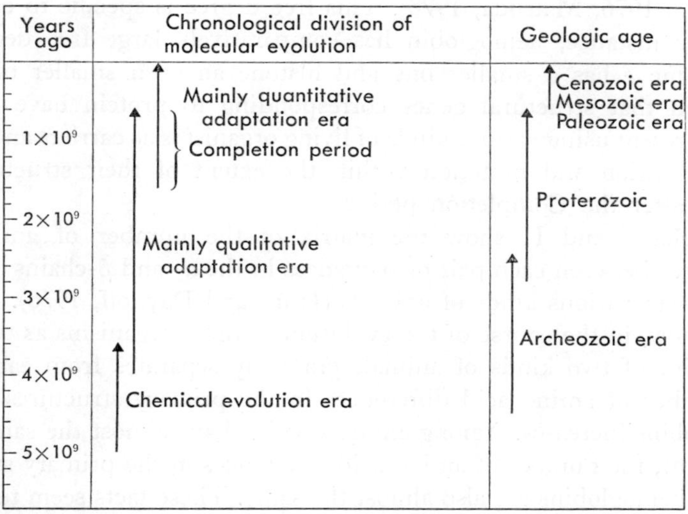

Fig. 1. Chronological division of molecular evolution and geologic age.

quantitative manner. In other words, in this era living organisms seem to have been evolving and adapting many kinds of regulation mechanisms of the biochemical reactions in their bodies to circumstances. For instance, an organism synthesizes various kinds of proteins and the synthesis of one of these proteins could be accelerated while another is inhibited. Moreover, one step in a metabolic reaction system could be accelerated while another is inhibited. These two phenomena are included in this quantitative adaptation. Of course it can be assumed that the qualitative adaptation previously described occurred partially in this era, too.

THE STRUCTURAL FREE DEGREES IN THE MOLECULAR EVOLUTION OF HEMOGLOBIN AND OTHER PROTEINS

As mentioned before, it is assumed that the fundamental structures and functions of structural genes corresponding to many proteins were formed in the Completion period. But why are there species-specific differences in the primary structures of many homologous proteins? I suppose that each structural gene corresponding to many proteins composing living organisms has varied to some degree even after it was formed in the Completion period. The degree of variability of the primary structures of proteins based upon such variability of the structural genes could be called "the structural free degree in the molecular evolution of protein" (Matsuda, 1976, Matsuda, 1978). This free degree is specific to each protein. For instance, hemoglobin has comparatively large free degree but cytochrome *c* has a smaller one and histone an even smaller one. It is presumed that structural genes corresponding to protein have survived till the present using various kinds of living organisms as carriers and repeating duplication and mutation within the extent of their structural free degrees after the Completion period.

Tables I and II show the matrix of the number of amino acid differences between each pair of sequences in the α- and β-chains of hemoglobin from various kinds of animals (Hunt and Dayhoff, 1976). As these tables show, in the course of the evolution of these organisms as the divergent point of two kinds of animals gradually separates from each other, the number of amino acid differences in the primary structures of these hemoglobins increases. Among animals which have almost the same divergent point, the number of amino acid differences in the primary structures of their hemoglobins are also almost the same. These facts seem to support the neutral theory which proposes that the chief cause of molecular evolu-

TABLE I
Matrix of the Numbers of Amino Acid Differences between Each Pair of Sequences in the α-Chains of Vertebrate Hemoglobins

	Human	Mouse	Rabbit	Dog	Bovine	Horse	Sheep	Gray kangaroo	Echidna	Chicken	Carp
Human	0										
Mouse	18	0									
Rabbit	25	27	0								
Dog	23	25	28	0							
Bovine	17	19	25	28	0						
Horse	17	24	23	26	17	0					
Sheep	21	23	30	30	12	19	0				
Gray kangaroo	27	33	37	33	26	28	30	0			
Echidna	37	43	49	42	43	41	45	49	0		
Chicken	35	40	43	44	38	39	43	41	48	0	
Carp	68	69	71	67	65	66	70	71	75	72	0

TABLE II
Matrix of the Numbers of Amino Acid Differences between Each Pair of Sequences in the β-Chains of Vertebrate Hemoglobins

	Human (β)	Mouse	Rabbit	Dog	Bovine	Horse	Gray kangaroo	Echidna	Human (γ)	Chicken	Frog
Human (β)	0										
Mouse	26	0									
Rabbit	14	28	0								
Dog	15	29	21	0							
Bovine	24	37	29	27	0						
Horse	24	35	24	29	28	0					
Gray kangaroo	38	45	37	35	43	42	0				
Echidna	31	37	31	31	37	34	39	0			
Human (γ)	39	46	40	40	39	40	43	40	0		
Chicken	45	49	48	42	54	47	47	42	44	0	
Frog	67	65	68	65	68	67	73	65	69	71	0

tion is random fixation of selective neutral mutants. But I think this neutral theory can be applied within the extent of the structural free degree of the protein. Genes which mutated beyond this extent seem to have undergone natural selection and vanished from the organism kingdom.

As we know from detailed studies of the conformations of myoglobin, hemoglobin, cytochrome *c*, and so on, it is conceivable that each protein has a minimum essential primary structure to express its regular tertiary structure and its characteristic function. Species specificity in the primary structure of each homologous protein is admitted only within the extent of the structural free degree peculiar to each protein, and therefore the species specificity seems to have a progressively smaller meaning in the evolution of living organisms. As I have mentioned in the preceding section, it is assumed that after the Completion of genes quantitative regulation of the metabolic systems of organisms had a remarkable influence on individual evolution at the molecular level.

CHANGES IN THE STRUCTURAL FREE DEGREE IN THE MOLECULAR EVOLUTION OF PRIMATE HEMOGLOBINS

In the preceding section it was assumed that each protein has a characteristic structural free degree. But is this free degree constant? Since 1962, we have been doing comparative studies of the primary structures of various kinds of primate hemoglobins. Tables III and IV show the matrix of the number of amino acid differences between each pair of sequences in the α- and β-chains of hemoglobins from various kinds of primates (Matsuda, 1976).

Primates may be divided into four broad classes, the Prosimiae, Ceboidea (new world monkeys), Cercopithecoidea (old world monkeys), and Hominoidea (apes and human). Slow loris can be included in Prosimiae, spider monkey and capuchin monkey in Ceboidea, hanuman langur, rhesus monkey, Japanese monkey, and Savannah monkey in Cercopithecoidea, and human in Hominoidea. It is said that primates separated from other mammals 70–80 million years ago in the middle of the Paleocene, a little after that Prosimiae separated from other primates about 60 million years ago, and Ceboidea and Cercopithecoidea separated from Hominoidea 35–40 million years ago from the latter period of the Eocene to the former period of the Oligocene.

As Tables III and IV show, as the divergent points separate from Hominoidea to Cercopithecoidea, Ceboidea, and Prosimiae, the number

TABLE III
Matrix of the Numbers of Amino Acid Differences between Each Pair of Sequences in the α-Chains of Primate Hemoglobins

	Human	Hanuman langur	Rhesus monkey	Japanese monkey	Savannah monkey	Spider monkey	Capuchin monkey	Slow loris
Human	0							
Hanuman langur	3	0						
Rhesus monkey	4	5	0					
Japanese monkey	4	5	0	0				
Savannah monkey	5	4	1	1	0			
Spider monkey	5	6	5	5	6	0		
Capuchin monkey	6	7	8	8	9	3	0	
Slow loris	9	10	11	11	11	7	8	0

TABLE IV
Matrix of the Numbers of Amino Acid Differences between Each Pair of Sequences in the β-Chains of Primate Hemoglobins

	Human	Hanuman langur	Rhesus monkey	Japanese monkey	Savannah monkey	Spider monkey	Capuchin monkey	Slow loris
Human	0							
Hanuman langur	4	0						
Rhesus monkey	8	5	0					
Japanese monkey	7	4	1	0				
Savannah monkey	6	3	3	2	0			
Spider monkey	6	4	9	8	7	0		
Capuchin monkey	8	8	9	8	8	8	0	
Slow loris	11	10	12	11	12	12	14	0

of the differences in the amino acid sequences increases. The divergent points are the same in the almost similar number of their differences. Consequently it seems that neutral theory can be roughly applied to these primate hemoglobins also. Then the value was calculated by dividing the numbers of differences of amino acid sequences between the α- and β-chains of human hemoglobin and those of each of the other primates and mammals by the time from the divergent point where those primates and mammals parted from humans to the present. The values calculated by this simple method could be called "the separation rate." Table V shows the separation rates of the α- and β-chains of primate and mammalian hemoglobins from the α- and β-chains of human hemoglobin.

In Table V the tendency is observed that the separation rates of the α- and β-chains of primate hemoglobins become slower in comparison with those of the α- and β-chains of other mammalian hemoglobins. What does this mean? I suppose that the structural free degree of the α- and β-chains of primate hemoglobins are gradually becoming smaller. In other words, it is assumed that primate hemoglobins carried out orthoevolution and qualitative adaptation by decreasing the free degree of the primary structures. Of course, as mentioned previously the more important factor in molecular evolution after the Completion period seems to be quantitative adaptation. But for the case of these primate hemo-

TABLE V
The Rates of Separation in the α- and β-Chains of Primate and Other Mammalian Hemoglobins from Human Hemoglobin

Species	T (years)	D_α	R_α	D_β	R_β
Primates					
Orangutan	20×10^6	2	0.10	2	0.10
Hanuman langur	35×10^6	3	0.09	4	0.11
Rhesus monkey	35×10^6	4	0.11	8	0.23
Japanese monkey	35×10^6	4	0.11	7	0.20
Savannah monkey	35×10^6	5	0.14	6	0.17
Spider monkey	40×10^6	5	0.13	6	0.15
Capuchin monkey	40×10^6	6	0.15	8	0.20
Slow loris	60×10^6	9	0.15	11	0.18
Other mammals					
Mouse	80×10^6	18	0.23	26	0.33
Rabbit	80×10^6	25	0.31	14	0.18
Dog	80×10^6	23	0.29	15	0.19
Bovine	80×10^6	17	0.21	24	0.30
Horse	80×10^6	17	0.21	24	0.30

T, the time from the divergent point where primates and mammals parted from humans to the present; D_α, the numbers of the differences of amino acid sequences between the α-chain of human hemoglobin and that of each other primate and mammal; D_β, the numbers of the differences of amino acid sequences between the β-chain of human hemoglobin and that of each other primate and mammal; R_α, R_β, the rates of separation in the α- and β-chains of primate and other mammalian hemoglobins from human hemoglobin.

globins, it is assumed that qualitative adaptation aslo occurred partially after the Completion period. These phenomena may have occurred not only in primate hemoglobins but also in other animal hemoglobins. To investigate these facts, detailed comparative studies of the primary structures of closely related animal hemoglobins must be made.

Why did the above-mentioned decrease in the free degree of the primary structure of protein occur? It is possible that it is an adaptation against all changes, for instance, a change in external circumstances such as a drop in temperature or in internal circumstances such as the production of an allosteric effector. At first these orthogenetic qualitative adaptations will act advantageously for the individuals, but if this orthogenesis advances too far, the structural free degree of the protein will decrease remarkably and the structural gene controlling it will vanish. It is assumed that the more the living organisms have proteins which have a tendency to decrease their structural free degrees, the more they become sensitive in terms of extreme changes in external circumstances. In the history of the evolution of living organisms, many kinds which carried out orthoevolutions finally died out and the same is conceivable at the molecular level.

SUMMARY

It has been assumed that the fundamental structures of constructed compounds of almost all living organisms had been established before the various organisms separated in the process of the evolution. Before and after the completion of the structural genes, the aspect of the evolution of living organisms at the molecular level might differ remarkably. In the period before the completion of genes, living organisms adapted themselves to circumstances at the molecular level, generally underwent natural selection, continued orthoevolution, and the fundamental structures and functions of many genes in them seem to have been established in a qualitative manner. On the other hand, in the period from the completion of genes to the present, living organisms seem to have been evolving and adapting regulation mechanisms for many kinds of biochemical reactions in their bodies to their circumstances in a quantitative manner.

It is supposed that each structural gene corresponding to the many proteins, composing living organisms has varied to some degree even after it was established in the Completion period. The degree of variability of the primary structures of proteins based upon such variability ofstr uctural genes is called "the structural free degree in the molecular evolution of protein" (Matsuda, 1976, 1978). The free degree is specific for each protein. Furthermore one can suppose that the structural free degrees of the α- and β-chains of primate hemoglobins are gradually becoming smaller. In other words, it is assumed that primate hemoglobins have carried out qualitative adaptation at the molecular level by decreasing their structural free degrees.

REFERENCES

Barghoorn, E. S. and Schopf, J. W. (1966). Microorganisms 3 billion years old from the precambrian of South Africa (Fig Tree). *Science* **152**, 758–763.

Braunitzer, G., Gehring-Müller, R., Hilschmann, H., Hilse, K., Hobom, G., Rudloff, V., and Wittman-Liebold, B. (1961). Die Konstitution des normalen adulten Humanhämoglobins. *Hoppe-Seyler's Z. Physiol. Chem.* **325**, 283–286.

Braunitzer, G. and Matsuda, G. (1961). Die Analyse der Tryptischen Peptide des Pferdehämoglobins. *Hoppe-Seyler's Z. Physiol. Chem.* **324**, 91–93.

Eglinton, G. and Calvin, M. (1967). Chemical fossils. *Sci. Am.* **216**, No. 1, 32–43.

Goodman, M., Moore, W., and Matsuda, G. (1975). Darwinian evolution in the genealogy of hemoglobin. *Nature* **253**, 603–608.

Hunt, L. T. and Dayhoff, M. O. (1976). Globins. In *Atlas of Protein Sequence and Structure* (ed. M. O. Dayhoff), Vol. 5, Suppl. 2, pp. 191–195. National Biomedical Research Foundation, Georgetown University Medical Center, Washington, D.C.

Kimura, M. (1968). Evolutionary rate at the molecular level. *Nature* **217**, 624–626.

King, J. L. and Jukes, T. H. (1969). Non-Darwinian evolution. *Science* **164**, 788–798.

Matsuda, G. (1976). Evolution of the primary structures of primate and other vertebrate hemoglobins. In *Molecular Anthropology* (eds. M. Goodman, R. E. Tashian, and J. H. Tashian), pp. 223–237. Plenum Publ., New York.

Matsuda, G. (1978). Molecular evolution of hemoglobins. In *Evolution of Protein Molecules* (eds. H. Matsubara and T. Yamanaka), pp. 175–185. Japan Sci. Soc. Press, Tokyo.

Matsuda, G., Maita, T., and Ota, H. (1968). The primary structures of primate hemoglobins and their phylogenetic implications. In *Proceedings of VIIIth International Congress of Anthropological and Ethnological Sciences*, Vol. 1, pp. 376–383. Science Council of Japan, Tokyo.

Schopf, J. W. (1967). Antiquity and evolution of precambrian life. In *McGraw-Hill Yearbook of Science and Technology*, pp. 47–55. McGraw-Hill Book Comp., New York.

Zuckerkandl, E. and Pauling, L. (1962). Molecular disease, evolution and genic heterogeneity. In *Horizons in Biochemistry* (eds. M. Kasha and B. Pullman), pp. 189–225. Academic Press, New York and London.

13

Molecular Evolution, Protein Polymorphism and the Neutral Theory (1982) (ed. M. Kimura), pp. 299–312.
Japan Scientific Societies Press, Tokyo / Springer-Verlag, Berlin

Evolutionary Relationship between [2Fe-2S] Ferredoxins and an Ancestral Ferredoxin

TOMITAKE TSUKIHARA,*[1] YUKITERU KATSUBE,*[1] TOSHIHARU HASE,*[2] KEISHIRO WADA,*[2] AND HIROSHI MATSUBARA*[2]

Faculty of Engineering, Tottori University, Tottori 680[1] and Department of Biology, Faculty of Science, Osaka University, Toyonaka, Osaka 560,*[2] Japan*

The name "ferredoxin" was originally given to the eight-iron protein from a nonphotosynthetic fermentative bacterium, *Clostridium pasteurianum* (Mortenson *et al.*, 1962). It functioned in the phosphoroclastic reaction system as an electron carrier and showed a very low redox potential. Tagawa and Arnon (1962) found that the bacterial ferredoxin and chloroplast factors were interchangeable in a photoreduction system of chloroplasts to reduce $NADP^+$ and proposed calling those electron carriers ferredoxin. Later Matsubara *et al.* (1967) determined the primary structure of spinach ferredoxin and found it to be homologous to the bacterial ferredoxins. The presence of a common ancestral form was proposed (Matsubara *et al.*, 1969) and supported (Keresztes-Nazy *et al.*, 1969; Schwartz and Dayhoff, 1978). It is now recognized that ferredoxins are distributed in a wide variety of organisms and classified into several groups according to their chromophore structures, such as [2Fe-2S], [4Fe-4S], and [8Fe-8S] and further, a new type of ferredoxin with a [7Fe-7S] cluster has recently been found in microorganisms (Stout *et al.*, 1980; Huynh *et al.*, 1980). A commonality is recognized in the primary structures of these ferredoxins, and a phylogenetic tree has been constructed to organize all bacterial and chloroplast-type ferredoxins (Matsubara *et al.*, 1980).

We have recently determined the tertiary structure of a [2Fe-2S] ferredoxin from a blue-green algae, *Spirulina platensis*, and discussed the relationship between the primary structures of chloroplast-type ferredoxins and the tertiary structure in terms of molecular evolution (Fukuyama *et al.*, 1980). In this paper we reestablish the evolutionary relationship among ferredoxins by comparing the amino acid sequences and the tertiary structure of the *S. platensis* ferredoxin. The structure of halobacterial ferredoxin is also predicted.

STRUCTURAL BASIS FOR THE ANCESTRAL FERREDOXIN AND ALIGNMENT OF [2Fe-2S] FERREDOXIN SEQUENCES

The repetitive structure of [8Fe-8S] ferredoxin composed of 55 residues has been found not only in the amino acid sequence (Tanaka *et al.*, 1964; Matsubara *et al.*, 1969) but also in the tertiary structure (Adman *et al.*, 1973; Rossmann and Argos, 1978). The ancestral forms of [8Fe-8S] ferredoxin were deduced as shown below in (a) (Matsubara *et al.*, 1969) and (b) (Schwartz and Dayhoff, 1978) based on a comparison of several bacterial and chloroplast-type ferredoxins.

(a) $\overset{1}{\text{A}}$—Y—V—I—D—A—D—S—C—I—D—C—G—A—C—A—A—E—
C—P—V—G—A—I—N—Q—G—D—$\overset{29}{\text{S}}$

(b) A—Y—V—I— —A—D—S—C—I—D—C—G—A—C—A—A—E—
C—P—V—G—A—I—S—Q—G—D—S.

These ancestral forms show no practical difference and have 27 identical sites out of 29. Form (a) will be used for the following discussion.

So far 28 amino acid sequences of [2Fe-2S] ferredoxin are known (Matsubara *et al.*, 1980; Takrui and Boulter, 1979a, b). Out of 28 sequences, 26 are the chloroplast type from plants and algae consisting of 93–99 residues and two others are from halobacteria with 128 residues. These are aligned as shown in Fig. 1. Halobacterial ferredoxins function as electron carriers in the photoreduction of cytochrome *c* catalyzed by ferredoxin-supplemented chloroplasts (Werber *et al.*, 1980). They are neither efficient in catalyzing the photo-induced electron transfer to $NADP^+$ nor do they form complexes with ferredoxin-$NADP^+$ reductase (Werber *et al.*, 1980), indicating a clear difference in the tertiary structures between

```
(1)  L. glauca          —AFKVKLLT—PDG—PKEFECPDDVYILDQAEELGIDLPYSCRAGS
(2)  S. oleracea        AAYKVTLVT—PTG—NVEFQCPDDVYILDAAEEEGIDLPYSCRAGS
(3)  M. sativa          ASYKVKLVT—PEG—TQEFECPDDVYILDHAEEEGIVLPYSCRAGS
(4)  S. nigra           ASYKVKLIT—PDG—PQEFECPDDVYILEHAEELGIDIPYSCRAGS
(5)  T. aestivum        ATYKVKLVT—PEG—EVELEVPDDVYILDQAEEEGIDLPYSCRAGS
(6)  C. esculenta       ATYKVKLVT—PSG—QQEFQCPDDVYILDQAEEVGIDLPYSCRAGS
(7)  P. americana I     ATYKVTLVT—PSG—TQTIDCPDDTYVLDAAEEAGLDLPYSCRAGS
(8)  P. americana II   AASYKVTFVT—PSG—TNTITCPADTYVLDAAEESGLDLPYSCRAGA
(9)  P. esculenta I     ATYKVTLVT—PSG—TQTIDCPDDTYVLDAAEEAGLDLPYSCRAGS
(10) P. esculenta II   AASYKVTFVT—PSG—TKTITCPADTYVLDAAEDTGLDLPYSCRAGA
(11) G. japonica        AIFKVKFLT—PDG—ERTIEVPDDKFILDAGEEAGLDLPYSCRAGA
(12) E. telmateia I     —AYKTVLKT—PSG—EFTLDVPEGTTILDAAEEAGYDLPFSCRAGA
(13) E. telmateia II    —AYKVTLKT—PDG—DITFDVEPGERLIDIASEKA—DLPLSCQAGA
(14) E. arvense I       —AYKTVLKT—PSG—EFTLDVPEGTTILDAAEEAGYDLPFSCRAGA
(15) E. arvense II      —AYKVTLKT—PDG—DITFDVEPGERLIDIGSEKA—DLPLSCQAGA
(16) S. quadricauda     ATYKVTLKT—PSG—DQTIECPDDTYILDAAEEAGLDLPYSCRAGA
(17) D. salina I        —SYMVTLKT—PSG—EEKVEVSPDSYILDAAEEAGVDLPYSCRAGS
(18) D. salina II       —AYKVTLKT—PSG—DQTIEVSPDAYILDAAEEAGLDLPYSCRAGA
(19) P. umbilicalis     ADYKIHLVSKEEGIDVTFDCSEDTYILDAAEEEGIELPYSCRAGA
(20) C. caldarium       ASYKIHLVNKDQGIDETIECPDDQYILDAAEEQGLDLPYSCRAGA
(21) S. platensis       ATYKVTLINEAEGINETIDCDDDTYILDAAEEAGLDLPYSCRAGA
(22) S. maxima          ATYKVTLISEAEGINETIDCDDDTYILDAAEEAGLDLPYSCRAGA
(23) N. muscorum I      ATFKVTLINEAEGTKHEIEVPDDEYILDAAEEEGYDLPFSCRAGA
(24) M. laminosus       ATYKVTLINEAEGLNKTIEVPDDQYILDAAEEAGIDLPYSCRAGA
(25) A. sacrum I        ASYKVTLKT—PDG—DNVITVPDDEYILDVAEEEGLDLPYSCRAGA
(26) A. sacrum II       ATYKVTLINEEEGINAILEVADDQTILDAGEEAGLDLPSSCRAGS
(27) H. halobium        LFEKAADAGLDGEDYGTMEVAEGEYILEAAEAQGYDWPFSCRAGA
(28) H. Dead Sea        VFGEASDMDLDDEDYGSLEVNEGEYILEAAEAQGYDWPFSCRAGA

(1)  CCSCAGKLVEGDL—DQSDQSFLDDEQIEEGWV—LTCAAYPRSDVVIETHKEEELTG
(2)  CSSCAGKLKTGSL—NQDDQSFLDDDQIDEGWV—LTCAAYPVSDVTIETHKEEELTA
(3)  CSSCAGKVAAGEV—NQSDGSFLDDDQIEEGWV—LTCVAYAKSDVTIETHKEEELTA
(4)  CSSCAGKLVAGSV—BZSBZSFLDDEZIEEGWV—LTCVAYPKSDVT(I,E,T,H)KEEELTA
(5)  CSSCAGKLVSGEI—DQSDQSFLDDDQMEAGWV—LTCHAYPKSDIVIETHKEEELTA
(6)  CSSCAGKVKVGDV—DQSDGSFLDDEQIGEGWV—LTCVAYPVSDGTIETHKEEELTA
(7)  CSSCTGKVTAGTV—DQEDQSFLDDDQIEAGFV—LTCVAFPKGDVTIETHKEEDIV—
(8)  CSSCAGKVTAGAV—NQEDGSFLEEEQMEAGWV—LTCVAYPTSDVTIETHKEEDLTA
(9)  CSSCTGKVTAGTV—DQEDQSFLDDDQIEAGFV—LTCVAYPKGDVTIETHKEEDIA—
(10) CSSCAGKVTAGSV—NQEDGSFLDDEQMEAGWV—LTCVAYPTSDVTIETHKEEDLSA
(11) CSSCTGKLLDGRV—DQSEQSFLDDDQMAEGFV—LTCVAYPAGDITIETHAEEKL——
(12) CSSCLGKVVSGSV—DQSEGSFLDDGQMEEGFV—LTCIAIPESDLVIETHKEEELF—
(13) CSTCLGKIVSGTV—DQSEGSFLDDEQIEQGYV—LTCIAIPESDVVIETHKEDEL——
(14) CSSCLGKVVSGSV—DESEGSFLDDGQMEEGFV—LTCIAIPESDLVIETHKEEELF—
(15) CSTCLGKIVSGTV—DQSEGSFLDDEQIEQGYV—LTCIAIPESDVVIETHKEDEL——
(16) CSSCAGKVEAGTV—DQSDQSFLDDSQMDGGFV—LTCVAYPTSDCTIATHKEEDLF—
(17) CSSCAGKVESGTV—DQSDQSFLDDDQM(D,S,G,F)V—LTCVAYATSDCTIVTHQEENLY—
(18) CSSCAGKVEAGTI—DQSDQSFLDDDQQGRGFV—LTCVAYATSDCTISTHQEESLY—
(19) CSTCAGKVTEGTV—DQSDQSFLDDEQMLKGYV—LTCIAYPESDCTILTHVEQELY—
(20) CSTCAGKLLEGEV—DQSDQSFLDDDQVKAGFV—LTCVAYPTSNATILTHQEESLY—
(21) CSTCAGTITSGTI—DQSDQSFLDDDQIEAGYV—LTCVAYPTSDCTIKTHQEEGLY—
(22) CSTCAGKITSGSI—DQSDQSFLDDDQIEAGYV—LTCVAYPTSDCTIQTHQEEGLY—
(23) CSTCAGKLVSGTV—DQSDQSFLDDDQIEAGYV—LTCVAYPTSDVVIQTHKEEDLY—
(24) CSTCAGKLISGTV—DQSDQSFLDDDQIEAGYV—LTCVAYPTSDCVIETHKEEELY—
(25) CSTCAGKLVSGPAPD—EDQSFLDDDQIQAGYI—LTCVAYPTGDCVIETHKEEALY—
(26) CSTCAGKLVSGAAPNQDDQAFLDDDQLAAGWV—MTCVAYPTGDCTIMTHQESEVL—
(27) CANCASIVKEGEI—DMDMQQILSDEEVEEKDVRLTCIGSPAADEVKIVYNAKHLDY
(28) CANCAAIVLEGDI—DMDMQQILSDEEVEDKNVRLTCIGSPDADEVKIVYNAKHLDY
```

Fig. 1. Amino acid sequences of [2Fe-2S] ferredoxins. Gaps shown by — are inserted to make the most probably alignment. In halobacterial ferredoxins 22 amino- and 6 carboxyl-terminal are deleted from this figure.

halobacterial and chloroplast-type ferredoxins in spite of their sequence similarity.

TERTIARY STRUCTURE OF *S. PLATENSIS* FERREDOXIN

The crystal structure of *S. platensis* ferredoxin has recently been revealed at 2.5 Å resolution (Fukuyama *et al.*, 1980). The polypeptide chain of ferredoxin folds mainly in a β-barrel-like structure as shown in

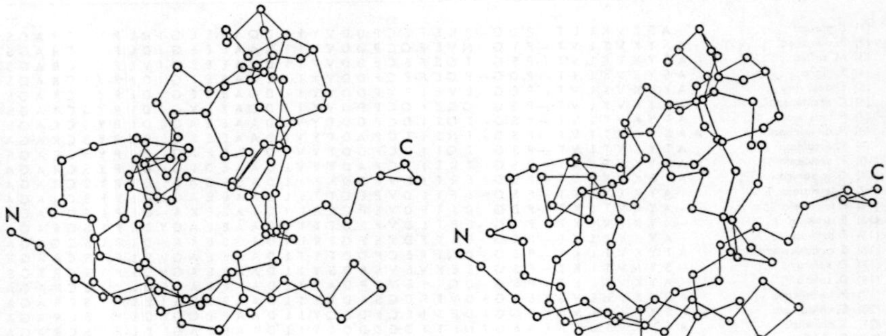

Fig. 2. Main chain fold of *S. platensis* ferredoxin is shown by a stereo pair of α-carbon. Most of the chain takes part in construction of a pseudo-β-barrel at the lower part of the molecule. The [2Fe-2S] cluster is located in the loop region outside of the barrel.

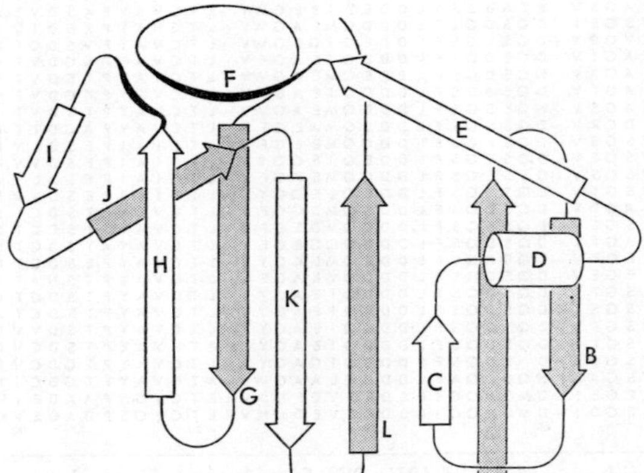

Fig. 3. A schematic drawing of *S. platensis* ferredoxin. The symbols A to L refer to the polypeptide fragments given in Table I. The bottom of the molecule consists of shadowed ones.

Fig. 2 with a stereo pair of C_α atoms. The main chain backbone is also schematically represented in Fig. 3. Polypeptide fragments are termed A, B, . . . , and L along the sequence (Table I). The fold exhibits a right-handed cross-over connection as found in most of the β-sheet proteins. Assignment of pseudo-β-sheet topology was carried out from a hydrogen-bonding scheme or extended conformation of the main chain. Secondary structures are shown in Fig. 4. There are hydrogen-bonding networks

TABLE I
MBDCs within Plant-type Ferredoxins and Those between Plant-type and Halobacterial-type Ferredoxins

Segment	A	B	C	D	E	F
Residues	1–11	12–21	22–24	25–32	33–40	41–49
MBDC (A)	0.47	0.90	0.80	0.36	0.36	0.13
MBDC (B)	1.55	1.26	1.05	0.44	0.68	0.27
Segment	G	H	I	J	K	L
Residues	50–58	59–66	67–73	74–81	82–88	89–101
MBDC (A)	0.57	0.32	0.33	0.36	0.41	0.51
MBDC (B)	0.92	1.28	0.74	0.73	0.79	1.36

(A), comparison between plant-type ferredoxins. (B), comparison between plant-type ferredoxin and halobacterial-type ferredoxin.

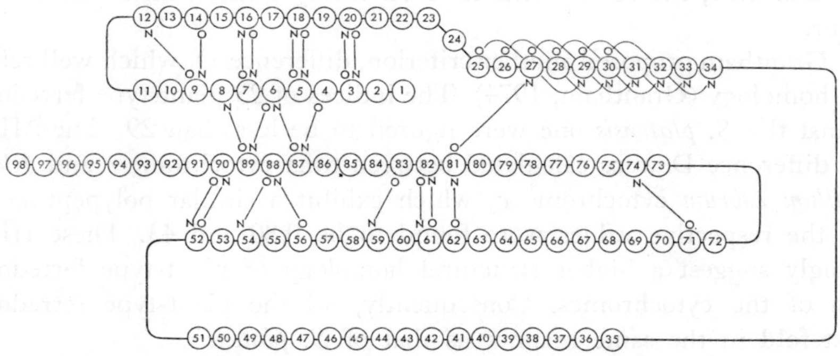

Fig. 4. Secondary structure of *S. platensis* ferredoxin. Hydrogen bonds between main chain atoms are shown by thin lines. The sequence numbers from 1 to 98 are given without a gap. The numbering from 59 to 76 is shifted by one residue from the common alignment in Fig. 1, and that from 78 to 98 by two residues.

among polypeptide fragments B, A, K-L, and G. The pseudo-β-barrel consists of fragments A, B, C, G, H, J, K, and L. The [2Fe-2S] cluster is at the top of the molecule outside of the barrel. Out of four intrinsic cysteinyl residues three are in fragment F and the other is at the end of J. Fragment F consisting of nine residues folds in a loop conformation. An α-helix of D and a successive extended chain of E are provided for the functionally important loop.

MUTABILITY AND TERTIARY STRUCTURE OF PLANT- AND HALOBACTERIAL-TYPE FERREDOXINS

In the cases of oxygen-carrying heme proteins (Perutz *et al.*, 1960), NAD^+-binding proteins, and polysaccharide-binding proteins (Rossmann and Argos, 1978) it has been pointed out that folding similarity is found among proteins with functional similarity. A high conservation of molecular folding has also been observed during the divergent evolution of protein sequences in several cases (Rossmann and Argos, 1978; Dickerson, 1980). An electron density map of *Aphanothece sacrum* (a bluegreen algae) ferredoxin at 3.9 Å resolution revealed that the protein folded in the same manner as in the *S. platensis* ferredoxin. The score of the minimum base difference per codon (MBDC) between these ferredoxins is 0.39. Since the MBDC score of any pair of plant-type ferredoxins (26 proteins) is less than 0.74, the ferredoxins have obviously come from a common ancestor.

Grantham proposed another criterion, difference D, which well reflects this homology (Grantham, 1974). The indices of 25 plant-type ferredoxins against the *S. platensis* one were figured to be less than 29. The MBDC and difference D were calculated between tuna cytochrome *c* and *Rhodospirillum rubrum* cytochrome c_2 which exhibit a similar polypeptide fold and the respective values were found to be 0.80 and 43. These criteria strongly suggest a higher structural homology of plant-type ferredoxins than of the cytochromes. Consequently, all the plant-type ferredoxins must fold in the same manner as that of *S. platensis*.

The MBDCs were calculated among chloroplast-type ferredoxins for each position and fragment along the polypeptide chain and the average score for each fragment is listed in Table I. The mutability is not uniform in all over the fragments and positions. Those in fragments B and C are intensive as compared with those in the others and a high homology is observed in the primary structures of fragments D, E, F, H, I, J, and K during evolution. These polypeptide fragments, except for J, are located on one side of the β-barrel. The average MBDC score for each position is plotted against the distance between its C_α atom and the center of the [Fe-S] cluster as shown in Fig. 5. The polypeptide chain of fragment F containing intrinsic cysteinyl residues folds as a loop where the main chain itself is generally unstable. Some of the conservative residues must be important for preserving the active center conformation. The other con-

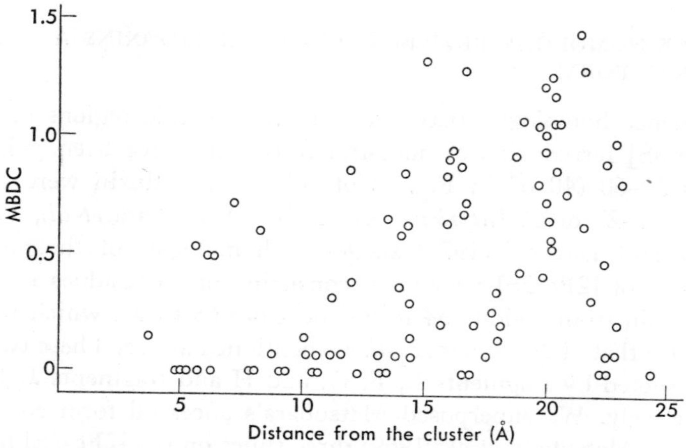

Fig. 5. Average MBDC scores among 26 plant-type ferredoxins are plotted for each position along the sequence against the distance between the center of the cluster and each C_α atom.

servative residues may play key roles in intermolecular interaction with other proteins such as ferredoxin-$NADP^+$ reductase.

The average MBDCs between chloroplast-type and halobacterial-type ferredoxins are also shown in Table I for each fragment. The overall score for fragments D through K is 0.74. This value is lower than the score between tuna cytochrome *c* and *R. rubrum* cytochrome c_2 which exhibit a common folding pattern. It is inferred that the two types of ferredoxin fold in a very similar manner at least in fragments D through K.

The score for fragment F among chloroplast-type ferredoxins is 0.13 and that between chloroplast-type and halobacterial-type ferredoxins is 0.23, which are very low and not unexpected because fragment F together with a part of J forms clusters which have the same physico-chemical properties. Low scores are also observed for helical fragment D, suggesting that this region may be important for maintaining the loop structure of fragment F. Fragments H, I, J, and K have low scores among chloroplast-type, suggesting that the molecular surface of chloroplast-type ferredoxin constructed by these fragments (upper left of molecule in Fig. 2) is considerably different from the corresponding region of halobacterial-type ferredoxin, and that this region may take part in the interaction with ferredoxin-$NADP^+$ reductase.

SEQUENCE HOMOLOGY BETWEEN [2Fe-2S] FERREDOXINS AND ANCESTRAL FORM

Sequence homologies between particular peptide regions in [2Fe-2S] and [8Fe-8S] ferredoxins or ancestral ferredoxin have been pointed out. Residues 38–60 (40–63 in Fig. 1) of spinach ferredoxin were compared with those 7–27 of *C. butyricum* ferredoxin (Matsubara *et al.*, 1967), and Schwartz and Dayhoff (1978) suggested homologies of the ancestor to two regions of [2Fe-2S] ferredoxin consisting of 28 residues in the polypeptide chain from residue 34 to 64 and from 65 to 94, which were compatible with that of Matsubara *et al.* as mentioned above. These two regions are constructed by fragments E, F, G, and H and fragments I, J, K, and L, respectively. We superposed Matsubara's ancestral form consisting of 29 residues (Matsubara *et al.*, 1969) three times on the [2Fe-2S] ferredoxin sequence. The first region is the chain from 1 to 31, the second, the chain from 33 to 64, and the third, the chain from 65 to 94. In order to make the alignment better, we modified the alignments proposed by the authors mentioned above as follows. The cysteinyl residues 41, 46, and 49 of the second region were made to correspond to cysteinyl residues 9, 12, and 15 of the ancestor, respectively, by introducing gaps in 10 and 11. Only one cysteinyl residue 81 in the last region of chloroplast-type ferredoxins is preserved and corresponds to the one at 15 of the ancestor. The alignment is summarized in Fig. 6.

The MBDCs between the 28 [2Fe-2S] ferredoxins and the ancestor

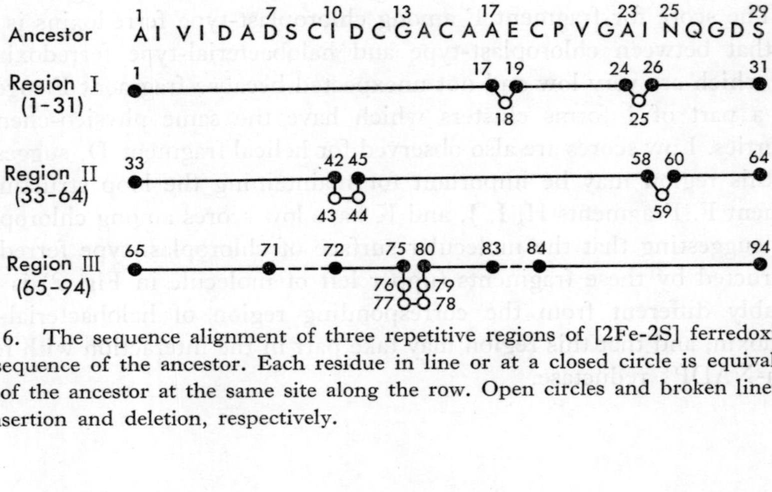

Fig. 6. The sequence alignment of three repetitive regions of [2Fe-2S] ferredoxin and the sequence of the ancestor. Each residue in line or at a closed circle is equivalent to that of the ancestor at the same site along the row. Open circles and broken lines refer to insertion and deletion, respectively.

TABLE II
Sequence Comparisons between [2Fe-2S] Ferredoxins and an Ancestral Form as Expressed in MBDCs

	I	II	III
(1)–(26)/ancestor	1.31 ± 0.06	0.95 ± 0.09	1.11 ± 0.07
(27), (28)/ancestor	1.02	1.25	1.21

I, II, and III refer to the compared segments in the ferredoxins and ancestor. I: [2Fe-2S] ferredoxin segments 1–17, 19–24, and 26–31 were compared with ancestral segments 1–17, 18–23, and 24–29, respectively, in a total of 29 residues. II: ferredoxin segments 33–42, 45–58, and 60–64 were compared with ancestral segments 1–10, 11–24, and 25–29, respectively, in a total of 29 residues. III: ferredoxin segments 65–71, 72–75, 80–83, and 84–94 were compared with ancestral segments 1–7, 10–13, 14–17, and 19–29, respectively, in a total of 26 residues.

were enumerated for each region as listed in Table II. Comparison of the first regions between the two halobacterial-type ferredoxins and the ancestor gave an average MBDC of 1.02, suggesting a measure of the possible evolutionary relationship between them. However, an average MBDC of 1.31 was scored between the 26 chloroplast-type ferredoxins and the ancestor. The MBDCs for the second and third regions between the chloroplast-type and ancestor were 0.95 and 1.11, respectively, while those between the halobacterial-type and ancestor were 1.25 and 1.21, respectively. The score in the third region of the chloroplast-type ferredoxin was slightly higher than that in the second. However, these scores, obtained from many sequence comparisons, are reliable enough and reasonably lower than those from a random comparison. From these considerations it may be concluded that both fragments E, F, G, and H, and fragments I, J, K, and L in the chloroplast-type ferredoxin have evolved from the ancestor.

The halobacterial-type ferredoxins are less divergent from the ancestor in fragments A through D than in the other regions and the chloroplast-type ferredoxins are less divergent from the ancestor in fragments E through H and I through L. As the mutational rate of each residue site is considered to be largely dependent on the importance of the site's function, the different pattern of divergency at different regions of the molecules would be attributable to the functional differences in chloroplast-type and halobacterial-type ferredoxins.

Now, let us compare the ancestor with the common ferredoxins derived from 26 chloroplast-type ferredoxins and 2 halobacterial-type ferredoxins, which are presented in Table III. The most frequently occurring residue

TABLE III
Common Sequence of Two [2Fe-2S] Ferredoxin Types and Their Similarity to the Ancestor

Common sequence derived from plant and algae ferredoxins	
Region I (1–32)	ATYKVTLVTEPSGIDQTIECPDDTYILDAAEE
Region II (33–64)	AGLDLPYSCRAGACSSCAGKVVSGTV—DQSDQ
Region III (65–94)	SFLDDDQIEAGXV—LTCVAYPTSDVTIETH
Common sequence derived from two halobacterial ferredoxins	
Region I (1–32)	XFXXAXDXXLDXEDYGXXEVXEGEYILEAAEA
Region II (33–64)	QGYDWPFSCRAGACANCAXIVXEGXI—DMDMQ
Region III (65–94)	QILSDEEVEXKXVRLTCIGSPXADEVKIVY

X refers to undetermined residues and — to the gap shown in Fig. 1.

TABLE IV
Common Sequence Comparisons between [2Fe-2S] Ferredoxins and the Ancestor Form as Expressed in MBDCs

Region	I (1–32)	II (33–64)	III (65–94)
MBDCs against the ancestor			
Chloroplast-type	1.24 (29)	0.86 (29)	1.04 (26)
Halobacterial-type	0.85 (20)	1.23 (26)	1.17 (24)
MBDCs between two types	1.00 (21)	0.68 (28)	0.88 (26)

Numbers in parentheses are number of sites compared.

at each site among the ferredoxins in each group, though only two samples are available of the halobacterial ferredoxins, was taken as the representative residue. As shown in Table IV, every MBDC between each region and the ancestor, on the basis of common sequences, was lower than the corresponding average value shown in Table II. This result strongly supports the concept that [2Fe-2S] ferredoxin was derived from a common ancestor by gene triplication and it has diverged to chloroplast-type and halobacterial-type ferredoxins.

DIVERGENT EVOLUTION OF AN ANCESTOR OF [2Fe-2S] FERREDOXINS

As mentioned above, [2Fe-2S] ferredoxin was generated by gene triplication of the ancestor followed by structural change of the cluster, that is, from [8Fe-8S] to [2Fe-2S]. Three regions in [2Fe-2S] ferredoxins, A-D, E-H, and I-L, have come from the ancestor.

Polypeptide folding is usually conservative during divergent evolution as in the cases of cytochrome *c* (Dickerson, 1980), globins, and NAD^+-binding proteins (Rossmann and Argos, 1978). Table V gives a comparison between the folds of the regions in the [2Fe-2S] ferredoxin and of the ancestor deduced from the [8Fe-8S] ferredoxin exhibiting a 2-fold symmetry. The average r.m.s. of distances were figured out by the least-squares method programmed by Nishikawa (personal communication, 1980) which minimized the sum of the squares of distances between equivalent atoms. The pairs of corresponding atoms were made as in the case of sequence comparison. The distances for three regions were 9.50, 5.81, and 8.82 Å, respectively. Though the value of 5.81 Å is smaller than the others, it is difficult to deduce the folding convergency from the distance itself. Structural comparisons within the [2Fe-2S] molecule were also carried out by the same method mentioned above. The average r.m.s. distance between G, H, and K, L were 2.5 Å for 17 pairs of equivalent atoms. This may indicate the evolutionary relationship between the two regions, but the reliability of this speculation is low, because the two short polypeptides in the β-barrels as in the present case sometimes show a high structural similarity to each other independent of their evolutionary origins. There is no clear structural repetition corresponding to gene triplication.

The polypeptide conformation of F containing three intrinsic cysteinyl residues seems to be stabilized by cluster coordination. In the cases of myoglobin and NAD^+-binding proteins the cofactors do not affect their folding, while cytochrome *c* protein folding is strongly affected by the heme group (Dickerson *et al.*, 1971) in a manner similar to the divergent molecules. Since the [2Fe-2S] and [8Fe-8S] ferredoxins have clusters different from each other in size and nature, their respective folds seem to be affected by the characteristic interaction between the cluster and the

TABLE V

Comparisons between Three Peptide Regions and Half-chain of an [8Fe-8S] Ferredoxin

	Average r.m.s. distance (Å)
The first region (A, B, C, and D)	9.50
The second region (E, F, G, and H)	5.81
The third region (I, J, K, and L)	8.82

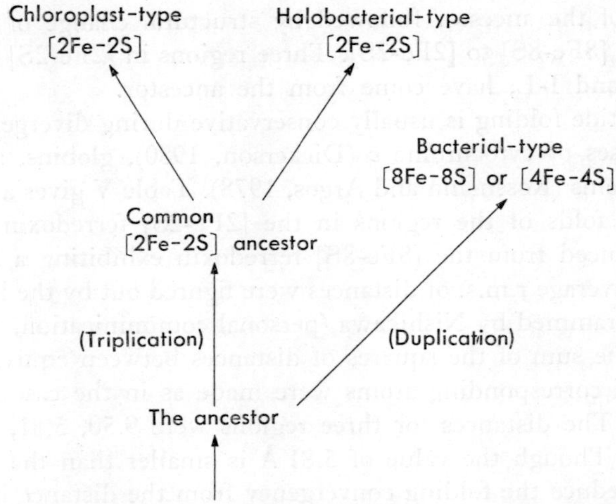

Fig. 7. Evolutionary process of [2Fe-2S] ferredoxins. A common [2Fe-2S] ancestor generated by triplication diverged to two types of [2Fe-2S] ferredoxins.

main chain. That is, gene triplication followed by structural change in the clusters has introduced a divergent evolution of the folding.

After gene triplication fragments A, B, and C in the first part of the chloroplast-type ferredoxin diverged more than those in the halobacterial-type ferredoxin. This difference may be attributed to their different physiological functions (Werber *et al.*, 1980). The fact that the first part except for fragment D in chloroplast-type ferredoxin shows a very high divergency in comparison with the others suggests that the first region in chloroplast-type ferredoxin, which is located far from the cluster, is physiologically less important. The second part containing the three intrinsic cysteinyl residues in chloroplast- and halobacterial-type ferredoxins shows a higher similarity than in the other regions, suggesting that the divergency is restrained to preserve the common properties for transferring electrons in the two groups of ferredoxins.

The overall evolutionary process is illustrated in Fig. 7.

SUMMARY

Ferredoxins are classified into several groups according to their chromophore structures, such as [2Fe-2S], [4Fe-4S], [8Fe-8S], and [7Fe-7S]. So far 28 amino acid sequences of [2Fe-2S] ferredoxin are known.

Out of 28 proteins, 26 from plants and algae are called chloroplast-type ferredoxin, and two others are from halobacteria. The common sequences of the chloroplast-type and the halobacterial ferredoxins, respectively, have been established. The common sequence of chloroplast-type protein is less divergent from the ancestral sequence than any chloroplast-type ferredoxins. The polypeptide of [2Fe-2S] ferredoxin is divided into three segments comprising 26 to 29 amino acid residues. Sequence comparison among the ancestral and common forms of the chloroplast-type and the halobacterial ferredoxins suggests that each of three polypeptide segments of [2Fe-2S] ferredoxins has come from the ancestral ferredoxin. The variation in amino acid sequence among [2Fe-2S] ferredoxins seems to be strongly related to the tertiary structure and function of these ferredoxins.

REFERENCES

Adman, E. T., Sieker, L. C., and Jensen, L. H. (1973). The structure of a bacterial ferredoxin. *J. Biol. Chem.* **248**, 3987–3996.

Dickerson, R. E. (1980). Evolution and gene transfer in purple photosynthetic bacteria. *Nature* **283**, 210–212.

Dickerson, R. E., Takano, T., Eisenberg, D., Kalai, O. B., Samon, L., Cooper, A., and Margoliash, E. (1971). Ferricytochrome *c*. I. General features of the horse and bovine proteins at 2.8 A resolution. *J. Biol. Chem.* **246**, 1511–1535.

Fukuyama, K., Hase, T., Matsumoto, S., Tsukihara, T., Katsube, Y., Tanaka, N., Kakudo, M., Wada, K., and Matsubara, H. (1980). Structure of *S. platensis* 2Fe-2S ferredoxin and evolution of chloroplast-type ferredoxins. *Nature* **286**, 522–524.

Grantham, R. (1974). Amino acid difference formula to help explain protein evolution. *Science* **185**, 862–864.

Huynh, B. H., Moura, J.J.G., Moura, I., Kent, T. A., LeGall, J., Xavier, A. V., and Munck, E. (1980). Evidence for a three-iron center in a ferredoxin from *Desulfovibrio gigas*. *J. Biol. Chem.* **255**, 3242–3244.

Keresztes-Nahy, S., Perini, F., and Margoliash, E. (1969). Primary structure of alfalfa ferredoxin. *J. Biol. Chem.* **244**, 981–995.

Matsubara, H., Hase, T., Wakabayashi, S., and Wada, K. (1980). Structure and evolution of chloroplast- and bacterial-type ferredoxins. In *Evolution of Protein Structure and Function* (eds. D. S. Sigmen and M.A.B. Brazier), pp. 245–266. Academic Press, New York.

Matsubara, H., Jukes, T. H., and Cantor, C. R. (1969). Structural and evolutionary relationships of ferredoxins. *Brookhaven Symp. Biol.* **21**, 201–211.

Matsubara, H., Sasaki, R. M., and Chain, R. K. (1967). The amino acid sequence of spinach ferredoxin. *Proc. Natl. Acad. Sci. U.S.* **57**, 439–445.

Mortenson, L. E., Valentine, R. C., and Carnahan, J. E. (1962). An electron transport factor from *Clostridium pasteurianum*. *Biochem. Biophys. Res. Commun.* **7**, 448–452.

Perutz, M. F., Rossmann, M. G., Cullis, A. F., Muirhead, H., Will, G., and North, A.C.T.

(1960). Structure of haemoglobin. A three dimensional fourier synthesis at 5.5 A resolution, obtained by X-ray analysis. *Nature* **185**, 416–422.

Rossmann, M. G. and Argos, P. (1978). The taxonomy of binding sites in proteins. *Mol. Cell. Biochem.* **21**, 161–182.

Schwartz, R. M. and Dayhoff, M. O. (1978). Other electron transport proteins. In *Atlas of Protein Sequence and Structure* (ed. M. O. Dayhoff), Vol. 5, pp. 45–48. National Biomedical Research Foundation, Washington, D. C.

Stout, C. D., Ghosh, D., Pattabhi, V., and Robbins, A. H. (1980). Iron-sulfur clusters in *Azotobacter* ferredoxin at 2.5 A resolution. *J. Biol. Chem.* **255**, 1797–1800.

Tagawa, K. and Arnon, D. I. (1962). Ferredoxins as electron carriers in photosynthesis and in the biological production and consumption of hydrogen gas. *Nature* **195**, 537–543.

Takruri, I.A.H. and Boulter, D. (1979a). The amino acid sequence of ferredoxin from *Triticum aestivum* (wheat). *Biochem. J.* **179**, 373–378.

Takruri, I.A.H. and Boulter, D. (1979b). The amino acid sequence of ferredoxin from *Sambucus nigra*. *Phytochemistry* **18**, 1481–1484.

Tanaka, M., Nakashima, T., Benson, A., Mower, H. F., and Yasunobu, K. T. (1964). The amino acid sequence of *Clostridium pasteurianum* ferredoxin. *Biochem. Biophys. Res. Commun.* **16**, 422–427.

Werber, M. M., Shahak, Y., and Avron, M. (1980). One-site reactivity of halobacterial 2Fe-ferredoxin as a plant ferredoxin substrate. *FEBS Lett.* **113**, 111–114.

14

Molecular Evolution, Protein Polymorphism and the Neutral Theory (1982) (ed. M. Kimura), pp. 315–329.
Japan Scientific Societies Press, Tokyo / Springer-Verlag, Berlin

A Coevolutionary Aspect Observed between Cytochrome Oxidase and Cytochrome *c*

TATEO YAMANAKA, YOSHIHIRO FUKUMORI, YUHSUKE KAMITA,*1 AND KEIKO FUJII*2

Department of Biology, Faculty of Science, Osaka University, Toyonaka 560, Japan

Aerobic organisms oxidize organic or inorganic substances with molecular oxygen to acquire energy which is necessary for their life processes. In these organisms, cytochrome oxidases play an important role in the reduction of molecular oxygen, and in general, cytochrome *c* acts as the direct electron donor for the oxidases.

In previous studies (Yamanaka and Okunuki, 1964, 1968; Yamanaka, 1967, 1972, 1973, 1975), we have revealed that there is a distinct specificity in the reactions of cytochrome *c* with certain redox enzymes; *e.g.*, eukaryotic cytochromes *c* react rapidly with cow cytochrome oxidase but react poorly with *Pseudomonas aeruginosa* nitrite reductase (=*P. aeruginosa* cytochrome oxidase). Many bacterial cytochromes *c* react rapidly with the *P. aeruginosa* enzyme but react poorly with the cow oxidase, while a few are known which react relatively rapidly with the cow oxidase and react slowly with the *P. aeruginosa* enzyme. Some bacterial cytochromes *c* are known which

*1 Present address: Institute for Biological Science, Sumitomo Chemical Co., Ltd., Takarazuka, Hyogo 665.
*2 Present address: Shionogi Research Laboratory, Osaka 553.

react poorly with both the enzymes. On the basis of the occurrence of cytochromes *c* with various enzymatic properties as mentioned above, we have predicted that cytochrome oxidase might have coevolved with cytochrome *c* during the evolution of the organisms so that oxidases with different specificities for the cytochrome might exist (Yamanaka, 1972, 1973; Yamanaka and Fukumori, 1978).

Recently, we have obtained pure or partially purified preparations of cytochrome oxidases derived from several bacteria and determined their reactivities with cytochromes *c* isolated from various organisms (Yamanaka and Fukumori, 1978; Yamanaka *et al.*, 1979b; Yamanaka and Fujii, 1980; Yamanaka *et al.*, 1981). As we had predicted, the specificities for the cytochrome *c* of these oxidases differ from those of the *P. aeruginosa* and cow enzymes. That is each oxidase shows a distinct specificity for cytochrome *c* and the maximal activity is observed for each enzyme in the reaction with cytochromes *c* which are derived from organisms in an evolutionary stage similar to the host organisms of the oxidase.

In the present investigation, we further developed our previous study (Yamanaka *et al.*, 1979b) and show evidence which suggests that cytochrome *c* and cytochrome oxidase might have coevolved.

CYTOCHROME OXIDASES

At present, cytochrome oxidases derived from several kinds of organisms are available. The oxidases used in the present study were purified from *P. aeruginosa* (Yamanaka, 1963, 1964), *Nitrosomonas europaea* (Erickson *et al.*, 1972), *Thiobacillus novellus* (Yamanaka and Fujii, 1980), *Paracoccus denitrificans* (Ludwig and Schatz, 1980), *Nitrobacter agilis* (Yamanaka *et al.*, 1979a, 1981), and the cow (Okunuki, 1966). The *P. aeruginosa* enzyme is nitrite reductase and acts also as a cytochrome oxidase (Yamanaka, 1964). Therefore, in the present investigation, it was used as a cytochrome oxidase. The enzyme is a cytochrome *cd*-type oxidase, while the other five enzymes are cytochrome aa_3-type oxidases. The enzymes except for the *N. europaea* and *P. denitrificans* oxidases were highly purified as previously described. The *P. denitrificans* oxidase was partially purified according to a method similar to that used for the purification of the *T. novellus* oxidase. The enzyme preparation used showed absorption peaks at 415, 517, and 552 nm besides 442 and 605 nm when it was reduced by $Na_2S_2O_4$. Therefore, it was still contaminated with cytochrome c_1, although Ludwig and Schatz (1980) have obtained the enzyme in a pure state.

Nitrosomonas europaea cytochrome oxidase purified by Erickson *et al.* (1972) was kindly supplied by Dr. A. B. Hooper. We also obtained a cruder preparation of it using a method similar to that used for the purification of the *T. novellus* oxidase. So far as has been tested with a few kinds of cytochromes *c*, the specificity for the cytochrome *c* of the *N. europaea* oxidase did not differ between the highly purified and the cruder preparations. The reactivity for cytochrome *c* of the enzyme was determined with the cruder preparation in the present study.

1. *Pseudomonas aeruginosa Cytochrome Oxidase*

Although the biological role of this enzyme is to reduce nitrite to nitric oxide, it has the general properties of a cytochrome oxidase (Yamanaka, 1964; Yamanaka *et al.*, 1963). The enzyme molecule is composed of one peptide, and has one molecule each of haem *c* and haem *d*, but does not possess Cu (Yamanaka, 1964, 1966). Its molecular weight is 67,700 (Nagata *et al.*, 1970). The enzyme shows a very complicated absorption spectrum; the reduced form shows absorption peaks at 418, 460, 523, 549, 554, and 625 nm at pH 5.7 (Yamanaka, 1963, 1966).

2. *Thiobacillus novellus Cytochrome Oxidase*

This enzyme resembles mitochondrial cytochrome oxidase as regards the absorption spectrum and prosthetic groups. It shows absorption peaks at 602 and 442 nm in the reduced state, and has 1 mol of haem *a* and 1 g atom of Cu per 55,000 g (Yamanaka and Fujii, 1980). The enzyme molecule is composed of two kinds of subunits of 32,000 and 23,000 daltons (Yamanaka *et al.*, 1979a).

3. *Nitrobacter agilis Cytochrome Oxidase*

This enzyme is very similar to the *T. novellus* enzyme in various properties. It shows absorption peaks at 606 and 442 nm in the reduced state. The enzyme contains 2 mol of haem *a* and 2 g atoms of Cu per 82,000 g (Yamanaka *et al.*, 1981). The enzyme is composed of two kinds of subunits of 40,000 and 27,000 daltons (Yamanaka *et al.*, 1979a, 1981).

4. *Cow Cytochrome Oxidase*

This enzyme shows absorption peaks at 443 and 605 nm in the reduced state, and has 1 mol of haem *a* and 1 g atom of Cu per 90,000 g (Yonetani, 1963; Okunuki, 1966; Caughey *et al.*, 1976; Malmström, 1979). The enzyme molecule is composed of 6–7 kinds of subunits (see Table I).

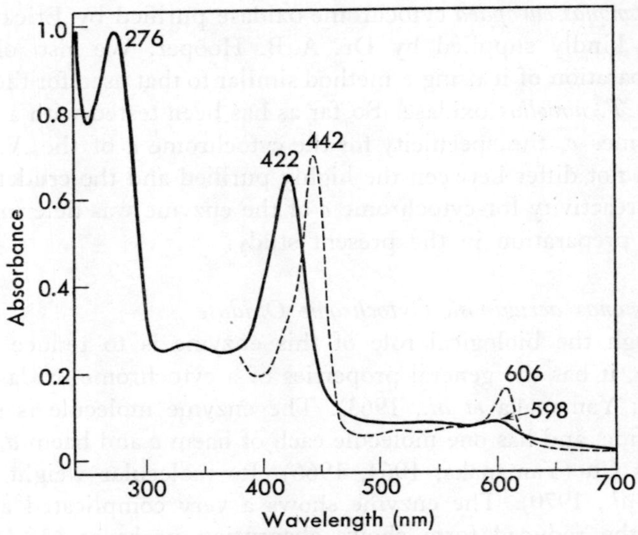

Fig. 1. Absorption spectrum of *N. agilis* cytochrome oxidase. The enzyme was dissolved in 20 mM Tris-HCl buffer, pH 8.0, containing 1% Emasol 1130 at a concentration of 7.2 μM. —— oxidized; ----- reduced with $Na_2S_2O_4$ (Yamanaka *et al.*, 1981).

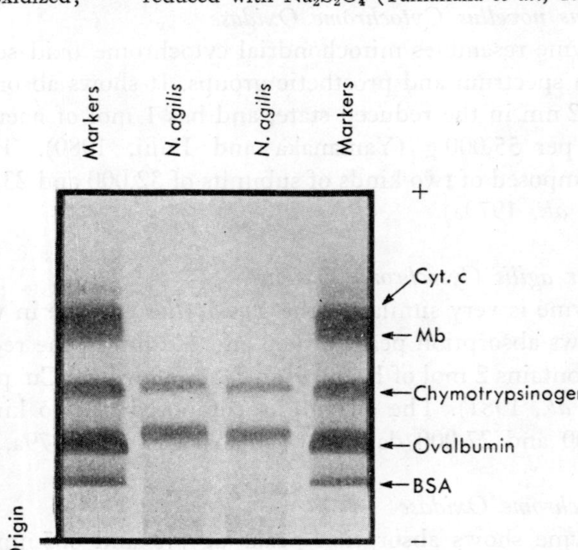

Fig. 2. SDS-gel electrophoretic patterns of *N. agilis* cytochrome oxidase. The gel was produced by polymerization of 10% acrylamide. The electrophoresis was performed in 0.1 M Tris-acetate, pH 8.0, in the presence of 0.1% SDS (Yamanaka *et al.*, 1981).

TABLE I
Comparison of Molecular Weights of the Subunits between Bacterial and Eukaryotic Cytochrome Oxidases[a]

Subunit		Apparent molecular weight				
		T. novellus[c]	*N. agilis*[b,c]	*P. denitrificans*[d]	*S. cerevisiae*[e]	Cow[f]
Mitochondrial	I	32,000	40,000	45,000	40,000	35,400
	II	23,000	27,000	28,000	33,000	24,000
	III				22,000	21,000
Cytoplasmic	IV				14,500	16,800
	V				12,700	12,400
	VI				12,700	8,200
	VII				4,600	4,400

[a] Cytochrome aa_3-type or cytochrome *a*-type bacterial enzymes are listed. Besides the enzymes listed here, bacterial cytochrome oxidases have been purified which possess haems *a* and *c* (Sone and Kagawa, 1979; Fee *et al.*, 1980; Hon-nami and Oshima, 1980).
[b] Although the molecular weights of the two subunits were deterimined to be 51,000 and 31,000, respectively, by Ferguson plot, those listed here were determined by electrophoresis at one gel concentration (10%) like the cases of the other enzymes listed.
[c] Yamanaka *et al.* (1979a).
[d] Ludwig and Schatz (1980).
[e] Poyton and Schatz (1975).
[f] Downer *et al.* (1976).

The larger three subunits are synthesized in the mitochondrion, while the smaller four subunits are synthesized on the cytoplasmic ribosomes (Mason and Schatz, 1973; Bernstein *et al.*, 1978; Hare *et al.*, 1980).

CYTOCHROMES *C*

Cytochromes *c* purified from the following organisms were used as electron donors for the oxidases: *P. aeruginosa* (Horio *et al.*, 1960), *N. europaea* (Yamanaka and Shinra, 1974), *Chlorobium limicola* f. *thiosulfatophilum* (Meyer *et al.*, 1968; Yamanaka and Okunuki, 1968), *Rhodospirillum rubrum* (Horio and Kamen, 1961), *T. novellus* (Yamanaka *et al.*, 1971), *P. denitrificans* (Scholes *et al.*, 1971), *N. agilis* (Ketchum *et al.*, 1969), *Candida krusei* (Yamanaka *et al.*, 1964), *Saccharomyces oviformis* (Motonaga *et al.*, 1965), tuna (Hagihara *et al.*, 1958), bonito (Hagihara *et al.*, 1958), the cow (Hagihara *et al.*, 1958), and human (Matsubara and Yasunobu, 1961). Horse cytochrome *c* was purchased from Sigma Chemical Company (U.S.A.). *Rhodospirillum rubrum* cytochrome *c* (cytochrome c_2) and bonito cytochrome *c* were kindly supplied by Dr. Horio and Dr. Kakudo, respectively.

REACTIVITIES OF CYTOCHROMES *C* WITH CYTOCHROME OXIDASES

The reactivities of cytochromes *c* with the oxidases described above were determined spectrophotometrically; the decrease in the absorbance at the α peak of each cytochrome *c* was followed with time when cytochrome oxidase was aerobically added to ferrocytochrome *c*. The relative reactivity was obtained on the basis of the molecular activity per min which was calculated from the initial reaction rate. The relative reactivities with various cytochromes *c* of these six oxidases are shown in Fig. 3.

The *P. aeruginosa* enzyme reacted rapidly with *P. aeruginosa* cytochrome *c* and reacted relatively rapidly with cytochrome *c*'s derived from *N. europaea* and *C. limicola* f. *thiosulfatophilum*, while it scarcely reacted with eukaryotic cytochromes *c*. The enzyme reacted only slowly with *T. novellus*, *P. denitrificans*, and *R. rubrum* cytochromes *c*.

The *N. europaea* oxidase reacted most rapidly with *P. aeruginosa* cytochrome *c* among the enzyme possessing haem *a* tested. It reacted rapidly with *S. oviformis* and *C. krusei* cytochromes *c*, while it reacted relatively slowly with the other eukaryotic cytochromes *c* tested. It seemed curious that the enzyme reacted fairly slowly with its native cytochrome *c*. The enzyme reacted very poorly with cytochromes *c* derived from *R. rubrum* and *T. novellus*.

The *T. novellus* oxidase reacted very poorly with cytochromes *c* derived from *P. aeruginosa*, *N. europaea*, and *R. rubrum*, while it reacted rapidly with cytochromes *c* derived from *C. krusei*, tuna, and bonito in addition to its native cytochrome *c*. The enzyme reacted relatively slowly with horse and cow cytochromes *c*, while its reactivity with human cytochrome *c* was distinctly higher than that with horse and cow cytochromes *c*. Although the reactivity of the enzyme with *S. oviformis* cytochrome *c* was very low in 10 mM phosphate buffer, it was maximal in 40 mM buffer. In the buffer at the latter concentration the reactivity was comparable to those of tuna and bonito cytochromes *c* under the same experimental conditions. The reactivity with cytochromes *c* decreased with increase in the salt concentration or ionic strength with all cytochromes *c* tested but *S. oviformis* and *C. krusei* cytochromes *c* (Yamanaka and Fujii, 1980). Eukaryotic cytochromes *c* which react rapidly with the *T. novellus* enzyme

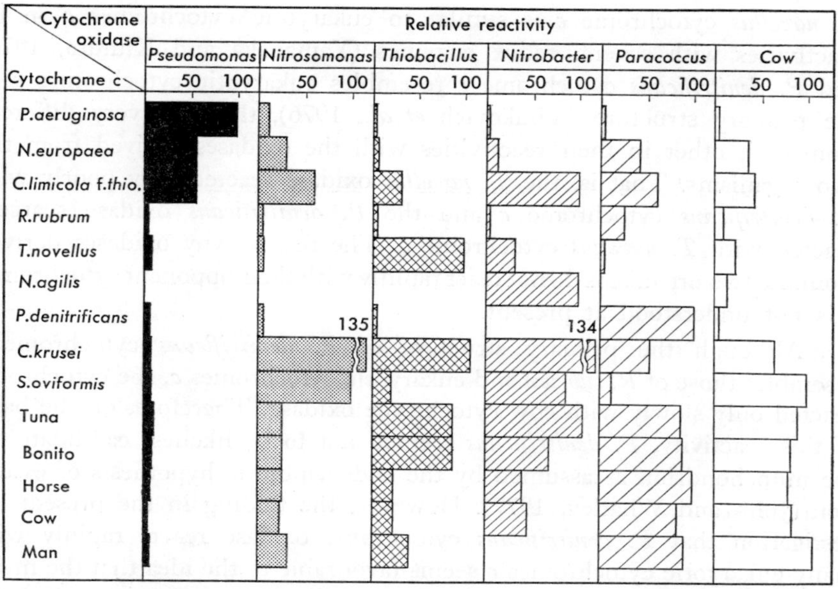

Fig. 3. Specificities for cyrtochrome *c* of cytochrome oxidases derived from *P. aeruginosa, N. europaea, T. novellus, N. agilis, P. denitrificans*, and the cow. The reactions catalyzed by these enzymes were carried out in 40 mM phosphate buffer, pH 6.5 (*P. aeruginosa* and cow enzymes) or in 10 mM phosphate buffer, pH 5.5 (*N. europaea, T. novellus*, and *P. denitrificans* oxidases), and at pH 6.0 (*N. agilis* oxidase). The reactivities of cytochromes *c* were expressed as relative values; the molecular activity (mol of cytochrome *c* oxidized per mol of enzyme) per min was taken as 100% which was observed when each enzyme reacted with its native cytochrome *c* except for *N. europaea* and cow enzymes. In the cases of the latter two enzymes, their respective reactivities with *S. oviformis* cytochrome *c* were taken as 100%.

seem to have a tyrosine residue at the 46th position in their primary structures.

The *N. agilis* enzyme reacted rapidly with cytochromes *c* derived from *C. limicola* f. *thiosulfatophilum, C. krusei, S. oviformis*, and tuna besides its native cytochrome *c*, while its reactivity with horse and cow cytochromes *c* was relatively low.

The *P. denitrificans* enzyme reacted rapidly with its native cytochrome *c* and all the eukaryotic cytochromes *c* tested except for *S. oviformis* cytochrome *c*. Further, the enzyme reacted relatively rapidly with *N. europaea, C. limicola* f. *thiosulfatophilum*, and *R. rubrum* cytochromes *c*. Although

T. novellus cytochrome *c* is similar to eukaryotic cytochromes *c* in its reactivities with several redox enzymes (Yamanaka and Kimura, 1974) and *P. denitrificans* cytochrome *c* resembles eukaryotic cytochromes *c* in the primary structure (Timkovich *et al.*, 1976), they are very different from each other in their reactivities with the oxidases derived from the two organisms. That is, the *T. novellus* oxidase reacted very poorly with *P. denitrificans* cytochrome *c* and the *P. denitrificans* oxidase scarcely reacted with *T. novellus* cytochrome *c*. The reason why oxidases derived from the two organisms do not react rapidly with their opponent cytochromes *c* is not understood at present.

Although the primary structure of *P. denitrificans* cytochrome *c* resembles those of *R. rubrum* and eukaryotic cytochromes *c*, the cytochrome reacted only slowly with cow cytochrome oxidase. Therefore, on the basis of this reactivity, *P. denitrificans* seemed not to be likeliest candidate for the mitochondrion as assumed by the endosymbiotic hypothesis of evolution (John and Whatley, 1977). However, the finding in the present investigation that *P. denitrificans* cytochrome oxidase reacts rapidly with many eukaryotic cytochromes *c* seems favourable to the idea that the mitochondrion may have been evolutionarily derived from the bacterium.

Cow cytochrome oxidase reacted rapidly with eukaryotic cytochromes *c* but scarcely reacted with *P. aeruginosa* and *P. denitrificans* cytochromes *c*. The enzyme reacted relatively rapidly with *T. novellus* and *C. limicola* f. *thiosulfatophilum* cytochromes *c*. *Chlorobium limicola* f. *thiosulfatophilum* cytochrome *c* seems to be a very particular cytochrome since it reacts relatively rapidly with any of the six oxidases used in the present investigation. Further, it seems important that the spatial structures of *P. aeruginosa* and *C. limicola* f. *thiosulfatophilum* cytochromes *c* resemble each other (Timkovich *et al.*, 1976; Korszun and Salemme, 1977) but their reactivities with some of the six oxidases tested differed greatly from each other.

In previous studies (Yamanaka, 1973, 1975), we reported that the cow enzyme does not react with *N. europaea* or *P. aeruginosa* cytochrome *c*. However, we found in the present investigation that the reactivity with the enzyme of these bacterial cytochromes *c* depends to some extent on the state of the oxidase preparation. In previous studies, we used enzyme preparations, which were dissolved in 40 mM phosphate buffer, pH 7.4, containing 0.25% Emasol 1130, and stored in a frozen state at −20°C until use. In the present investigation, the enzyme preparation was dissolved in 20 mM Tris-HCl buffer, pH 8.0, containing 1% Tween 20, and stored in a frozen state at liquid nitrogen temperature until use. The enzyme preparation used in the present investigation reacted with *C. limicola* f. *thiosulfatophilum* cytochrome *c* more rapidly than the previous preparation and reacted relatively rapidly with *N. europaea* cytochrome *c*. Further, it was surprising that this enzyme preparation reacted at all with *P. aeruginosa* cytochrome *c* even though the rate was very low.

When the oxidase preparation used in the present investigation was kept for several days in the frozen state at −20°C, its reactivity with the bacterial cytochromes *c* decreased more rapidly than that with eukaryotic cytochromes *c*. This will explain why the previous preparation which reacted rapidly with eukaryotic cytochromes *c* did not react or reacted very poorly with some of bacterial cytochromes *c*.

COEVOLUTION OF CYTOCHROME OXIDASE WITH CYTOCHROME *C*

The results presented in Fig. 3 show that host organisms whose cytochromes *c* react rapidly with cytochrome oxidases shift from primitive to higher as the host organisms of the oxidases used vary from primitive to higher. This relationship can be diagrammatically expressed as shown in Fig. 4. That is, these observations mentioned above suggest that cytochrome oxidase might have coevolved with cytochrome *c*. This will verify our idea that cytochrome *c* and cytochrome oxidase might have coevolved and the degree of "fitness" between both the proteins reached the highest point in eukaryotes (Yamanaka and Fukumori, 1978; Yamanaka *et al.*, 1979b).

Cytochrome *c* (as previously defined (Yamanaka, 1972, 1973)) which reacts rapidly with the *P. aeruginosa* enzyme but reacts poorly with the eukaryotic enzyme could have evolved to eukaryotic cytochrome *c* which reacts rapidly with the eukaryotic enzyme but reacts poorly with the *P. aeruginosa* enzyme. The evolutionary change in the cytochrome *c* molecule could be caused by the appearance of different cytochrome oxidases during the evolution of organisms (Yamanaka, 1966, 1972).

In general, the cytochrome *cd*-type enzyme was originally nitrite reductase which also acts as a cytochrome oxidase (Yamanaka, 1963, 1964). An enzyme of this type could have functioned exclusively as a nitrite reductase under the conditions where molecular oxygen was not available. After molecular oxygen appeared due to photosynthesis of blue-green algae, the enzyme might have acted as a cytochrome oxidase. However, as the enzyme is extremely labile in the presence of hydrogen peroxide (Yamanaka *et al.*, 1963, 1964), cytochrome oxidase must have changed to the enzyme which is more stable against peroxide. That is, enzymes

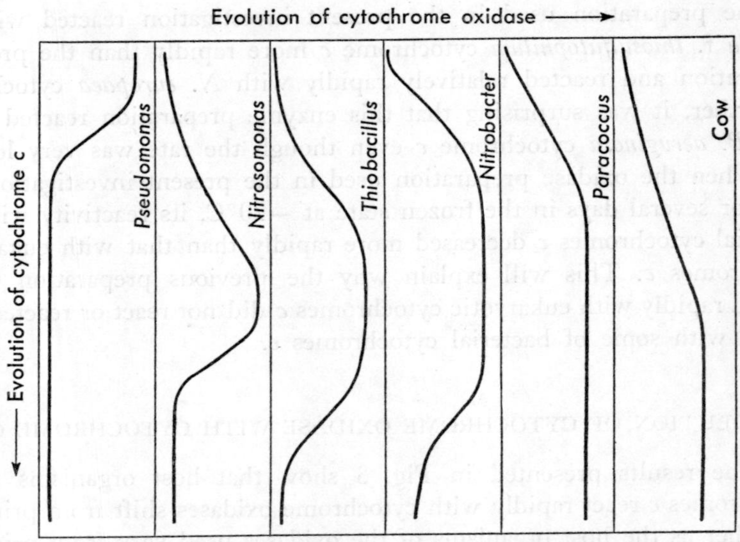

Fig. 4. A diagrammatical presentation of the evolutional relationship between cytochrome oxidase and cytochrome *c*.

having haem *a*, especially the cytochrome aa_3-type oxidase, appeared. The protein moiety of the cytochrome *cd*-type enzyme does not fit haem *a* (Yamanaka and Okunuki, 1963). Therefore, the protein moiety should have changed to the proteins which are adapted for haem *a*; various kinds of cytochrome aa_3-type oxidases can thus exist. In the present study, the occurrence of cytochrome oxidases with various specificities for cytochrome *c* has been revealed.

As Fig. 3 shows, in general the cytochrome aa_3-type enzymes of bacteria react with bacterial cytochromes *c* more rapidly than the eukaryotic enzyme does. Further, the former enzymes react with yeast and tuna cytochromes *c* more rapidly than with cow and horse cytochromes *c*. These findings suggest that the bacterial enzymes might be intermediates during evolution. This may be supported by the structural features of the bacterial enzymes.

STRUCTURAL CONSIDERATION OF EVOLUTION OF CYTOCHROME OXIDASE

As mentioned above, the molecules of bacterial enzymes are composed of two kinds of subunits (Yamanaka *et al.*, 1979a; Ludwig and Schatz,

1980), while the molecules of the eukaryotic enzymes are composed of 6–7 kinds of subunits (*e.g.*, Malmström, 1979). Further, the larger subunits of the eukaryotic enzymes are biosynthesized in the mitochondrion, while the other subunits are biosynthesized on the cytoplasmic ribosomes (Mason and Schatz, 1973; Bernstein *et al.*, 1978; Hare *et al.*, 1980). The subunits of the bacterial enzymes are similar to the subunits of the eukaryotic enzymes biosynthesized in the mitochondrion with respect to their molecular weight (Table I).

These facts suggest that the eukaryotic enzymes might have been evolutionarily completed by adding cytoplasmic subunits to the mitochondrial subunits. Thus, it has been reported that all haem *a* and Cu of the eukaryotic enzymes are located in the larger subunits which are biosynthesized in the mitochondrion (Winter *et al.*, 1980), although there are some researchers who report that haem *a* and Cu are distributed both in the mitochondrial and cytoplasmic subunits (see, *e.g.*, Malmström, 1979; King *et al.*, 1979; Winter *et al.*, 1980).

What was the advantage for the enzymes possessing haem *a* to have changed from the two-subunit molecule to the seven-subunit molecule? The bacterial enzyme is considerably sensitive to the salt concentration or ionic strength (Yamanaka and Fujii, 1980; Yamanaka *et al.*, 1981). Thus, the oxidation rate of ferrocytochrome *c* catalyzed by the *N. agilis* enzyme in 100 mM phosphate buffer is 1–2% of that in 10 mM buffer (Fig. 5). A similar phenomenon has been observed with the *T. novellus* enzyme. The dependency on ionic strength of the activity of the eukaryotic enzymes differs from that of the bacterial enzymes; the oxidation of ferrocytocrhome *c* catalyzed by the cow enzyme shows a maximal rate at a certain ionic strength (Smith *et al.*, 1973). However, at present we do not know whether or not the difference in the sensitivity to ionic strength between the eukaryotic and bacterial enzymes is of evolutionary significance. This should be elucidated in future. In any case, elucidation of the evolutionary relationship between the bacterial and eukaryotic cytochrome aa_3-type enzymes seems to give a clue to elucidation of the evolutionary relationship between prokaryotes and eukaryotes.

SUMMARY

The specificity for cytochrome *c* of cytochrome oxidases was determined with the enzymes purified from six organisms: *P. aeruginosa*, *N. europaea*, *T. novellus*, *N. agilis*, *P. denitrificans*, and the cow.

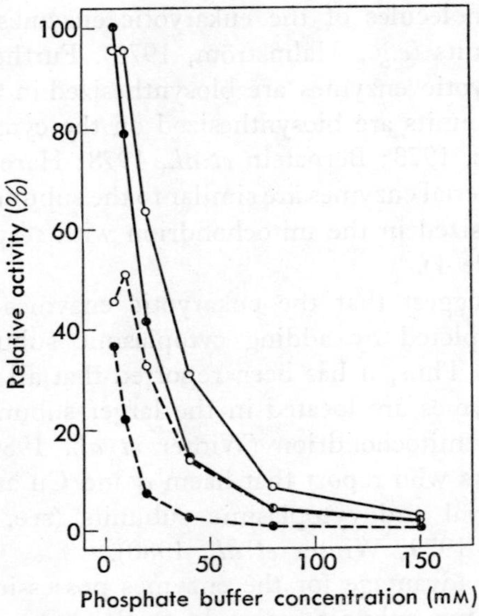

Fig. 5. Dependency on salt concentration of oxidation rates of ferrocytochromes *c* catalyzed by *N. agilis* cytochrome oxidase. The reactions were performed at pH 6.0 and at 20°C. The concentrations of ferrocytochromes *c* were around 30 μM. —○— *C. krusei*; —●— tuna; ---○--- *S. oviformis*; ---●--- horse.

Each cytochrome oxidase reacts most rapidly with cytochromes *c* of organisms which are thought to be near to the host organism of the oxidase in an evolutionary sense. Therefore, the host organisms of cytochromes *c* which react rapidly with each oxidase shift from primitive to higher as the oxidase used varies from the *P. aeruginosa* enzyme to the cow enzyme *via* the *N. europaea*, *T. novellus*, *N. agilis*, and *P. denitrificans* enzymes. The findings suggest that cytochrome oxidase might have co-evolved with cytochrome *c*. The evolution of cytochrome oxidase has also been discussed from the viewpoint of its structure and function relationship.

Acknowledgments

The authors wish to thank Professor H. Matsubara for valuable discussions during the course of this work. They are also grateful to Drs. T. Horio and M. Kakudo (Institute for Protein Research, Osaka University) for their generosity in supplying *R. rubrum* cytochrome c_2 and bonito cytochrome *c*, respectively.

REFERENCES

Bernstein, J. D., Bucher, J. R., and Penniall, R. (1978). Origin of mitochondrial enzymes. V. The polypeptide character and the biosynthesis of rat liver cytochrome *c* oxidase polypeptides by mitochondria. *J. Bioenerg. Biomembr.* **10**, 59–74.

Caughey, W. S., Wallace, W. J., Volpe, J. A., and Yoshikawa, S. (1976). In *The Enzymes* (ed. P. D. Boyer), Vol. XIII, pp. 299–344. Academic Press, New York.

Downer, N. W., Robinson, N. C., and Capaldi, R. A. (1976). Characterization of a seventh different subunit of beef heart cytochrome *c* oxidase. Similarities between the beef heart enzyme and that from other species. *Biochemistry* **15**, 2930–2936.

Erickson, R. H., Hooper, A. B., and Terry, K. (1972). Solubilization and purification of cytochrome a_1 from *Nitrosomonas*. *Biochim. Biophys. Acta* **283**, 155–166.

Fee, J. A., Choc, M. E., Findling, K. L., Lorence, R., and Yoshida, T. (1980). Properties of a copper-containing cytochrome c_1aa_3 complex: A terminal oxidase of the extreme thermophile *Thermus thermophilus* HB8. *Proc. Natl. Acad. Sci. U.S.* **77**, 147–151.

Hagihara, B., Morikawa, I., Tagawa, K., and Okunuki, K. (1958). Crystalline animal cytochrome *c*. In *Biochemical Preparations* (ed. C. S. Vestling), Vol. 6, pp. 1–7. Wiley, New York.

Hare, J. F., Ching, E., and Attardi, G. (1980). Isolation, subunit composition, and site of synthesis of human cytochrome *c* oxidase. *Biochemistry* **19**, 2023–2030.

Hon-nami, K. and Oshima, T. (1980). Cytochrome oxidase from an extreme thermophile *Thermus thermophilus* HB8. *Biochem. Biophys. Res. Commun.* **92**, 1023–1029.

Horio, T., Higashi, T., Sasagawa, M., Kusai, K., Nakai, M., and Okunuki, K. (1960). Preparation of crystalline *Pseudomonas* cytochrome *c*-551 and its general properties. *Biochem. J.* **77**, 194–201.

Horio, T. and Kamen, M. D. (1961). Preparation and properties of three pure crystalline bacterial haem proteins. *Biochim. Biophys. Acta* **48**, 266–286.

John, P. and Whatley, F. R. (1977). *Paracoccus denitrificans* and the evolutionary origin of the mitochondrion. *Nature* **254**, 495–498.

Ketchum, P. A., Sanders, H. K., Gryder, J. W., and Nason, A. (1969). Characterization of cytochrome *c* from *Nitrobacter agilis*. *Biochim. Biophys. Acta* **189**, 360–365.

King, T. E., Orii, Y., Chance, B., and Okunuki, K. (eds.) (1979). *Cytochrome Oxidase*. Elsevier/North-Holland, Amsterdam.

Korszun, Z. R. and Salemme, F. R. (1977). Structure of cytochrome *c*-555 of *Chlorobium thiosulfatophilum*: Primitive low-potential cytochrome *c*. *Proc. Natl. Acad. Sci. U.S.* **74**, 5244–5247.

Ludwig, B. and Schatz, G. (1980). A two-subunit cytochrome *c* oxidase (cytochrome aa_3) from *Paracoccus denitrificans*. *Proc. Natl. Acad. Sci. U.S.* **77**, 196–200.

Malmström, B. G. (1979). Cytochrome *c* oxidase. Structure and catalytic activity. *Biochim. Biophys. Acta* **549**, 281–303.

Mason, T. L. and Schatz, G. (1973). Cytochrome *c* oxidase from baker's yeast. II. Site of translation of the protein components. *J. Biol. Chem.* **248**, 1355–1360.

Matsubara, H. and Yasunobu, K. T. (1961). Comparative biochemical studies. I. Purification and crystallization of human heart cytochrome *c*. *J. Biol. Chem.* **236**, 1701–1705.

Meyer, T. E., Bartsch, R. G., Cusanovich, M. A., and Mathewson, J. H. (1968). The cytochromes of *Chlorobium thiosulfatophilum*. *Biochim. Biophys. Acta* **153**, 854–861.

Motonaga, K., Misaka, E., Nakajima, E., Ueda, S., and Nakanishi, K. (1965). Structure of yeast cytochrome *c*. I. Monomer and dimer interconversion. *J. Biochem.* **57**, 22–28.

Nagata, Y., Yamanaka, T., and Okunuki, K. (1970). Amino acid composotion and N-terminus of *Pseudomonas* cytochrome oxidase (=*Pseudomonas aeruginosa* nitrite reductase). *Biochim. Biophys. Acta* **221**, 668–671.

Okunuki, K. (1966). Cytochromes and cytochrome oxidase. In *Comprehensive Biochemistry* (eds. M. Florkin and E. H. Stotz), Vol. 14, pp. 232–308. Elsevier, Amsterdam.

Poyton, R. O. and Schatz, G. (1975). Cytochrome *c* oxidase from baker's yeast. III. Physical characterization of isolated subunits and chemical evidence for two different classes of polypeptides. *J. Biol. Chem.* **250**, 752–761.

Scholes, P. B.,McLain, G., and Smith, L. (1971). Purification and properties of a *c*-type cytochrome from *Micrococcus denitrificans*. *Biochemistry* **10**, 2072–2076.

Smith, L., Nava, M. E., and Margoliash, E. (1973). The relationship of the amino acid sequence and of substances bound to cytochrome *c* to its reaction with cytochrome *c* oxidase. In *Oxidases and Related Redox Systems* (eds. T. E. King, H. S. Mason, and M. Morrison), Vol. 2, pp. 629–638. Univ. Park Press, Baltimore.

Sone, N., Ohyama, T., and Kagawa, Y. (1979). Thermostable single-band cytochrome oxidase. *FEBS Lett.* **106**, 39–42.

Timkovich, R., Dickerson, R. E., and Margoliash, E. (1976). Amino acid sequence of *Paracoccus denitrificans* cytochrome *c*-550. *J. Biol. Chem.* **251**, 2197–2206.

Winter, D. B., Bruyninckx, W. J., Foulke, F. G., Grinich, N. P., and Mason, H. S. (1980). Location of heme *a* on subunits I and II and Cu on subunit II of cytochrome *c* oxidase. *J. Biol. Chem.* **255**, 11408–11414.

Yamanaka, T. (1963). Studies on *Pseudomonas* cytochrome oxidase. *Annu. Rep. Sci. Works, Fac. Sci., Osaka Univ.* **11**, 77–115.

Yamanaka, T. (1964). Identity of *Pseudomonas* cytochrome oxidase with *Pseudomonas* nitrite reductase. *Nature* **204**, 253–255.

Yamanaka, T. (1966). *Pseudomonas* cytochrome oxidase. In *Biochemistry of Copper* (eds. J. Peisach, P. Aisen, and W. E. Blumberg), pp. 275–292. Academic Press, New York.

Yamanaka, T. (1966). Evolution based on comparative biochemical studies of cytochrome *c*. *Annu. Rep. Biol. Works, Fac. Sci., Osaka Univ.* **14**, 47–71.

Yamanaka, T. (1967). Cytochrome *c* and evolution. *Nature*, **213**, 1183–1186.

Yamanaka, T. (1972). A comparative study on reactivity of cytochromes *c* with yeast cytochrome *c* peroxidase. In *Aspects of Cellular and Molecular Physiology* (ed. K. Hamaguchi), pp. 75–87. Japan Sci. Soc. Press, Tokyo.

Yamanaka, T. (1973). Cytochrome *c* and evolution of the energy acquiring system. *Space Life Sci.* **4**, 490–504.

Yamanaka, T. (1975). A comparative study on the redox reactions of cytochromes *c* with certain enzymes. *J. Biochem.* **77**, 493–499.

Yamanaka, T. and Fujii, K. (1980). Cytochrome *a*-type terminal oxidase derived from *Thiobacillus novellus*. Molecular and enzymatic properties. *Biochim. Biophys. Acta* **591**, 53–62.

Yamanaka, T., Fujii, K., and Kamita, Y. (1979a). Subunits of cytochrome *a*-type terminal oxidase derived from *Thiobacillus novellus* and *Nitrobacter agilis*. *J. Biochem.* **86**, 821–824.

Yamanaka, T. and Fukumori, Y. (1978). An evolutionary aspect of the reactivity of cytochrome *c* with cytochrome oxidase. In *Evolution of Protein Molecules* (eds. H. Matsubara and T. Yamanaka), pp. 387–399. Japan Sci. Soc. Press, Tokyo.

Yamanaka, T., Fukumori, Y., and Fujii, K. (1979b). Some evolutionary aspects of cytochrome oxidase. In *Cytochrome Oxidase* (eds. T. E. King, Y. Orii, B. Chance, and K. Okunuki). Elsevier/North-Holland, Amsterdam.

Yamanaka, T., Kamita, Y., and Fukumori, Y. (1981). Molecular and enzymatic properties of "cytochrome aa_3"-type terminal oxidase derived from *Nitrobacter agilis*. *J. Biochem.* **89**, 265–273.

Yamanaka, T., Kijimoto, S., and Okunuki, K. (1963). Biological significance of *Pseudomonas* cytochrome oxidase in *Pseudomonas aeruginosa*. *J. Biochem.* **53**, 416–421.

Yamanaka, T. and Kimura, K. (1974). Eukaryotic cytochrome *c*-like properties of cytochrome *c*-550 (*Thiobacillus novellus*). *FEBS Lett.* **48**, 253–255.

Yamanaka, T., Mizushima, H., Katano, H., and Okunuki, K. (1964). Comparison of three yeast cytochrome *c*'s purified from *Saccharomyces cerevisiae*, *Kloeckera* sp. and *Candida krusei*. *Biochim. Biophys. Acta* **85**, 11–17.

Yamanaka, T. and Okunuki, K. (1963). Reconstitution of *Pseudomonas* cytochrome oxidase from haem a_2 and its protein moiety and some properties of the reconstituted enzyme. *Biochem. Z.* **338**, 62–72.

Yamanaka, T. and Okunuki, K. (1964). Comparative biochemistry of cytochrome *c*. *J. Biol. Chem.* **239**, 1813–1817.

Yamanaka, T. and Okunuki, K. (1968). Comparison of *Chlorobium thiosulfatophilum* cytochrome *c*-555 with *c*-type cytochromes derived from algae and nonsulphur purple bacteria. *J. Biochem.* **63**, 341–346.

Yamanaka, T. and Okunuki, K. (1968). Comparative studies on reactivities of cytochrome *c* with cytochrome oxidases. In *Structure and Function of Cytochromes* (eds. K. Okunuki, M. D. Kamen, and I. Sekuzu), pp. 390–403. Univ. Tokyo Press, Tokyo.

Yamanaka, T., Ota, A., and Okunuki, K. (1961). Activation of *Pseudomonas* cytochrome oxidase by catalase. *J. Biochem.* **49**, 414–420.

Yamanaka, T. and Shinra, M. (1974). Cytochrome *c*-552 and cytochrome *c*-554 derived from *Nitrosomonas europaea*: Purification, properties and their function in hydroxylamine oxidation. *J. Biochem.* **75**, 1265–1273.

Yamanaka, T., Takenami, S., Akiyama, N., and Okunuki, K. (1971). Purification and properties of cytochrome *c*-550 and cytochrome *c*-551 derived from facultative chemoautotroph, *Thiobacillus novellus*. *J. Biochem.* **70**, 349–358.

Yonetani, T. (1963). The *a*-type cytochromes. In *The Enzymes* (eds. P. D. Boyer, H. Lardy, and K. Myrbäck), pp. 41–79. Academic Press, New York.

15

Molecular Evolution, Protein Polymorphism and the Neutral Theory (1982) (ed. M. Kimura), pp. 333–346.
Japan Scientific Societies Press, Tokyo / Springer-Verlag, Berlin

Reconstruction of Evolutionary Processes of Legume Double-headed Proteinase Inhibitors by Chemical-enzymatic Methods

SHOJI ODANI*[1] AND TOKUJI IKENAKA*[2]

Department of Biochemistry, Niigata University School of Medicine, Niigata 951,[1] *and Department of Chemistry, Osaka University, College of Science, Toyonaka 560,*[2] *Japan*

Protein proteinase inhibitors have been found in, and isolated from a large number of the Leguminosae plants in the past thirty years. They have been classified into two types, the Kunitz type and Bowman-Birk type, according to their molecular weights and cystine contents. The latter type, of which the soybean Bowman-Birk inhibitor is a typical representative, is characterized by a low molecular weight (less than 10,000) and high cystine content (about 20 mol %). Inhibitors of this type are of particular interest, because they can bind two proteinases simultaneously and exist as sets of numerous inhibitor variants within a single species.

Structural analyses of these double-headed inhibitors revealed the presence of internally homologous sequences, each of which contains a reactive site for a proteinase, suggesting their evolution from a common single-headed ancestor. Similar multi-headed inhibitors are frequently found in nature, but what advantage(s) has been gained by multiplying the ancestral single-headed structure remains unknown. We tried to split the double-headed inhibitor into single-headed inhibitors, expecting that comparison of the properties of the resulting monovalent inhibitors and of the parent inhibitor would afford some clues to this problem.

Isoinhibitors of legume plants have closely related amino acid sequences, but their inhibitory properties vary widely. Examination of sequence data indicated several prominent features. The reactive site P_1 residues (notation by Schechter and Berger, 1967), which primarily determine the specificity of the inhibitors, are quite variable. As pointed out by Laskowski and his coworkers (1974), it is rather exceptional in a homologous group of functional proteins for critical residues at the active sites not to be conserved but to be substituted "radically." On the other hand, reactive site P_1' serine residues, which form scissile bonds with P_1 residues, appear to be strictly conserved. In a homologous family of proteins, constant residues are thought to be critical for function and/or conformation, and to have been conserved up to the present time. However, sequence data are always limited and it is not possible to rule out the possibility that a different amino acid will appear at a constant site when more sequence data became available. This problem could be solved by replacing the critical residue by the other 19 protein amino acids and examining the properties of the resulting *mutants*. We have used chemical-enzymatic mutation methods (Laskowski, Jr. *et al.*, 1974; Kowalski and Laskowski, Jr., 1976) to examine whether or not the highly variable P_1 residue of the double-headed inhibitors accepts amino acids without any restriction, and whether the strictly conserved P_1' serine residue can be replaced by any other residue.

AMINO ACID SEQUENCE AND EVOLUTION OF LEGUME DOUBLE-HEADED PROTEINASE INHIBITORS

Figure 1 shows the known amino acid sequences of lequme double-headed inhibitors (Odani *et al.*, 1971; Tan and Stevens, 1971; Wilson and Laskowski, Sr., 1975; Odani and Ikenaka, 1977, 1978a; Yoshikawa *et al.*, 1979; Ishikawa *et al.*, 1979; Joubert *et al.*, 1979). The sequences clearly indicate a high degree of homology and internal sequence duplications (domains 1 and 2*). Although many amino acids seemed to be conserved in all the molecules, the critical reactive site P_1 residues themselves varied considerably. The P_1 amino acids in domain 1 of the inhibitors are lysine, alanine, and arginine, and those in domain 2 are leucine, phenylalanine,

* The term "domain" may not be appropriate to describe these disulfide-linked peptide moieties which are not arranged linearly in the amino acid sequence.

```
                  D D E S S K P C C D Q C A C T K* S N P P Q C R C S D M   BBI
      S D H S S S D D E S S K P C C D L C M C T A* S M P P Q C H C A D I   C-II
  S D Q S S S Y D D D E Y S K P C C D L C M C T R* S M P P Q C S C E D I   D-II
S G H H E H S T D Z P S Z S S K P C(C,B)H(C,A/L,C)T K* S I P P Q C R C T/S D L/F   LBI-IV   Domain 1
                  Z P S Z S S P P C C B I C V C T A* S I P P Q C(I,C,T,B,V)   GBI-II
S G H H D E T T D E P S E S S K P C C D Q C S C T K* S M P P K C R C S D I   ABI-I
  D H H H S T D E P S E S S K P C C D E C A C T K* S I P P Q C R C T D V   Mal DE-3
    H E H S S D E S S E S S K P C C D L C T C T K* S I P P Q C H C N D M   Mal DE-4

                  R L N S C H S A C K S C I C A L* S Y P A Q C F C V D I   BBI
                  R L N S C H S A C D R C A C T R* S M P G Q C R C L D T   C-II
                  R L N S C H S D C K S C M C T R* S Q P G Q C R C L D T   D-II
                  R L D S C H S A C K S C I C T L/F* S I P A Q C V(C,T,B)I   LBI-IV   Domain 2
                  R L B S C H S A C K S C M C T R* S M P G K C R C L B T   GBI-II
                  R L N S C H S A C K S C A C T Y* S I P A K C F C T D I   ABI-I
                  R L N S C H S A C S S C V C T F* S I P A Q C V C V D M   Mal DE-3
                  R L N S C H S A C K S C I C A L* S E P A Q C F C V D T   Mal DE-4

T D F C Y E P C K P S E D D K E N          Soybean Bowman-Birk inhibitor
T D F C Y K P C K S S D E D D D            Soybean inhibitor C-II
N D F C Y K P C K S R D D                  Soybean inhibitor D-II
B D F C Y E P C K S S H S D D D N N N      Limabean inhibitor IV,IV'
T B Y C Y K S C K S B S G Z B B            Garden bean inhibitor II'
N D F C Y E P C K S S R D D D W D N        Adzuki bean inhibitor I
K D F C Y A P C K S S H D D                Macrotyloma inhibitor DE-3
T D F C Y K S C H N N A E K D              Macrotyloma inhibitor DE-4
```

Fig. 1. Amino acid sequences of legume double-headed proteinase inhibitors. The sequence of the adzuki bean inhibitor differs somewhat from that reported by Yoshikawa *et al.* (1979). The reactive site P_1 residues are marked with asterisks.

tyrosine, and arginine. The nature of these P_1 residues is in accord with the substrate specificity of the proteinases inhibited. It is also seen from the figure that all the P_1' positions are occupied by serine residues.*

The double-headed inhibitors are currently thought to have evolved from a single-headed common ancestor by internal gene duplication. It is expected that the first double-headed inhibitor had the same residues at both reactive sites, just like soybean inhibitor D-II (Fig. 1). Therefore, D-II may be regarded as a survivor of an archetype double-headed inhibitor, the original ancestor being an inhibitor for trypsin or trypsin-like enzymes and having an arginyl-serine at the reactive site P_1-P_1' residues.

* Recently, the presence of an arginine residue in the P_1' position of the first domain of a peanut Bowman-Birk-type proteinase inhibitor was found by Norioka and Ikenaka (unpublished data). However, the true reactive site of the inhibitor has not yet been determined.

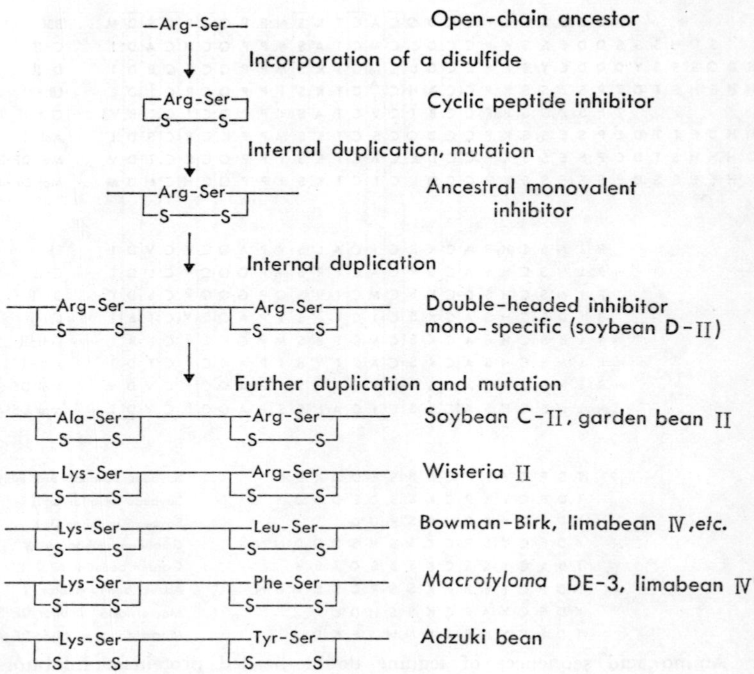

Fig. 2. A possible evolutionary process of legume double-headed proteinase inhibitors. Only reactive site residues are shown.

One possible, but highly speculative, evolutionary process in legume double-headed inhibitors is illustrated in Fig. 2.

ACTIVE FRAGMENTS OF SOYBEAN BOWMAN-BIRK INHIBITOR— A POSSIBLE MODEL OF THE HYPOTHETICAL ANCESTOR

The disposition of the seven disulfide bonds in the Bowman-Birk inhibitor (Odani and Ikenaka, 1973a) also showed the essentially identical arrangements of the disulfides in two domains (Fig. 3). We have succeeded in splitting the inhibitor into two single-headed inhibitors by treatment with cyanogen bromide and pepsin as shown in the figure (Odani and Ikenaka, 1973b, 1978b). This finding that the separated domains themselves show inhibitory activity strongly supports the idea of the existence of a single-headed ancestor. The ancestor might not have been very different from these monovalent fragments, and thus this experiment

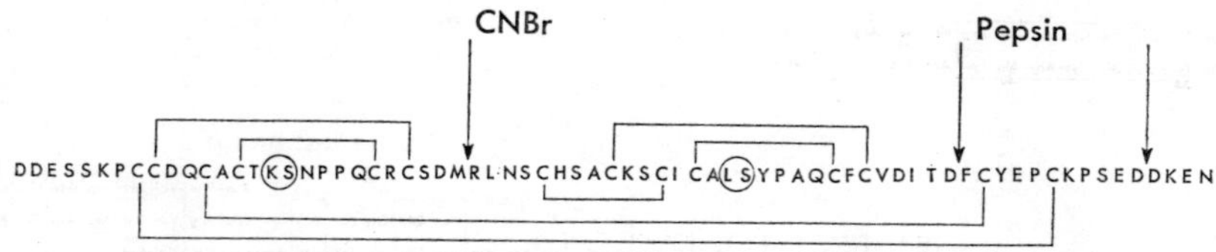

Fig. 3. Topological structure of the soybean Bowman-Birk inhibitor showing the disposition of 7 disulfide bonds. Reactive site P_1 and P_1' residues are shown in circles. Sites of action of cyanogen bromide and pepsin are indicated by vertical arrows.

may correspond to a reversal of the evolutionary process. In contrast to the native inhibitor which is very stable, the separated domains were rather unstable under extreme conditions (*e.g.*, incubation in 0.1 N NaOH at 25°C) and were gradually inactivated upon incubation with excess proteinases (Odani and Ikenaka, 1978b). Thus the ancestral monovalent inhibitor may have become more stable and effective as an inhibitor by doubling its sequence, possibly by additional interactions between the two halves to form a rigid structure around the two sites, protecting them from proteolytic attack.

Recently Izumiya and his coworkers (Nishino *et al.*, 1975; Terada *et al.*, 1980) chemically synthesized the reactive site nona peptide loop of the Bowman-Birk inhibitor and its analogs, which showed weak, but definite activity against proteinases. It is also possible that this smallest reactive site loop duplicated and that its sequence changed to form a more effective structure, like the separated domains described here, without forming a second reactive site. Internal sequence repetitions, which did not lead to the evolution of multi-headed inhibitors, have been recognized for soybean trypsin inhibitor (Kunitz) (Koide and Ikenaka, 1973) and bovine pancreatic inhibitor (Kunitz) (Huber *et al.*, 1970) (Fig. 4). The latter example is particularly indicative, because one of the repeated regions contains the reactive site of the inhibitor. Izumiya and his coworkers (Terada *et al.*, 1980) also reported that the open-chain derivatives, peptides I_h and II_d, of the reactive site loops (*i.e.*, reduced and S-carboxamidomethylated forms of the nonapeptide loop) still retained considerable activity against proteinases. Consequently, it is possible that the original ancestor was a small peptide fragment like these open-chain derivatives. Szilágyi *et al.* (1978) examined the sequence of the Bowman-Birk inhibitor and also suggested the evolution of this inhibitor by multiplication of a small segment, which initially had no disulfide bond. These speculations are summarized in Fig. 2.

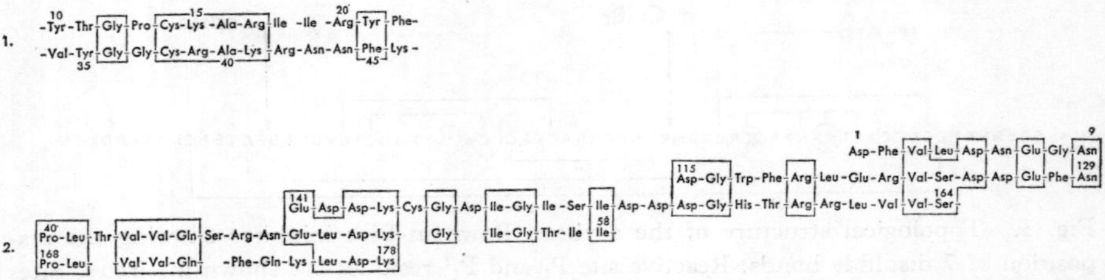

Fig. 4. Internal sequence repetitions found for : 1) bovine pancreatic trypsin inhibitor (Kunitz) and for 2) soybean trypsin inhibitor (Kunitz). Identical and highly conservative residues are shown in boxes.

CHEMICAL-ENZYMATIC MUTATION AT THE REACTIVE SITE P_1 RESIDUE THAT DETERMINES THE SPECIFICITY OF INHIBITORS

The inhibitor specificity of a proteinase inhibitor is primarily determined by the nature of its reactive site P_1 residue, although the contributions of other residues and "leaky specificity" are frequently recognized. Laskowski and his coworkers (1974) converted soybean trypsin inhibitor (Kunitz) to a strong chymotrypsin inhibitor by replacing its trypsin reactive site P_1 arginine residue (Arg^{63}) by a tryptophan residue. Their procedure included limited proteolysis of the P_1-P_1' bond by trypsin, removal of Arg^{63} with carboxypeptidase B, and addition of tryptophan to the P_1 position by the reverse action of carboxypeptidase A in the presence of a high concentration of tryptophan and α-chymotrypsin. In their procedure, the reverse (synthetic) reaction of the carboxypeptidase was driven by removal of newly formed Trp^{63}-inhibitor from the system as a tight complex with chymotrypsin. Although the procedure is of much interest from the protein chemical and enzymological viewpoints, it seems to have an inherent limitation that arises from the use of specific enzymes. This method can only be used to introduce a very limited number of kinds of amino acids into the P_1 position unless a suitable enzyme system is found. To overcome this limiation, we tried to add amino acids by chemical methods and found that condensation with water-soluble carbodiimide was effective. This procedure is summarized in Fig. 5.

The carboxyl groups of soybean Bowman-Birk inhibitor were first blocked by coupling taurine with water-soluble carbodiimide (Carraway *et al.*, 1972). This carboxyl-protected inhibitor still retained full inhibitory activity and was cleaved at the chymotrypsin-reactive site Leu^{43}-Ser^{44} (P_1-P_1') bond by limited proteolysis and Leu^{43} was removed by carboxy-

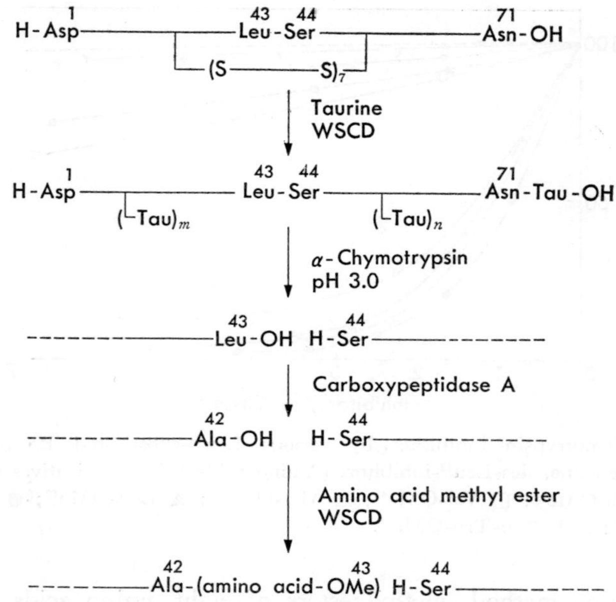

Fig. 5. Chemical-enzymatic mutation at the chymotryspin reactive site P_1 Leu[43] of the soybean Bowman-Birk inhibitor. WSCD, watersoluble carbodiimide or 1-ethyl-3-(3-dimethylaminopropyl)-carbodiimide.

peptidase A. An appropriate amino acid methyl ester was then introduced into the P_1 position by the carbodiimide. Amino acid analysis of the products indicated an essentially quantitative incorporation of amino acid methyl esters. The efficiency of this procedure was evaluated by introducing the methyl ester of the original leucine. The resulting [Leu-OMe[43]] derivative showed restoration of about 50% of the inhibitory activity of the native inhibitor. This may be due to some side reactions, such as racemization at Ala[42] or modification of other residues. However, the use of racemization-suppressing agents (*i.e.*, N-hydroxysuccinimide or 1-hydroxybenzotriazole) did not improve the recovery of activity significantly. We are now examining the optinum conditions for this reaction. This method can be used to introduce essentially all protein (and nonprotein, if desired) amino acids, because legume inhibitors are quite stable in trifluoroacetic acid and the side chain amino-group of lysine and the ω-carboxyl groups of aspartic or glutamic acids can be blocked by the trifluoroacetic acid-sensitive *tert*-butoxy and *tert*-butyl groups, respectively.

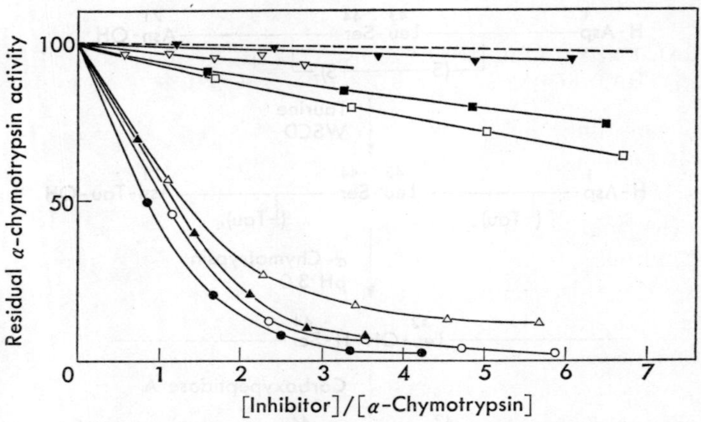

Fig. 6. α-Chymotrypsin inhibition by various Leu[43]-substituted Bowman-Birk inhibitors. Broken line, des-Leu[43]-inhibitor. [Amino acid-OMe[43]] derivatives are: ▼ Gly-OMe[43]; ■ Ala-OMe[43]; □ Val-OMe[43]; △ Met-OMe[43]; ▲ Leu-OMe[43]; ● Phe-OMe[43]; ○ Trp-OMe[43]; and ▽ D-Trp-OMe[43].

Using this method, methyl esters of eight amino acids, Gly, Ala, Val, Met, Leu, Phe, Trp, and D-Trp, were introduced into the P_1 position. The modes of inhibition of bovine α-chymotrypsin by these derivatives are shown in Fig. 6. It is seen that the [Leu-OMe[43]], [Phe-OMe[43]], and [Trp-OMe[43]] derivatives were equally effective for the inhibition of chymotrypsin. The [Met-OMe[43]] inhibitor showed somewhat weaker interaction with the proteinase than the former three. The [Val-OMe[43]], [Ala-OMe[43]], and [D-Trp-OMe[43]] derivatives had much lower affinities to α-chymotrypsin, and the [Gly-OMe[43]] inhibitor was essentially inactive. These observations were in good agreement with the substrate specificity of chymotrypsin, and we can predict the presence of a chymotrypsin inhibitor having a tryptophan residue at the reactive site and an absence of inhibitors with Val, Ala, or Gly at this position in nature.

Though experimental conditions should be improved for quantitative restoration of the activity, the present procedure provides a new approach to investigations of the evolutionary processes of functional protein molecules.

CHEMICAL-ENZYMATIC MUTATION AT THE REACTIVE SITE P_1' RESIDUE THAT FORMS A SCISSILE BOND WITH THE P_1 RESIDUE

By examining the nature of the P_1' residues of various homologous groups of proteinase inhibitors, Kowalski and Laskowski (1976) suggested the presence of three families of inhibitors; that is, those in which P_1' residues are highly variable (family 1), conservative (family 2), and invariant (family 3). Figure 1 shows that legume double-headed inhibitors belong to the third family and that their reactive site P_1' positions are invariably occupied by serine residues. This invariability suggests the importance of serine for the inhibitory activity. To test this, the P_1' serine of the chymotrypsin-reactive site (P_1=Leu^{43}) of the Bowman-Birk inhibitor was replaced by ohter amino acids and the inhibitory activities of the resulting "mutants" were investigated.

The procedure (Kowalski and Laskowski, Jr., 1976; Odani and Ikenaka, 1978c) used here is summarized in Fig. 7. First, the amino groups of the protein were successively blocked with O-methylisourea and cyanate. The resulting amino-protected inhibitor was then cleaved specifically at the chymotrypsin reactive site Leu^{43}-Ser^{44} bond by limited hydrolysis with chymotrypsin, and the newly formed amino-terminal Ser^{44} was removed by Edman degradation. Then, an appropriate amino acid was introduced into this position using the active ester method.

An α-chymotrypsin-modified Bowman-Birk inhibitor having a glycine residue instead of Ser^{44} was prepared using a different method, which was first applied by Dixon (1968) for the conversion of the amino-terminal serine of corticotropin to glycine by periodate oxidation followed by transamination. In a similar way, derivatives having a glycine residue at the trypsin reactive site P_1' position of the Bowman-Birk inhibitor (Ser^{17}) or of soybean inhibitor C-II (Ser^{50}) were prepared.

Chemical analyses of the products indicated essentially quantitative reactions in the protection of amino groups, removal of serine residues by Edman degradation, and condensation of amino acids *via* active esters. Periodate oxidation and the transamination reaction also proceeded quantitatively.

Using the activated ester method, six inhibitor derivatives having a Ser, Ala, Thr, Val, Leu, and Gly residue, respectively, at position 44 were prepared. The inhibitory activities of these derivatives against bovine α-chymotrypsin are summarized in Fig. 8. The [Ser^{44}] and [Ala^{44}] derivatives caused stoichiometric inhibition of chymotrypsin under the standard

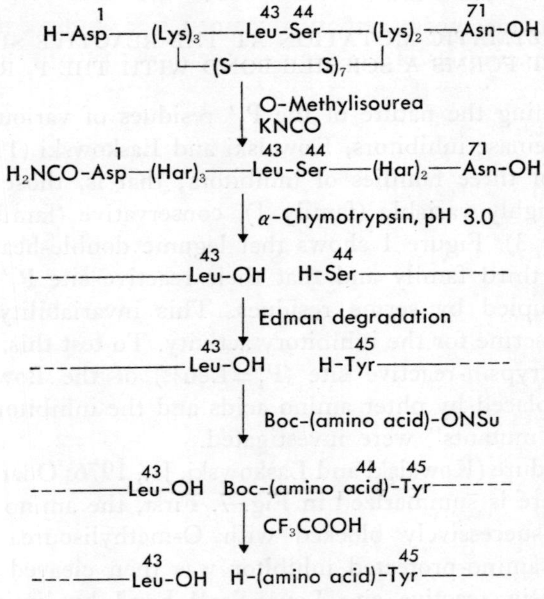

Fig. 7. Chemical-enzymatic mutation of chymotrypsin reactive site P_1' Ser^{44} of the Bowman-Birk inhibitor. Har, homoarginine (Odani and Ikenaka 1978c).

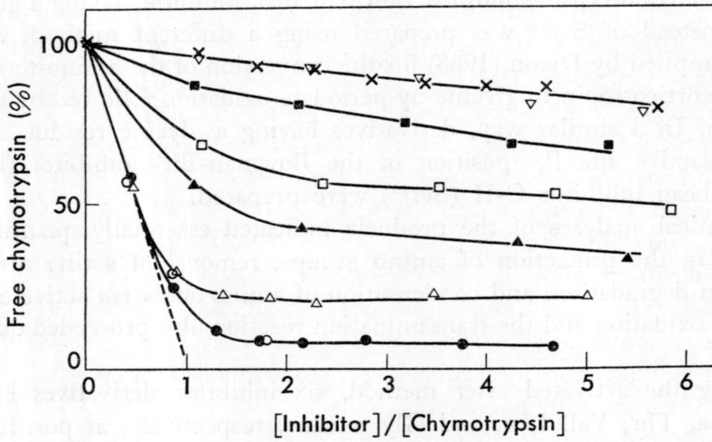

Fig. 8. Titration of α-chymotrypsin with various Ser^{44}-substituted Bowman-Birk inhibitors. × des-Ser^{44}-inhibitor. [Amino acid[44]] derivatives are: ● Ser^{44}; △ Ala^{44}; ▲ Thr^{44}; □ Val^{44}, ■ Leu^{44}; and ▽ Gly^{44}.

assay conditions (Odani and Ikenaka, 1978c), but with the [Ala^{44}] inhibitor considerable chymotrypsin activity remained free at the equivalent point. The inhibitory activities of the semi-synthetic derivatives decreased in the following order: [Ser^{44}], [Ala^{44}], [Thr^{44}], [Val^{44}], [Leu^{44}], and [Gly^{44}]. The [Gly^{44}] inhibitors prepared by the two different methods showed no significant activity against α-chymotrypsin. Similar transformation experiments on the trypsin reactive sites Lys^{16}-Ser^{17} in the first domain of the Bowman-Birk inhibitor and Arg^{49}-Ser^{50} in the second domain of soybean inhibitor C-II indicated that the derivatives were strong inhibitors of trypsin.

In the present study we mainly used the method developed by Kowalski and Laskowski, Jr. (1976). Because of the unusual stability of the double-headed inhibitors already mentioned, replacement of the P_1' serine residue was much easier than that of the P_1' isoleucine residue in the soybean trypsin inhibitor (Kunitz) that they studied. The other method, periodate oxidation followed by transamination, was particularly effective for obtaining glycine derivatives, since the reaction conditions are much milder than those in the activated ester method, and extra protection of the amino groups of the protein is not necessary.

None of the five amino acids incorporated into the P_1' position was as effective as serine for inhibition of α-chymotrypsin. Our preliminary experiments showed that phenylalanine and proline introduced into the P_1' position also restored only a little, if any, of the original activity. These findings are in contrast to the case of soybean trypsin inhibitor (kunitz) in which introduction of alanine, leucine, or glycine into the P_1' position fully restored the activity.

Unexpectedly, glycine was ineffective as the P_1' residue of the chymotrypsin reactive site. This observation, together with the considerable potency of alanine, suggests that the β-carbon atom is indispensable for inhibition of chymotrypsin. But this is not the case for trypsin-inhibitory domains, since inhibitors having a glycine residue at the trypsin reactive site P_1' positi on showed marked activity against bovine trypsin.

Another hydroxy amino acid, threonine, restored less activity than serine or even alanine, indicating severe hindrance of the side chain methyl group at this position for restoration of the inhibitory activity. It appears that amino acids with bulky side chains are inappropriate at the P_1' position, since the inhibitory activities of the [Thr^{44}], [Val^{44}], [Leu^{44}], and [Phe^{44}] derivatives decreased in this order. Although not all amino acids were tested, serine seemed to be the most effective for the P_1' position of the legume double-headed inhibitors. One possible role of proteinase inhibitors in plant seeds is to protect the seeds from insects and animals. Variants having amino acids other than serine as the P_1' residue, therefore, must have been eliminated by natural selection, because of their low proteinase-inhibiting capacity.

CONCLUSION

We were able to follow the evolutionary processes of legume double-headed inhibitors by chemical-enzymatic methods. The experiments presented here provide a new approach to studies of the evolutionary events and structure-function relationships of many functional proteins. It may also be possible to "evolve" from the present-day inhibitors novel proteinase inhibitors with entirely new specificities, such as inhibitors for post-proline and post-glutamic acid-cleaving serine proteinases, or for the highly differentiated proteinases of blood coagulation, fibrinolysis, and others, by combining the reactions described above.

SUMMARY

The evolutionary process of legume double-headed proteinase inhibitors was investigated chemically. Comparison of the amino acid sequences of a number of these inhibitors suggested the evolution of these inhibitors from an ancestral single-headed inhibitor with Arg-Ser at the reactive site for trypsin (or trypsin-like enzymes) by fused gene duplication. Characterization of artificially separated inhibitory domains of the Bowman-Birk inhibitor, which was regarded as a possible model of the hypothetical ancestor, suggested that duplication of an ancestral sequence resulted in the unusual stability of the present-day inhibitors.

Chemical-enzymatic methods to exchange P_1 and P_1' residues at the reactive sites were described. These methods were applied to the Bowman-Birk inhibitor and another soybean inhibitor. The primary specificity site (P_1) for chymotrypsin accepted several amino acids other than the original leucine with restoration of strong inhibitory activity, confirming the hypervariability of the P_1 residues. On the other hand, none of the amino acids tested for the ability to replace the P_1' serine residue had any ability to take the place of serine. This serine seems to have been strictly conserved throughout the evolution of legume double-headed inhibitors.

The possibility of applying the present methods to the preparation of artificial mutants of functional proteins was also suggested.

REFERENCES

Carraway, K. L. and Koshland Jr., D. E., (1972). Carbodiimide modification of proteins. *Methods Enzymol.* **25**, 616–623.

Dixon, H.B.F. (1968). The conversion of glyoxyloyl groups into glycyl groups and their formation from maleyl residues. *Biochem. J.* **107**, 124–126.

Huber, R., Kukla, D., Rühlmann, A., and Steigemann, W. (1970). The atomic structure of the basic trypsin inhibitor of bovine organs. In *Proceedings of International Research Congression Proteinase Inhibitors* (eds. H. Fritz and H. Tschesche), pp. 56–65. Walter de Gruyter, Berlin and New York.

Ishikawa, C., Nakamura, S., Watanabe, K., and Takahashi, K. (1979). Amino acid sequence of Adzuki bean proteinase inhibitor I. *FEBS Lett.* **99**, 97–100.

Joubert, F. J., Kruger, H., Townshend, and G. S. Botes, D. P. (1979). Purification, some properties and the complete primary structures of two protease inhibitors (DE-3 and DE-4) from *Macrotyloma axillare* seed. *Eur. J. Biochem.* **97**, 85–91.

Koide, T. and Ikenaka, T. (1973). Studies on soybean trypsin inhibitors 3. Amino acid sequence of the carboxyl-terminal region and the complete amino acid sequence of soybean trypsin inhibitor (Kunitz). *Eur. J. Biochem.* **32**, 417–431.

Kowalski, D. and Laskowski Jr., M., (1976). Chemical-enzymatic replacement of Ile^{64} in the reactive site of soybean trypsin inhibitor (Kunitz). *Biochemistry* **15**, 1300–1309.

Laskowski Jr., M., Kato, I., Leary, T. R., Schroede, J., and Sealock, R. W. (1974). Evolution of specificity of protein proteinase inhibitors. In *Bayer Symposium V: Proteinase Inhibitors* (eds. H. Fritz, H. Tschesche, L. J. Greene, and E. Truscheit), pp. 597–611. Springer-Verlag, Berlin, Heidelberg and New York.

Nishino, N., Aoyagi, H., Kato, T., and Izumiya, N. (1975). Synthesis and activity of nonapeptide fragments of soybean Bowman-Birk inhibitor. *Experientia* **31**, 410–412.

Odani, S., Koide, T., and Ikenaka, T. (1971). The amino acid sequence of Bowman-Birk soybean proteinase inhibitor. *Proc. Japan. Acad.* **47**, 621–624.

Odani, S. and Ikenaka, T. (1973a). Studies on soybean trypsin inhibitors VIII. Disulfide bridges in soybean Bowman-Birk proteinase inhibitor. *J. Biochem.* **74**, 697–715.

Odani, S. and Ikenaka, T. (1973b). Scission of soybean Bowman-Birk proteinase inhibitor into two small fragments having either trypsin or chymotrypsin inhibitory activity. *J. Biochem.* **74**, 857–860.

Odani, S. and Ikenaka, T. (1977). Studies on soybean trypsin inhibitors XI. Complete amino acid sequence of a soybean trypsin-chymotrypsin-elastase inhibitor, C-II. *J. Biochem.* **82**, 1523–1537.

Odani, S. and Ikenaka, T. (1978a). Studies on soybean trypsin inhibitors XII. Linear sequences of two soybean double-headed trypsin inhibitor D-II and E-I. *J. Biochem.* **83**, 737–745.

Odani, S. and Ikenaka, T. (1978b). Studies on soybean trypsin inhibitors XIII. Preparation and characterization of active fragments from Bowman-Birk proteinase inhibitor. *J. Biochem.* **83**, 747–753.

Odani, S. and Ikenaka, T. (1978c). Studies on soybean trypsin inhibitors XIV. Change of the inhibitory activity of Bowman-Birk inhibitor upon replacement of the α-chymotrypsin reactive site serine residue by other amino acids. *J. Biochem.* **84**, 1–9.

Schechter, I. and Berger, A. (1967). On the size of the active site in proteases. I. Papain. *Biochem. Biophys. Res. Commun.* **27**, 157–162.

Szilágyi, S., Szilágyiné, S., and Somogyi, E. (1978). Internal homology and evolution of Bowman-Birk soybean protease inhibitor molecule. *Acta Biol. Debrecina* **15**, 211–226.

Tan, C.G.L. and Stevens, F. C. (1971). Amino acid sequence of lima bean protease inhibitor component IV. 2. Isolation and sequence determination of the chymotryptic peptides and the complete amino acid sequence. *Eur. J. Biochem.* **18**, 515–523.

Terada, S., Sato, K., Kato, T., and Izumiya, N. (1980). Studies on the synthesis of proteinase inhibitors. II. Synthesis of cyclic nonapeptide fragments and analog related to the reactive sites of soybean Bowman-Birk inhibitor. *Int. J. Peptide Protein Res.* **15**, 441–454.

Wilson, K. A. and Laskowski Sr., M. (1975). The partial amino acid sequence of trypsin inhibitor II from garden bean, *Phasedus vulgaris*, with location of the trypsin and elastase reactive sites. *J. Biol. Chem.* **250**, 4261–4267.

Yoshikawa, M., Kiyohara, T., Iwasaki, T., Ishi, Y., and Kimura, N. (1979). Amino acid sequence of proteinase inhibitor II and II′ from Azuki beans. *Agric. Biol. Chem.* **43**, 787–796.

16

Molecular Evolution, Protein Polymorphism and the Neutral Theory (1982) (ea. M. Kimura), pp. 349–356.
Japan Scientific Societies Press, Tokyo / Springer-Verlag, Berlin

Molecular Mechanisms of Bacterial Adaptation to High Temperatures

TAIRO OSHIMA

Mitsubishi-Kasei Institute of Life Sciences, Machida 194, Japan

Thermophiles are interesting organisms for biochemists who are interested in the molecular aspects of evolution or adaptation to an environment. Enzymes, proteins, nucleic acids, and cell organella such as membranes and ribosomes of thermophiles are much more resistant to heat than the corresponding ones from mesophilic organisms. Studies of their unusual stabilities will give us insight into the molecular basis of microbial evolution.

The author and his colleagues have been studying the molecular basis of thermophily (Oshima, 1979; Oshima *et al.*, 1976). Although many important problems concerning thermophily remain unsolved, our results as well as those of a variety of reports from other groups suggest two general rules for molecular mechanisms: i) different tactics are used to make cell components thermostable; and ii) the structural changes which are important for improved stability are rather subtle. In this article, the author like would to first present some typical examples to explain these general rules and finally some preliminary findings concerning the biochemical genetics of thermophiles.

TACTICS

The tactics used to make thermophile proteins, nucleic acids, and cell organella thermostable are diverse, and different from molecular to molecule, and from organelle to organelle. The author (Oshima, 1979; Oshima *et al.*, 1976; Langworthy *et al.*, 1979) has proposed that the molecular mechanisms can be classified into three general types, the intrinsic, protector, and biochemical modification mechanisms.

Many enzymes and proteins of thermophiles seemed to be stabilized at high temperature by increased intracellular interactions caused by replacements of amino acid residues. The author would like to call this type of mechanism the intrinsic one. One of the most typical examples would be phosphoglycerate kinase from *Thermus thermophilus*. This enzyme is a simple protein without any prothetic group, and is stable at high temperatures without the presence of a substrate, metal ions such as Mg^{2+}, or other proteins (Nojima *et al.*, 1979). The catalytic properties of the thermophile phosphoglycerate kinase are similar to its mesophilic counterpart in many respects. A similar observation has been made for other enzymes from thermophiles (for instance, glyceraldehyde-3-phosphate dehydrogenase of *T. thermophilus* (Fujita *et al.*, 1976)). These facts suggest that the changes in chemical structure which confer upon the enzyme an inherent thermostability do not have so much effect on the molecular architecture of the active site.

Some enzymes from thermophiles are stabilized by binding with a stabilizing cofactor. For instance, a thermophile protease, thermolysin, is stable at high temperature only in the presence of Ca^{2+} ions (Weaver *et al.*, 1976). Another example of the protector mechanism is the protein synthesizing machinery of *T. thermophilus*. A polyamine has been reported to be the protector for the ribosome- messenger-aminoacyl tRNA ternary complex at high temperatures (Ohno-Iwashita *et al.*, 1975).

The third type of mechanism, biochemical modification, can be seen in tRNA from *T. thermophilus*. As will be described in the next section, the molecule's unusual stability is conferred in part by the modification of a ribothymidine residue. From this observation, one might imagine a thermophile enzyme which is stabilized by biochemical modification(s) on some amino acid residue(s). Up to now, however, no report has been made of a thermophile enzyme which is stabilized by the biochemical modification mechanism.

In many cases, a biopolymer or a cellular organelle is stabilized by a combination of two or three tactics. For instance, as will be described in the following section, tRNA from *Thermus* spp. is stabilized by a combination of the intrinsic mechanism by replacing a nucleotide base to another one, and the biochemical modification mechanism by which a base is thiolated.

If a thermophile biopolymer is stabilized only by the intrinsic mechanism, a mesophile which harbors the structural gene for the thermophile biopolymer could produce a heat resistant molecule at mesophilic temperature. Studies along this line will be described later.

SUBTLE ALTERATIONS

Many of the changes which confer upon the molecule an unusual thermostability may be subtle. The most convincing example is the structure-stability relationship of tRNA from *T. thermophilus*.

tRNA from the thermophile has a higher melting temperature than that from mesophilic organisms. Two structural features of the heat-stable tRNA were found to be related to heat resistance: i) a high G-C pair content in the base paired region. A physico-chemical study suggested a 90% G-C pair content for $tRNA_{mix}$ of the thermophile (Watanabe *et al.*, 1973); and ii) the presence of a modified base, 5-methyl-2-thiouridine (or 2-thio-ribothymidine) instead of unmodified ribothymidine in the T loop (Watanabe *et al.*, 1974). The implication of the thiolated base in the unusual stability was indicated by biochemical (Watanabe *et al.*, 1976), physico-chemical (Davanloo *et al.*, 1979; Watanabe *et al.*, 1979b), and organic chemical (Watanabe, 1980) studies.

tRNA specific for formyl methionine ($tRNA_{f1}^{Met}$) from the thermophile was isolated, sequenced, and compared with the chemical structure of the corresponding tRNA from *Escherichia coli* (Watanabe, 1979a). Only one base replacement was found; that is, a G-U pair (connected by two hydrogen bonds like the A-U pair) in *E. coli* tRNA was replaced by a G-C pair in the thermophile tRNA. The G-C pair content in the base paired region of the thermophile tRNA is 90% as suggested by the prior physicochemical study (Watanabe *et al.*, 1973).

Except for the base replacement, three modified bases were found in the thermophile $tRNA_{f1}^{Met}$: 2-thio-ribothymidine in the T loop, 1-methyladenosine in the T loop, and 2′-O-methylguanosine in the D loop. The latter two modifications might be neutral in terms of the unusual stability since the same or similar modifications are often found in eukaryote (mesophilic) tRNAs. Recently an enzyme which catalyzes the 2′-O-methylation of the specific guanosine residue in the D loop was isolated from the thermophile (Kumagai *et al.*, 1980). The melting temperature of *E. coli* tRNA methylated by the enzyme was identical to that of the unmodified tRNA. This experiment directly demonstrated that the methylation at the specific guanosine residue had no effect on the thermal stability.

The nucleotide sequence of *T. thermophilus* $tRNA_{f1}^{Met}$ is identical to that of *E. coli* $tRNA_f^{Met}$ except for the above-mentioned four residues (one replacement and three modifications). Thus the thermophile $tRNA_f^{Met}$ is so far the only molecule of thermophiles whose unusual stability is fully explained by changes in chemical structure. The molecule is stabilized by an extra intramolecular hydrogen bond formed by replacing a uridyl residue with a cytydyl residue and increased stacking interaction brought about by thiolating a ribothymidine residue. These changes are quite subtle compared to the sum of the intramolecular hydrogen bonds (roughly 60), or the number of atoms composing the molecule, one (oxygen) of which is replaced by a sulfur in the thermophile tRNA. It might be of interest to point out that these changes are concentrated in a part quite

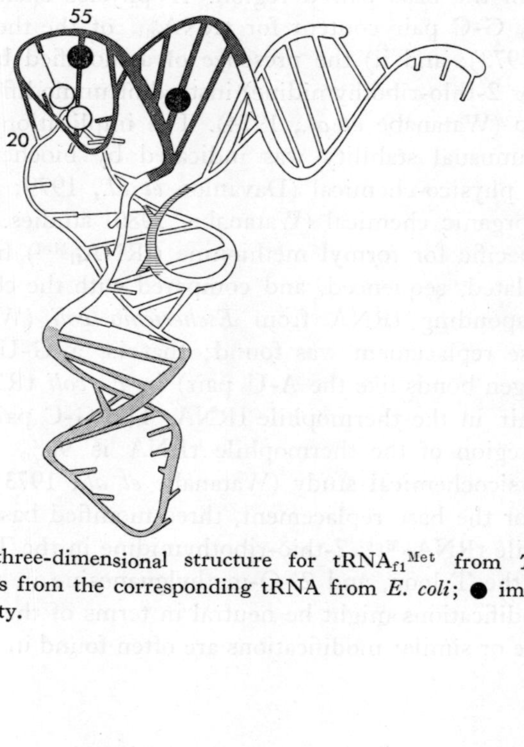

Fig. 1. Proposed three-dimensional structure for $tRNA_{f1}^{Met}$ from *T. thermophilus*. ○ structural changes from the corresponding tRNA from *E. coli*; ● important changes for improved stability.

far from the anti-codon loop and amino acid accepting terminal, since both sites are of direct importance to the function of tRNA (see Fig. 1).

GENETIC STUDIES

Studies on the genetic control of thermophily will be important for understanding the adaptation mechanisms of thermophilic bacteria to hot environments. However, few studies in this field have been carried out and only four groups have so far reported the transformation of thermophile genes.

Lindsay and Creaser (1975) and Mojica *et al.* (1977) reported exciting results which suggest that the transfer of a few genes causes the change from a mesophilic to thermophilic nature. They carried out DNA-mediated transformation experiments using *Bacillus subtilis* as the recipient and DNA preparations from thermophiles of *Bacillus* spp., and isolated transformants which were able to grow at higher temperatures. Lindsay and Creaser (1975) suggested that the regions of the origin of DNA replication and ribosomal and tRNA functions are highly conservative, and that the replacement of these regions by thermophile DNA sequences may have wide pleiotropic effects.

To avoid the possibility of contamination of highly stable endospores produced by the thermophilic *Bacilli*, the author, in cooperation with Drs. S. Shinomiya and K. Yamane, studied the transformation of an α-amylase gene between a mesophilic bacteria and a non-spore-forming thermophile (Shinomiya *et al.*, 1980). A *B. subtilis* transformant producing heat-stable α-amylase was isolated by treating with the thermophile DNA. The transformant did not grow at higher temperature; the maximum growth temperature of the transformant was identical to that of the recipient. Enzymes other than α-amylase of the transformant were not resistant to heat. Shinomiya *et al.* (submitted) found that the structural gene for the thermostable α-amylase is incorporated in the *B. subtilis* chromosome at a different site from that of the *B. subtilis* α-amylase gene.

Recently we have cloned a thermophile gene into *E. coli* (Tanaka *et al.*, 1981). A recombinant plasmid which contains 3-isopropylmalate dehydrogenase gene from an extreme thermophile, *T. thermophilus*, was made from a *E. coli* plasmid pBR322 and the thermophile DNA, and transferred into an *E. coli* mutant which lacks the dehydrogenase. *Escherichia coli* cells carrying the thermophile gene produced heat-stable 3-isopropylma-

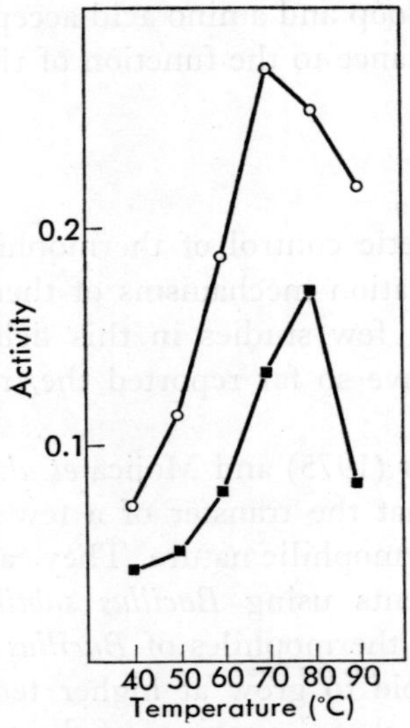

Fig. 2. Heat stability of 3-isopropylmalate dehydrogenase produced by *T. thermophilus* (■) and by *E. coli* carrying the gene coded for the enzyme from the thermophile (○). The *E. coli* enzyme has no activity above 60°C (taken from Tanaka *et al.*, 1981).

late dehydrogenase (Fig. 2). Nagahari *et al.* (1980) have also done a similar experiment using DNA from a different thermophilic strain.

These cloning studies as well as the DNA-mediated transformation carried by Shinomiya *et al.* (1980) indicate that the thermal stabilities of these enzymes are ultimately coded in their structural genes and transferable in mesophilic cells by genetic transformations. The stabilities of proteins coded by the mesophilic genes of the recipients seemed not to be affected by the transformations. These studies suggested that the thermophily of a cell is coded for by many genes.

Finally the author would like to point out an interesting observation made by Nagahari *et al.* (1980) in respect to the evolution of the thermophiles: the thermal stability of 3-isopropylmalate dehydrogenase produced by *E. coli* carrying the recombinant plasmid was lowered to some extent after the cells were kept growing at 37°C. This observation suggests that the chemical structure of a gene (DNA) is only stable in an appropriate environment. However, their study was a preliminary one, and more critical experiments should be done befofe a final conclusion is drawn.

SUMMARY

Biopolymers and cell organella from thermophiles are resistant to heat. The chemical changes which are important to the improved stability are subtle compared to the whole structure of cell constituents from thermophiles. The author proposes classifying the molecular tactics into three major groups: intrinsic, protector, and biochemical modification mechanisms. A thermophile cell constituent is often stabilized by a combination of two or more mechanisms. For instance, tRNA from an extreme thermophile, *T. thermophilus*, has a higher melting temperature than the corresponding tRNA from mesophilic organisms. The unusual stability is conferred by a combination of two mechanisms: 1) an intrinsic one by which the G-C pair content of the tRNA is increased and thus a few intramolecular hydrogen bonds are added to the molecule; and 2) a biochemical modification by which the thymine base in the T loop is thiolated, giving rise to a more tightly folded structure. Biochemical genetics could also be a promising method for understanding adaptations to temperature extremes. Genetic transformation experiments using thermophile DNA are briefly reviewed.

REFERENCES

Davanloo, P., Sprinzl, M., Watanabe, K., Albani, M., and Kersten, H. (1979). Role of ribothymidine in the thermal stability of transfer RNA as monitored by proton magnetic resonance. *Nucleic Acid. Res.* **6**, 1571–1581.

Fujita, S. C., Oshima, T., and Imahori, K. (1976). Purification and properties of D-glyceraldehyde-3-phosphate dehydrogenase from an extreme thermophile, *Thermus thermophilus* strain HB8. *Eur. J. Biochem.* **64**, 57–68.

Kumagai, I., Watanabe, K., and Oshima, T. (1980). Thermally induced biosynthesis of 2′-O-methylguanosine in tRNA from an extreme thermophile, *Thermus thermophilus* HB27. *Proc. Natl. Acad. Sci. U.S.* **77**, 1922–1926.

Langworthy, T. A., Brock, T. D., Castenholz, R. W., Esser, A. F., Johnson, E. J., Oshima, T., Tsuboi, M., Zeikus, J. G., and Zuber, H. (1979). Life at high temperatures (Group report). In *Strategies of Microbial Life in Extreme Environments* (ed. M. Shilo), pp. 489–502. Verlag Chemie, Weinheim and New York.

Lindsay, J. A. and Creaser, E. H. (1975). Enzyme thermostability is transformable property between *Bacillus* spp. *Nature* **255**, 650–652.

Ohno-Iwashita, Y., Oshima, T., and Imahori, K. (1975). *In vitro* protein synthesis at elevated temperature by an extract of an extreme thermophile. *Arch. Biochem. Biophys.* **171**, 490–499.

Oshima, T. (1979). Molecular basis for unusual thermostabilities of cell constituents from an extreme thermophile, *Thermus thermophilus*. In *Strategies of Microbial Life in*

Extreme Environments (ed. M. Shilo), pp. 455–469. Verlag Chemie, Weinheim and New York.

Oshima, T., Sakaki, Y., Wakayama, N., Watanabe, K., Ohashi, Z., and Nishimura, S. (1976). Biochemical studies on an extreme thermophile *Thermus thermophilus*: Thermal stabilities of cell constituents and a bacteriophage. In *Enzymes and Proteins from Thermophilic Microorganisms* (Experientia Suppl. 26) (ed. H. Zuber), pp. 317–331. Birkhäuser-Verlag, Basel.

Mojica, T., Cheung, D., and Friedman, S. M. (1977). Effect of transformation to thermophily on ribosome structure and function. Abstr. the Annual Meeting of American Society for Microbiology, 70 (I-94).

Nagahari, K., Koshikawa, T., and Sakaguchi, K. (1980). Cloning and expression of the leucine gene from *Thermus thermophilus* in *Escherichia coli*. *Gene* **10**, 137–145.

Nojima, H., Oshima, T., and Noda, H. (1979). Purification and Properties of phosphoglycerate kinase from *Thermus thermophilus* strain HB8. *J. Biochem.* **85**, 1509–1517.

Shinomiya, S., Oshima, T., and Yamane, K. *Bacillus subtilis* transformant producing thermostable α-amylase by a deoxyribonucleic acid from a thermophilic bacterium: Enzymological and genetic studies (submitted).

Shinomiya, S., Yamane, K., and Oshima, T. (1980). Isolation of a *Bacillus subtilis* transformant producing thermostable α-amylase by DNA from a thermophilic bacterium. *Biochem. Biophys. Res. Commun.* **96**, 175–179.

Tanaka, T., Kawano, N., and Oshima, T. (1981). Cloning of 3-isopropylmalate dehydrogenase gene of an extreme thermophile and partial purification of the gene product. *J. Biochem.* **89**, 677–682.

Watanabe, K. (1980). Reactions of 2-thioribothymidine and 4-thiouridine with hydrogen peroxide in transfer ribonucleic acids from *Thermus thermophilus* and *Escherichia coli* as studied by circular dichroism. *Biochemistry* **19**, 5542–5549.

Watanabe, K., Kuchino, Y., Yamaizumi, Z., Kato, M., Oshima, T., and Nishimura, S. (1979a). Nucleotide sequence of formylmethionine tRNA from an extreme thermophile, *Thermus thermophilus* HB8. *J. Biochem.* **86**, 893–905.

Watanabe, K., Oshima, T., Saneyoshi, M., and Nishimura, S. (1974). Replacement of ribothymidine by 5-methyl-2-thiouridine in sequence GTψC in tRNA of an extreme thermophile. *FEBS Lett.* **43**, 59–63.

Watanabe, K., Shinma, M., Oshima, T., and Nishimura, S. (1976). Heat-induced stability of tRNA from an extreme thermophile, *Thermus thermophilus*. *Biochem. Biophys. Res. Commun.* **72**, 1137–1144.

Watanabe, K., Yokoyama, S., Hanske, F., Kasai, H., and Miyazawa, T. (1979b). CD and NMR studies on the conformational thermostability of 2-thio-ribothymidine found in TΨC loop of thermophile tRNA. *Biochem. Biophys. Res. Commun.* **91**, 671–677.

Watanabe, K., Seno, T., Nishimura, S., Oshima, T., and Imahori, K. (1973). The study of circular dichroic difference spectra applied to polynucleotides and ribonucleic acids. *Polymer J.* **4**, 539–552.

Weaver, L. H., Kester, W. R., Ten Eyck, L. F., and Matthews, B. W. (1976). The structure and stability of thermolysin. In *Enzymes and Proteins from Thermophilic Microorganisms*(Experientia Suppl. 26) (ed. H. Zuber), pp. 31–39. Birkhäuser-Verlag, Basel.

Contributors

Fujii, K.	Shionogi Research Laboratory, Osaka 553, Japan
Fukumori, Y.	Department of Biology, Faculty of Science, Osaka University, Toyonaka, Osaka 560, Japan
Gojobori, T.	Center for Demographic and Population Genetics, The University of Texas at Houston, Houston, Texas 77025, U.S.A.
Hase, T.	Department of Biology, Faculty of Science, Osaka University, Toyonaka, Osaka 560, Japan
Ichinose, M.	Department of Biology, Kyushu University, Fukuoka 812, Japan
Ikenaka, T.	Department of Chemistry, Osaka University, College of Science, Toyonaka, Osaka 560, Japan
Kamita, Y.	Institute for Biological Science, Sumitomo Chemical Co., Ltd., Takarazuka, Hyogo 665, Japan
Katsube, Y.	Faculty of Engineering, Tottori University, Tottori 680, Japan
Kimura, M.	National Institute of Genetics, Mishima, Shizuoka 411, Japan
Kusakabe, S.	Department of Biology, Kyushu University, Fukuoka 812, Japan
Maruyama, T.	National Institute of Genetics, Mishima, Shizuoka 411, Japan

Matsubara, H.	Department of Biology, Faculty of Science, Osaka University, Toyonaka, Osaka 560, Japan
Matsuda, G.	Department of Biochemistry, Nagasaki University School of Medicine, Nagasaki 852, Japan
Matsuda, M.	Department of Biology, Kyushu University, Fukuoka 812, Japan
Miyata, T.	Department of Biology, Faculty of Science, Kyushu University, Fukuoka 812, Japan
Mukai, T.	Department of Biology, Kyushu University, Fukuoka 812, Japan
Odani, S.	Department of Biochemistry, Niigata University School of Medicine, Niigata 951, Japan
Ohta, T.	National Institute of Genetics, Mishima, Shizuoka 411, Japan
Omoto, K.	Department of Anthropology, Faculty of Science, The University of Tokyo, Tokyo 113, Japan
Oshima, T.	Mitsubishi-Kasei Institute of Life Sciences, Machida, Tokyo 194, Japan
Soeda, E.	National Institute of Genetics, Mishima, Shizuoka 411, Japan
Tachida, H.	Department of Biology, Kyushu University, Fukuoka 812, Japan
Takahata, N.	National Institute of Genetics, Mishima, Shizuoka 411, Japan
Tateno, Y.	Institute of Physical and Chemical Research, Wako, Saitama 351, Japan
Tsukihara, T.	Faculty of Engineering, Tottori University, Tottori 680, Japan
Wada, K.	Department of Biology, Faculty of Science, Osaka University, Toyonaka, Osaka 560, Japan
Yamaguchi, O.	Department of Biology, Kyushu University, Fukuoka 812, Japan
Yamanaka, T.	Department of Biology, Faculty of Science, Osaka University, Toyonaka, Osaka 560, Japan
Yamazaki, T.	Laboratory of Population Genetics, Faculty of Science, Kushu University, Fukuoka 812, Japan
Yoshimaru, H.	Department of Biology, Kyushu University, Fukuoka 812, Japan

Subject Index